FUTURE TRENDS
IN MICROELECTRONICS

FUTURE TRENDS
IN MICROELECTRONICS
Journey into the Unknown

Edited by

SERGE LURYI
JIMMY XU
ALEXANDER ZASLAVSKY

IEEE PRESS

WILEY

Published by John Wiley & Sons, Inc., Hoboken, New Jersey
Published simultaneously in Canada

For general information on our other products and services or for technical support, please contact our Customer Care Department within the United States at (800) 762-2974, outside the United States at (317) 572-3993 or fax (317) 572-4002.

Wiley also publishes its books in a variety of electronic formats. Some content that appears in print may not be available in electronic formats. For more information about Wiley products, visit our web site at www.wiley.com.

Library of Congress Cataloging-in-Publication Data:

Names: Luryi, Serge, editor. | Xu, Jimmy, editor. | Zaslavsky, Alex, 1963-
   editor.
Title: Future trends in microelectronics. Journey into the unknown / edited
   by Serge Luryi, Jimmy Xu, Alexander Zaslavsky.
Description: Hoboken, New Jersey : John Wiley & Sons, 2016. | Includes
   bibliographical references and index.
Identifiers: LCCN 2016022488 (print) | LCCN 2016025093 (ebook) | ISBN
   9781119069119 (cloth) | ISBN 9781119069171 (pdf) | ISBN 9781119069188
   (epub)
Subjects: LCSH: Microelectronics–Technological innovations. |
   Nanotechnology–Technological innovations. | Semiconductors–Technological
   innovations.
Classification: LCC TK7874 .F887 2016 (print) | LCC TK7874 (ebook) | DDC
   621.381–dc23
LC record available at https://lccn.loc.gov/2016022488

Printed in the United States of America

10 9 8 7 6 5 4 3 2 1

Contents

I   FUTURE OF DIGITAL SILICON

II   NEW MATERIALS AND NEW PHYSICS

III MICROELECTRONICS IN HEALTH, ENERGY HARVESTING, AND COMMUNICATIONS

List of Contributors

G. Ardila, *IMEP-LAHC/Minatec, CNRS-Grenoble INP, UJF, 38016 Grenoble, France*

Francis Balestra, *IMEP-LAHC, Minatec, Grenoble-Alpes University, 38016 Grenoble Cedex 1, France*

L. A. M. Barea, *Department of Electrical Engineering, UFSCAR, São Carlos, SP 13565-905, Brazil*

J. P. H. Benschop, *ASML, Veldhoven, The Netherlands*

D. A. Borton, *School of Engineering and the Brown Institute for Brain Science, Brown University, Providence, RI 02912, USA*

C. Schulte-Braucks, *Peter Grünberg Institut-9 and JARA-FIT, Forschungszentrum Jülich GmbH, 52425 Jülich, Germany*

D. Buca, *Peter Grünberg Institut-9 and JARA-FIT, Forschungszentrum Jülich GmbH, 52425 Jülich, Germany*

S. Burger, *Zuse Institute Berlin (ZIB), Takustraße 7, 14195 Berlin, Germany and JCMwave GmbH, Bolivarallee 22, 14050 Berlin, Germany*

A.V. Butenko, *Department of Physics and Faculty of Engineering, Jack and Pearl Institute of Advanced Technology, Bar Ilan University, Ramat Gan 52900, Israel*

Kyung-Eun Byun, *Device Laboratory, Samsung Advanced Institute of Technology, Suwon 443-803, South Korea*

Yeonchoo Cho, *Device Laboratory, Samsung Advanced Institute of Technology, Suwon 443-803, South Korea*

J. K. Choi, *Memory R&D Division, SK Hynix, Icheon-si, Gyeonggi-do 467-701, South Korea*

W. Crooijmans, *Philips Research, High Tech Campus 34, Eindhoven, The Netherlands*

S. Datta, *Department of Electrical Engineering, The Pennsylvania State University, University Park, PA 16802, USA and Department of Electrical Engineering, University of Notre Dame, Notre Dame, IN 46556, USA*

E. Dentoni Litta, *School of Information and Communication Technology, KTH Royal Institute of Technology, 16440 Kista, Sweden*

M. Dini, *Department of Electrical, Electronic and Information Engineering "G. Marconi" and Advanced Research Center on Electronic Systems "E. De Castro", University of Bologna, Cesena, Via Venezia 52, 47521 Cesena FC, Italy*

N. von den Driesch, *Peter Grünberg Institut-9 and JARA-FIT, Forschungszentrum Jülich GmbH, 52425 Jülich, Germany*

M. I. Dyakonov, *Laboratoire Charles Coulomb, Université Montpellier, CNRS, Montpellier, France*

M. Filippi, *Department of Electrical, Electronic and Information Engineering "G. Marconi" and Advanced Research Center on Electronic Systems "E. De Castro", University of Bologna, Cesena, Via Venezia 52, 47521 Cesena FC, Italy*

N. C. Frateschi, *"Gleb Wataghin" Physics Institute, University of Campinas, Campinas, SP 13083-859, Brazil*

F. Gamiz, *Department of Electronics, CITIC-UGR, University of Granada, 18071 Granada, Spain*

R. Geiger, *Laboratory for Micro- and Nanotechnology, Paul Scherrer Institut, 5232 Villigen PSI, Switzerland*

M. Goldstein, *Raymond and Beverly Sackler School of Physics and Astronomy, Tel-Aviv University, Tel Aviv 69978, Israel*

Detlev Grützmacher, *Peter Grünberg Institut-9 and JARA-FIT, Forschungszentrum Jülich GmbH, 52425 Jülich, Germany*

A. Haran, *Department of Physics and Faculty of Engineering, Jack and Pearl Institute of Advanced Technology, Bar Ilan University, Ramat Gan 52900, Israel*

J. M. Hartmann, *Université Grenoble Alpes and CEA-LETI/MINATEC, 38054 Grenoble, France*

P.-E. Hellström, *School of Information and Communication Technology, KTH Royal Institute of Technology, 16440 Kista, Sweden*

B. H. W. Hendriks, *Philips Research, High Tech Campus 34, Eindhoven, The Netherlands*

Jinseong Heo, *Device Laboratory, Samsung Advanced Institute of Technology, Suwon 443-803, South Korea*

R. Hevroni, *Raymond and Beverly Sackler School of Physics and Astronomy, Tel-Aviv University, Tel Aviv 69978, Israel*

R. Hinchet, *IMEP-LAHC/Minatec, CNRS-Grenoble INP, UJF, 38016 Grenoble, France*

Sungwoo Hwang, *Device Laboratory, Samsung Advanced Institute of Technology, Suwon 443-803, South Korea*

Z. Ikonic, *Institute of Microwaves and Photonics, University of Leeds, Leeds LS2 9JT, United Kingdom*

K.-S. Im, *School of Electronics Engineering, Kyungpook National University, 80, Daehak-ro, Buk-gu, Daegu, South Korea*

Gitae Jeong, *Semiconductor Business Division, Samsung Electronics Co. Ltd., Giheung, Gyeonggi-do, South Korea*

Y.-W. Jo, *School of Electronics Engineering, Kyungpook National University, 80, Daehak-ro, Buk-gu, Daegu, South Korea*

Yu. Kaganovskii, *Department of Physics and Faculty of Engineering, Jack and Pearl Institute of Advanced Technology, Bar Ilan University, Ramat Gan 52900, Israel*

M. Karpovski, *Raymond and Beverly Sackler School of Physics and Astronomy, Tel-Aviv University, Tel Aviv 69978, Israel*

M. Kaveh, *Department of Physics and Faculty of Engineering, Jack and Pearl Institute of Advanced Technology, Bar Ilan University, Ramat Gan 52900, Israel*

M. J. Kelly, *Department of Engineering, Centre for Advanced Photonics and Electronics, University of Cambridge, 9 JJ Thomson Avenue, CB3 0FA Cambridge, United Kingdom and MacDiarmid Institute for Advanced Materials and Nanotechnology, Victoria University of Wellington, Wellington 6140, New Zealand*

Kinam Kim, *Semiconductor Business Division, Samsung Electronics Co. Ltd., Giheung, Gyeonggi-do, South Korea*

E. Kogan, *Department of Physics and Faculty of Engineering, Jack and Pearl Institute of Advanced Technology, Bar Ilan University, Ramat Gan 52900, Israel*

J.-R. Kropp, *VI Systems GmbH, Hardenbergstraße 7, 10623 Berlin, Germany*

N. Ledentsov Jr., *VI Systems GmbH, Hardenbergstraße 7, 10623 Berlin, Germany*

N. N. Ledentsov, *VI Systems GmbH, Hardenbergstraße 7, 10623 Berlin, Germany*

J.-H. Lee, *School of Electronics Engineering, Kyungpook National University, 80, Daehak-ro, Buk-gu, Daegu, South Korea*

Min-Hyun Lee, *Device Laboratory, Samsung Advanced Institute of Technology, Suwon 443-803, South Korea*

S. Lenk, *Peter Grünberg Institut-9 and JARA-FIT, Forschungszentrum Jülich GmbH, 52425 Jülich, Germany*

Y. Li, *Department of Electrical Engineering, SUNY at Buffalo, Buffalo, NY 14260, USA*

Serge Luryi, *Department of Electrical and Computer Engineering, Stony Brook University, Stony Brook, NY 11794, USA*

A. Makarov, *Institute for Microelectronics, TU Wien, 1040 Vienna, Austria*

S. Mantl, *Peter Grünberg Institut-9 and JARA-FIT, Forschungszentrum Jülich GmbH, 52425 Jülich, Germany*

C. Marquez, *Department of Electronics, CITIC-UGR, University of Granada, 18071 Granada, Spain*

A. Michard, *IMEP-LAHC/Minatec, CNRS-Grenoble INP, UJF, 38016 Grenoble, France*

D. Mioni, *Philips Research, High Tech Campus 34, Eindhoven, The Netherlands*

V. Mitin, *Department of Electrical Engineering, SUNY at Buffalo, Buffalo, NY 14260, USA*

L. Montès, *IMEP-LAHC/Minatec, CNRS-Grenoble INP, UJF, 38016 Grenoble, France*

M. Mouis, *IMEP-LAHC/Minatec, CNRS-Grenoble INP, UJF, 38016 Grenoble, France*

G. Mussler, *Peter Grünberg Institut-9 and JARA-FIT, Forschungszentrum Jülich GmbH, 52425 Jülich, Germany*

D. Naveh, *Department of Physics and Faculty of Engineering, Jack and Pearl Institute of Advanced Technology, Bar Ilan University, Ramat Gan 52900, Israel*

Jean-Pierre Nozières, *eVaderis, Minatec Entreprises BHT, 7 Parvis Louis Néel, 38040 Grenoble, France and Spintec, Bat. 1005, 17 rue des Martyrs, 38054 Grenoble, France*

M. Östling, *School of Information and Communication Technology, KTH Royal Institute of Technology, 16440 Kista, Sweden*

A. Palevski, *Raymond and Beverly Sackler School of Physics and Astronomy, Tel-Aviv University, Tel Aviv 69978, Israel*

A. Parihar, *School of Electrical and Computer Engineering, Georgia Institute of Technology, Atlanta, GA 30332, USA*

Seongjun Park, *Device Laboratory, Samsung Advanced Institute of Technology, Suwon 443-803, South Korea*

A. Raychowdhury, *School of Electrical and Computer Engineering, Georgia Institute of Technology, Atlanta, GA 30332, USA*

G. F. M. Rezende, *"Gleb Wataghin" Physics Institute, University of Campinas, Campinas, SP 13083-859, Brazil*

V. Richter, *Department of Physics and Faculty of Engineering, Jack and Pearl Institute of Advanced Technology, Bar Ilan University, Ramat Gan 52900, Israel*

N. Rodriguez, *Department of Electronics, CITIC-UGR, University of Granada, 18071 Granada, Spain*

A. Romani, *Department of Electrical, Electronic and Information Engineering "G. Marconi" and Advanced Research Center on Electronic Systems "E. De Castro", University of Bologna, Cesena, Via Venezia 52, 47521 Cesena FC, Italy*

R. J. Ruiz, *Department of Electronics, CITIC-UGR, University of Granada, 18071 Granada, Spain*

K. Sablon, *U.S. Army Research Laboratory, Adelphi, MD 20783, USA*

E. Sangiorgi, *Department of Electrical, Electronic and Information Engineering "G. Marconi" and Advanced Research Center on Electronic Systems "E. De Castro", University of Bologna, Cesena, Via Venezia 52, 47521 Cesena FC, Italy*

F. Schmidt, *Zuse Institute Berlin (ZIB), Takustraße 7, 14195 Berlin, Germany and JCMwave GmbH, Bolivarallee 22, 14050 Berlin, Germany*

E. Sela, *Raymond and Beverly Sackler School of Physics and Astronomy, Tel-Aviv University, Tel Aviv 69978, Israel*

S. Selberherr, *Institute for Microelectronics, TU Wien, 1040 Vienna, Austria*

A. Sergeev, *U.S. Army Research Laboratory, Adelphi, MD 20783, USA*

A. Sharoni, *Department of Physics and Faculty of Engineering, Jack and Pearl Institute of Advanced Technology, Bar Ilan University, Ramat Gan 52900, Israel*

V. A. Shchukin, *VI Systems GmbH, Hardenbergstraße 7, 10623 Berlin, Germany*

V. Shelukhin, *Raymond and Beverly Sackler School of Physics and Astronomy, Tel-Aviv University, Tel Aviv 69978, Israel*

I. Shlimak, *Department of Physics and Faculty of Engineering, Jack and Pearl Institute of Advanced Technology, Bar Ilan University, Ramat Gan 52900, Israel*

Hadas Shtrikman, *Department of Condensed Matter Physics, Weizmann Institute of Science, Rehovot 76100, Israel*

N. Shukla, *Department of Electrical Engineering, The Pennsylvania State University, University Park, PA 16802, USA and Department of Electrical Engineering, University of Notre Dame, Notre Dame, IN 46556, USA*

H. Sigg, *Laboratory for Micro- and Nanotechnology, Paul Scherrer Institut, 5232 Villigen PSI, Switzerland*

P. M. Solomon, *IBM, T. J. Watson Research Center, Yorktown Heights, N Y 10598, USA*

D.-H. Son, *School of Electronics Engineering, Kyungpook National University, 80, Daehak-ro, Buk-gu, Daegu, South Korea*

M. C. M. M. Souza, *"Gleb Wataghin" Physics Institute, University of Campinas, Campinas, SP 13083-859, Brazil*

Boris Spivak, *Department of Physics, University of Washington, Seattle, WA 98195, USA*

D. Stange, *Peter Grünberg Institut-9 and JARA-FIT, Forschungszentrum Jülich GmbH, 52425 Jülich, Germany*

T. Stoica, *Peter Grünberg Institut-9 and JARA-FIT, Forschungszentrum Jülich GmbH, 52425 Jülich, Germany*

V. Sverdlov, *Institute for Microelectronics, TU Wien, 1040 Vienna, Austria*

R. Tao, *IMEP-LAHC/Minatec, CNRS-Grenoble INP, UJF, 38016 Grenoble, France*

M. Tartagni, *Department of Electrical, Electronic and Information Engineering "G. Marconi" and Advanced Research Center on Electronic Systems "E. De Castro", University of Bologna, Cesena, Via Venezia 52, 47521 Cesena FC, Italy*

G. Thomain, *Department of Electrical Engineering, SUNY at Buffalo, Buffalo, NY 14260, USA*

H. van Houten, *Philips Research, High Tech Campus 34, Eindhoven, The Netherlands*

A. A. G. von Zuben, *"Gleb Wataghin" Physics Institute, University of Campinas, Campinas, SP 13083-859, Brazil*

G. S. Wiederhecker, *"Gleb Wataghin" Physics Institute, University of Campinas, Campinas, SP 13083-859, Brazil*

T. Windbacher, *Institute for Microelectronics, TU Wien, 1040 Vienna, Austria*

S. Wirths, *Peter Grünberg Institut-9 and JARA-FIT, Forschungszentrum Jülich GmbH, 52425 Jülich, Germany*

L. Wolfson, *Department of Physics and Faculty of Engineering, Jack and Pearl Institute of Advanced Technology, Bar Ilan University, Ramat Gan 52900, Israel*

T. Yore, *Department of Electrical Engineering, SUNY at Buffalo, Buffalo, NY 14260, USA*

T. Zabel, *Laboratory for Micro- and Nanotechnology, Paul Scherrer Institut, 5232 Villigen PSI, Switzerland*

X. Zhang, *Department of Electrical Engineering, SUNY at Buffalo, Buffalo, NY 14260, USA*

E. Zion, *Department of Physics and Faculty of Engineering, Jack and Pearl Institute of Advanced Technology, Bar Ilan University, Ramat Gan 52900, Israel*

Preface

S. Luryi
Department of Electrical and Computer Engineering, Stony Brook University, Stony Brook, NY 11794, USA

J. M. Xu and A. Zaslavsky
School of Engineering, Brown University, Providence RI 02912, USA

This book is a brainchild of the eighth workshop in the *Future Trends in Microelectronics* series (FTM-8). The first of the FTM conferences, "*Reflections on the Road to Nanotechnology*," had gathered in 1995 on Ile de Bendor, a beautiful little French Mediterranean island.[1] The second FTM, "*Off the Beaten Path*," took place in 1998 on a larger island in the same area, Ile des Embiez.[2] Instead of going to a still larger island, the third FTM, "*The Nano Millennium*," went back to its origins on Ile de Bendor in 2001.[3] As if to compensate for small size of Bendor, the fourth FTM, "*The Nano, the Giga, the Ultra, and the Bio*," took place on the biggest French Mediterranean island of them all, Corsica.[4] Normally, the FTM workshops gather every 3 years; however, the FTM-4 was held 1 year ahead of the usual schedule, in the summer of 2003, as a one-time exception. Continuing its inexorable motion eastward, the fifth FTM workshop, "*Up the Nano Creek*," had convened on Crete, Greece, in June of 2006.[5] The inexorable motion was then interrupted to produce a semblance of a random walk in the Mediterranean and the FTM-6 "*Unmapped Roads*" went to the Italian island of Sardinia (June 2009).[6] Then, FTM-7, "*Into the Cross Currents*" returned to our earlier venue on Corsica (June 2012).[7] Finally, FTM-8 struck out toward new territories, jumping West all the way to the Spanish island of Mallorca.

The FTM workshops are relatively small gatherings (less than 100 people) by invitation only. If you, the reader, wish to be invited, please consider following a few simple steps outlined on the conference website. The FTM website at www.ece.stonybrook.edu/~serge/FTM.html contains links to all past and planned workshops, their programs, publications, sponsors, and participants. Our attendees have been an illustrious lot. Suffice it to say that among FTM participants, we find five Nobel laureates (Zhores Alferov, Herbert Kroemer, Horst Stormer, Klaus von Klitzing, and Harold Kroto) and countless others poised for a similar distinction. To be sure, high distinction is not a prerequisite for being invited to FTM, but the ability and desire to bring fresh ideas is. All participants of FTM-8 can be considered authors of this book, which in this sense is a collective treatise.

The main purpose of FTM workshops is to provide a forum for a free-spirited exchange of views, projections, and critiques of current and future directions, among the leading professionals in industry, academia, and government.

For better or worse, our civilization is destined to be based on electronics. Ever since the invention of the transistor and especially after the advent of integrated circuits, semiconductor devices have kept expanding their role in our lives. Electronic circuits entertain us and keep track of our money, they fight our wars and decipher the secret codes of life, and one day, perhaps, they will relieve us from the burden of thinking and making responsible decisions. Inasmuch as that day has not yet arrived, we have to fend for ourselves. The key to success is to have a clear vision of where we are heading. In the blinding light of a bright future, the FTM community has remained mindful of the fact that what controlled the past will still control the future – the basic principles of science. Thus, the trendy, red-hot projections of any given epoch deserve and require critical scrutiny.

Some degree of stability is of importance in these turbulent times and should be welcome. Thus, although the very term *"microelectronics"* has been generally rechristened *"nanoelectronics,"* we have stuck to the original title of the FTM workshop series.

The present volume contains a number of original papers, some of which were presented at FTM-8 in oral sessions, others as posters. From the point of view of the program committee, there is no difference between these types of contributions in weight or importance. There was, however, a difference in style and focus – and that was intentionally imposed by the organizers. All speakers were asked to focus on the presenter's views and projections of future directions, assessments or critiques of important new ideas/approaches, and *not* on their own achievements. This latter point is perhaps the most distinctive feature of FTM workshops. Indeed, we are asking scientists not to speak of their own work! This has proven to be successful, however, in eliciting powerful and frank exchange. The presenters are asked to be provocative and/or inspiring. Latest advances made and results obtained by the participants are to be presented in the form of posters and group discussions.

Each day of the workshop was concluded by an evening panel or poster session that attempted to further the debates on selected controversial issues connected to the theme of the day. Each such session was chaired by a moderator, who invited two or three attendees of his or her choice to lead with a position statement, with all other attendees serving as panelists. The debate was forcefully moderated and irrelevant digressions were cut off without mercy. Moderators were also assigned the hopeless task of forging a consensus on critical issues.

To accommodate these principles, the FTM takes a format that is less rigid than usual workshops to allow and encourage uninhibited exchanges and sometimes confrontations of different views. A central theme is designed together with the speakers for each day. Another traditional feature of FTM workshops is a highly informal vote by the participants on the relative importance of various fashionable current topics in modern electronics research. This tradition owes its origin to Horst Stormer, who composed the original set of questions and maintained the results over four conferences. These votes are perhaps too bold and irreverent for general publication, but they are carefully maintained and made available to every new generation of FTM participants. Another traditional vote concerned the best poster. The 2015 winning poster was by Michael Shur on "New ideas in smart lighting."

A joyful tradition of FTM meetings is the settling of scientific bets, a custom that dates back to the 1998 wager between Nikolai Ledentsov (pro) and Horst Stormer (con) about the putative future dominance of quantum dot-based lasers – a bet that Horst collected in 2004, at FTM-4. Another risky bet on the future dominance of SOI technology was adjudicated at the FTM-8 workshop. The precise statement of this bet (worth a six-magnum case of very good champagne) was the proposition that, by 2015, SOI would cover more than 35% of the complementary metal-oxide-semiconductor (CMOS) market, including memories, by value. This bet, proposed by Sorin Cristoloveanu, attracted three cons: Detlev Grützmacher, Dimitris Ioannou, and Enrico Sangiorgi. At FTM-8, Sorin Cristoloveanu conceded that his bet on SOI was premature – the vintage of the champagne remains to be determined.

Not every contribution presented at FTM-8 has made it into this book (not for the lack of persistence by the editors). Perhaps most sorely we miss the exciting contribution by Mihai Banu of Blue Danube Systems, Inc., in which he told us "How to increase the capacity of mobile wireless networks without changing anything (well, almost anything)!" Abstracts of his and all other presentations can be found at www.ece.stonybrook.edu/~serge/ARW-8/program.html.

The FTM meetings are known for the professional critiques – or even demolitions – of fashionable trends, that some may characterize as hype. The previous workshops had witnessed powerful assaults on quantum computing, molecular electronics, and spintronics, usually waged by the fearless Michel Dyakonov. It seems that by now most of the hype associated with some of these trends has dissipated and perhaps we can take some credit for the more balanced outlook that has emerged since. This time Michel waged no wars but gave a friendly overview of ubiquitous surface waves, a subject to which he has contributed so much that some of these waves bear his name.

We have grouped all contributions into three chapters: one dealing with the future of digital silicon technology, another with new materials and new physics, and still another with applications of microelectronics to health, energy harvesting, and communications. The breakdown could not be uniquely defined because some papers fit more than one category!

To produce a coherent collective treatise out of all of this, the interaction between FTM participants had begun well before their gathering at the workshop. All the proposed presentations were posted on the web in advance and could be subject to change up to the last minute to take into account peer criticism and suggestions. After the workshop is over, these materials (not all of which have made it into this book) remain on the web indefinitely, and the reader can peruse them starting at the www.ece.stonybrook.edu/~serge/FTM.html home page.

References

1. S. Luryi, J. M. Xu, and A. Zaslavsky, eds., *Future Trends in Microelectronics: Reflections on the Road to Nanotechnology*, NATO ASI Series E, Vol. **323**, Dordrecht: Kluwer Academic, 1996.

2. S. Luryi, J. M. Xu, and A. Zaslavsky, eds., *Future Trends in Microelectronics: The Road Ahead* New York: Wiley Interscience, 1999.

3. S. Luryi, J. M. Xu, and A. Zaslavsky, eds., *Future Trends in Microelectronics: The Nano Millennium*, New York: Wiley Interscience/IEEE Press, 2002.

4. S. Luryi, J. M. Xu, and A. Zaslavsky, eds., *Future Trends in Microelectronics: The Nano, The Giga, and The Ultra*, New York: Wiley Interscience/IEEE Press, 2004.

5. S. Luryi, J. M. Xu, and A. Zaslavsky, eds., *Future Trends in Microelectronics: Up the Nano Creek*, Hoboken, NJ: Wiley Interscience/IEEE Press, 2007.

6. S. Luryi, J. M. Xu, and A. Zaslavsky, eds., *Future Trends in Microelectronics: From Nanophotonics to Sensors to Energy*, Hoboken, NJ: Wiley Interscience/IEEE Press, 2010.

7. S. Luryi, J. M. Xu, and A. Zaslavsky, eds., *Future Trends in Microelectronics: Frontiers and Innovations*, Hoboken, NJ: Wiley Interscience/IEEE Press, 2013.

Acknowledgments

The 2015 FTM workshop on Mallorca and therefore this book were possible owing to support from the following:

- US National Science Foundation
- US Department of Defense: Army Research Office
- Industry: KoMiCo Technologies, Korea
- Academia: SUNY–Stony Brook, Brown University.

On behalf of all workshop attendees, sincere gratitude is expressed to the above organizations for their generous support and especially to the following individuals whose initiative was indispensable: William Clark, Sun-Q. Jeon, Y.-T. Kim, and Dimitris Pavlidis.

Finally, the organizers wish to thank all of the contributors to this volume and all the attendees for making the workshop a rousing success.

Part I

Future of Digital Silicon

We all have reasons to be concerned about the future of digital silicon. It looks bright to many, as long as the future is about the next-generation technology and products. Some might be even braver and willing to bet on the EUV lithography finally making its mark in the next wave of chips coming to the market. Unimaginable obstacles and uncertainties have never been short of supply in this space, unmatched by any seen in all fields of engineering except perhaps for the spirit of creativity, imagination, and determination, as artfully advocated by the contributors to this chapter.

Contributors

1.1 K. Kim and G. Jeong

1.2 J. P. H. Benschop

1.3 P. M. Solomon

1.4 M. Östling, E. Dentoni Litta, and P.-E. Hellström

1.5 F. Balestra

1.6 T. Windbacher, A. Makarov, V. Sverdlov, and S. Selberherr

1.7 J.-P. Nozières

1.8 M. J. Kelly

1.1

Prospects of Future Si Technologies in the Data-Driven World

Kinam Kim and Gitae Jeong
Semiconductor Business Division, Samsung Electronics Co. Ltd., Giheung, Gyeonggi-do, South Korea

1. Introduction

The evolution of human knowledge has been accelerated by storing and sharing information, which had been accomplished by using paper until twentieth century. In recent decades, magnetic tape and hard disk drives (HDDs) have taken over the role, as paper has been supplanted by semiconductor memory and storage.

Nowadays, it is expected that the amount of data will be doubling in every 2 years from 4.4 ZB in 2013 to 44 ZB in 2020.[1] The number of Internet-connected devices will increase from 0.5 billion in 2002 to 50 billion in 2020,[2] giving rise to the tremendous growth of Internet protocol (IP) traffic, which is expected to exceed 10^{21} bytes per year in 2016. The huge amount of information has a great impact on our daily lives, which can be filled with comfort, convenience, and safety by using and analyzing the so-called big data. That is, we live in data-driven world, where data-driven decision management,[3] agriculture,[4] health care,[5] and even data-driven journalism[6] are now a reality. It is noteworthy that we can store, share, and utilize the huge amount of data with the aid of silicon (Si) technology. This means that the novel Si technologies will be deployed to continuously enrich the data-driven world of the future.

In order to fully utilize big data, we need more accurate sensors, higher density memories, higher performance CPUs with low power consumption, and higher speed interconnection technologies. Higher density memories and faster CPUs with low power consumption will be enabled by the continuous scaling of Si technology, which is now heading to the 1X nm node. Higher speed interconnection can be achieved by parallel interconnection, such as through silicon vias (TSVs), and eventually by optical interconnection technologies. High-resolution complementary metal-oxide semiconductor image sensors (CISs) and advanced sensor technology like dynamic vision sensors (DVSs) employing global shutters will bring new valuable data to our data-driven world. Therefore, it is quite useful

Future Trends in Microelectronics: Journey into the Unknown, First Edition.
Edited by Serge Luryi, Jimmy Xu, and Alexander Zaslavsky.
© 2016 John Wiley & Sons, Inc. Published 2016 by John Wiley & Sons, Inc.

to review the bleeding-edge technologies and their future trends. In this chapter, we address the evolution of Si technology and predict its future directions.

2. Memory – DRAM

Dynamic random access memory (DRAM) has been successfully employed as a working memory for more than 40 years. The DRAM density has doubled every 18 months through technology innovations in the two key elements: the cell capacitor and the cell transistor.[7] In order to guarantee successful operation, the sensing voltage ΔV_{BL} should be higher than the sensing noise after the refresh time as shown in the following equation, where C_S is the cell storage capacitance, C_{BL} is the bit line (BL) loading capacitance, I_L is the leakage current of the storage node, t_{REF} is the refresh time, and ΔV_{MG} is the minimum signal sensing margin:

$$\Delta V_{BL} = \frac{C_S}{C_{BL} + C_S} \left[\frac{V_{CC}}{2} - \frac{\int_0^{t_{REF}} I_L(t)dt}{C_S} \right] \geq \Delta V_{MG}. \tag{1}$$

Scaling technologies have been developed to maintain C_S/C_{BL} ratio, suppress I_L, and minimize the signal sensing noise.[8] The capacitor's aspect ratio has been increased in every generation as shown in Fig. 1.[9] Innovative patterning

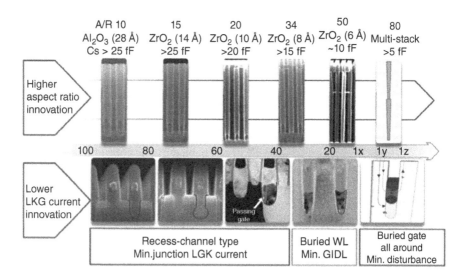

Figure 1. Key steps in the evolution of DRAM technology: (a) evolution of capacitor technology toward higher aspect ratio and lower equivalent oxide thickness; (b) innovation of cell access transistor structure for lower leakage current.

technology was developed to generate small and deep contact holes. In addition, equivalent oxide thickness (EOT) has been decreased by using the higher dielectric constant ZrO_2-based multilayered materials introduced at the 20 nm node. Further on, down at the 10 nm-class node, a novel stacking scheme will be required to conserve the integrity of the capacitor structure.

Cell transistor has evolved toward reducing the leakage current. The junction leakage current had been reduced by the recessed-channel cell transistor up to the 40 nm node. However, in the sub-40 nm regime, gate-induced-drain-leakage (GIDL) current due to the reduced overlap region between the gate and source/ drain (S/D) becomes the major problem, which can be solved by the buried word line (WL) scheme. Down to the 10 nm node, a gate-all-around (GAA) structure may have to be implemented to solve the intercell disturbance.

In DRAM cell capacitor technology, in spite of the increasing aspect ratio and decreasing EOT, cell capacitance has decreased from 25 fF to around 10 fF at the 20 nm node, reducing the signal sensing margin. Therefore, additional inno-vation is required to maintain the sensing voltage of Eq. (1). A key technology for fully compensating the small storage capacitance at the 20 nm node is to lower the bit line loading capacitance. The ratio of the bit line and storage capacitance is the key parameter for successful DRAM operation, since $\Delta V_{BL} \sim (1 + C_{BL}/C_S)$. It can be maintained at the same value as previous generations by introducing an air spacer around the bit line as shown in Fig. 2.

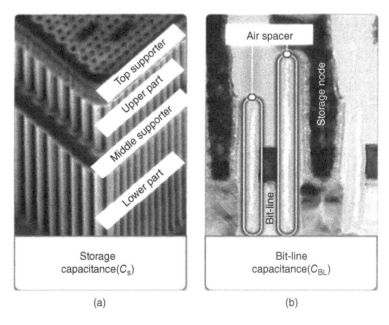

(a)   (b)

Figure 2. Twenty-nanometer DRAM can maintain the same sensing signal by lower-ing the bit line capacitance with an air spacer.

Fabricating the cell capacitor of 1X nm node DRAM is extremely challenging. The required aspect ratio will be near 80 and there is no space for dielectric material. Therefore, it will be necessary to achieve very high step coverage of the high-κ dielectric around the high aspect ratio stack and also develop thinner dielectric material with low enough leakage current. Greater than 95% step coverage will be required at the 10 nm node and the dielectric leakage current will need to be suppressed even at an EOT of 5 Å by workfunction engineering of the high-κ dielectric material.

At the device level, we expect that DRAM will be scaled down to 10 nm or less through the continuous innovations of cell storage capacitor and cell array transistor. Expected challenges at the 10 nm node are fabrication cost and manufacturability issues, such as overlay and uniformity. These issues will be resolved by novel integration schemes as well as improved process technology.

3. Memory – NAND

Over the last decade, physical dimensions of planar NAND flash memory have been successfully scaled down, so that even 1X nm technology node has been commercialized in recent years. NAND flash memory has played an important role in the data-driven world by providing higher bit density, lower cost, and smaller form factor indispensable in this mobile age.[9]

For NAND flash memory, the key scaling challenges involve overcoming the degradation of the V_T window, minimizing the interference coupling, and guaranteeing data retention for reliable device operation as shown in Fig. 3,[10, 11] while maintaining a wide process window for mass production. The V_T window for program and erase operation is decreased by short channel effects (SCEs), whereas the interference coupling increases since the electrostatic coupling becomes severe as the distance between adjacent cells is reduced. In addition, the number of stored electrons decreases rapidly as the device shrinks. Thus, only several tens of electrons are stored for 2X nm regime, which makes it difficult to satisfy the data retention requirements. At each node, those barriers have been overcome by significant technological breakthroughs in structures, materials, and intelligent operation algorithms implemented by circuits.

Up to the present technology node, the SCEs have been solved successfully by field-induced junction formation technology,[12] as well as continuous scaling of the tunnel oxide and oxide-nitride-oxide (ONO) layer thicknesses. The interference coupling has been suppressed by the timely adoption of the gate air-gap processes and innovative circuit technologies, including parallel programming, shadow programming, and extended error correction codes (ECCs).[13] However, the reduction in the number of stored electrons and the increases in the interference coupling are great obstacles to scaling beyond the 1X nm node. Moreover, uncontrollably small process windows pose another major challenge. For those

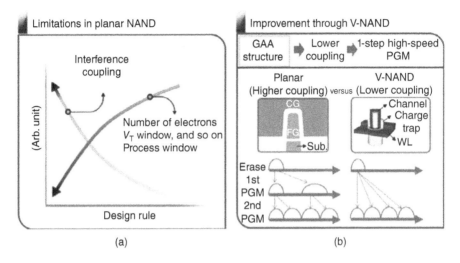

Figure 3. Scaling limitations for planar NAND flash memory (a) and an illustration of unit cell structures for planar and 3D V-NAND and corresponding programming schemes (b).

reasons, a paradigm shift from planar to 3D technology[7, 14, 15] may become necessary for further progress.

In terms of the memory cell itself, 3D vertical NAND (V-NAND) has three main differences compared to the planar NAND: a vertical channel with a GAA structure, a thin film transistor (TFT) with a poly-Si channel, and a charge trapping storage layer instead of the poly-Si floating gate, as shown in Fig. 3.[16] The GAA structure helps to suppress SCEs, as does the thin poly-Si channel, which leads to stronger gate control. The interference coupling along the word line (WL) direction is greatly reduced by increasing the distance between the adjacent cells, while the interference coupling along the bit line (BL) direction is eliminated by using a GAA structure in which each BL is completely shielded by the gate metal layer. The minimum feature size of the cell has been increased by about a factor of 10, resulting in an increase of the number of stored electrons.

Consequently, 3D V-NAND with negligible interference coupling allows single-step rather than dual-step programming, resulting in faster programming speed and lower power consumption, which makes it more competitive compared with planar NAND.[17, 18] In addition to the faster programming, 3D V-NAND with enhanced V_T window can push the limit of multilevel cell technology to 4 bits as shown in Fig. 4.

As illustrated in Fig. 4, the stacking of 3D cells in the vertical direction is another key advantage to overcome the small process windows beyond the 1X nm node. Since the physical design rule of the planar NAND should be decreased to achieve a higher bit density, the patterning process is getting more complex and difficult. The current double patterning technology (DPT)

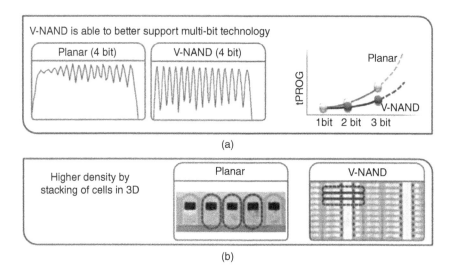

Figure 4. The V_T distributions of 4-bit cell operation and the programming speed (t_{PROG}) for planar and 3D V-NAND cells (a); vertical cross-sectional view of planar and 3D-VNAND cell strings (b).

has reached its limits and thus the more complex and expensive quadruple patterning technology (QPT) has become an essential process. On the other hand, 3D V-NAND relieves the photolithographic limit since the number of stacked layers determines the bit density.

Despite the advantages of 3D V-NAND over the conventional planar NAND, there was considerable doubt in the memory industry regarding mass production of 3D V-NAND because of several challenges, such as the fabrication of high-aspect ratio channel holes, the control of mechanical stress in thick cell stacks, the uniformity of dielectric layer deposition in deep contact holes, and so on. However, with advances in dry etching technology and materials have allowed 3D V-NAND to be commercialized, as shown in Fig. 5.[19] The 3D V-NAND with 32 stacked layers gives twice the density of a 16 nm planar NAND and, by increasing the number of stacked layers, it is expected that NAND flash memory scaling with 3D V-NAND technology will continue beyond 10 nm into the 1 Tb era.

4. Logic technology

Logic complementary metal-oxide semiconductor (CMOS) technology has successfully evolved over many decades, satisfying the market demand for high performance, low power, and low cost-per-function. This has been achieved through the "silicon shrinking technology" – the innovation of processes,

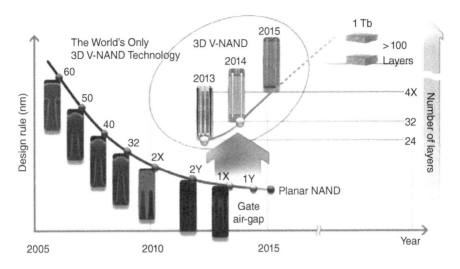

Figure 5. Samsung NAND flash memory technology outlook for planar NAND and 3D V-NAND devices.

materials, and device structures.[8] In a data-driven world, more emphasis will be placed on the power-constrained performance because only a limited amount of power can be available per chip. Total power dissipated by a CMOS circuit consists of two components, dynamic power P_{DYN} and static power P_{STAT}, as illustrated in the equation:

$$P_{TOT} = P_{DYN} + P_{STAT} = CV_{DD}^2 \times f \times \alpha + V_{DD} \times I_L, \qquad (2)$$

where $C, f,$ and α are total capacitance, operating frequency, and activity factor, respectively. Thus, for high performance with low power consumption, it is indispensable to scale the supply voltage V_{DD} without sacrificing the current drive I_{ON} or increasing the leakage current I_L.

The most critical challenges in conventional CMOS shrinkage down to 20 nm were the control of SCEs and the enhancement of carrier mobility. The SCEs have been suppressed by a thin gate oxide and shallow S/D junction depth. Both shorter gate length L_G and thinner gate oxide have continuously provided an increase in I_{ON} and a decrease in the total capacitance, thus faster switching speed despite a lower V_{DD}. At the same time, more CMOS transistors can be integrated onto a single chip for more functionality. However, the exponential increase in the gate leakage current puts a physical limitation on shrinking the SiO_2 gate oxide thickness to ~1.2 nm.[20] This limit has been overcome through the high-κ gate dielectric and metal gate (HKMG) process, shrinking the EOT below 1.0 nm without aggravating the gate leakage current.[21] On the other hand, carrier mobility can be enhanced by applying an appropriate mechanical stress to a silicon channel. Embedding SiGe into S/D region of PMOS and high stress SiN capping film over NMOS can provide much higher hole and electron mobilities at low V_{DD}.[22]

From the 14 nm node onward, it is extremely challenging to reduce V_T and its variability without sacrificing low subthreshold leakage current, which is key for low power consumption. Innovative device structures, such as 3D FinFETs or trigate devices, have successfully suppressed SCEs, and improved drain induced barrier lowering (DIBL) and subthreshold swing (SS) with much better control of channel potential. This made it possible to reduce V_{DD} and save over 30% of the dynamic power consumption, see Fig. 6.[23] Furthermore, a lightly doped channel provides the improved immunity to V_T variability thanks to suppressed random dopant fluctuation (RDF). However, there are some challenges to overcome before mass-production: low-cost patterning process, complexity in design optimization due to a quantized number of fins, and V_T tuning using gate workfunctions.[24] Cost-effective multiple patterning technology (MPT) may be an interim solution before extreme ultraviolet (EUV) lithography becomes cost-competitive.

For the 7 nm technology node and beyond, novel materials and device structures will be needed to meet market needs for low power consumption, low cost-per-function, small form-factor, and advanced functions, such as energy harvesting and communication – see Fig. 7. GAA or 1D nanowire (NW) transistors can provide a nearly ideal (60 mV/dec) electrostatic control over the channel potential.[25, 26] These device structures may be combined with alternative channel materials to enhance carrier mobility. Thus far, InGaAs and Ge are promising candidate materials for NMOS and PMOS, respectively.[27–29] However, there are many challenges to overcome, the main ones being the precise control of

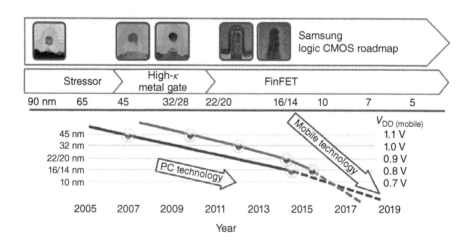

Figure 6. Samsung logic CMOS technology outlook for the supply voltage V_{DD}, innovative materials, and device structures.

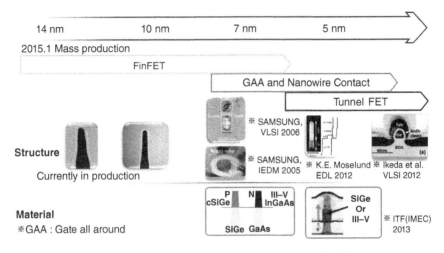

Figure 7. Device structures and materials for future logic CMOS technologies.

nanowire thickness and its interface for high performance and reliability, and the cost-effective integration of III-V channel materials onto a single silicon substrate. Also, tunneling field-effect transistors (TFETs) shows a great potential to provide an extremely low supply voltage operation with a steeper subthreshold swing of <60 mV/dec. However, its small on-current drive capability should be overcome; one of the promising approaches is the combination of perpendicular and parallel electric fields at the tunneling junction.[30, 31]

Logic CMOS technology is entering the sub-10 nm regime, emphasizing the low power consumption with high performance. Further downscaling will continue through the combination of innovative processes, materials, and device structures.

5. CMOS image sensors

Recently, CISs have gained widespread adoption by replacing charge-coupled device (CCDs). As the mobile market is rapidly expanding, it is very important to scale down the CIS pixel size for satisfying the needs of high resolution and small form factor. In the CIS development trend, the pixel size has been continuously shrunk toward 1.0 µm. The key technological challenge to going down to submicron pixels is to maintain a high signal-to-noise ratio (SNR) in the reduced pixel size.[32, 33] The SNR is mainly determined by pixel sensitivity and crosstalk between the pixels, which both degrade as the pixel size decreases. Therefore, technology innovation is focused on maintaining a large SNR with high pixel sensitivity and low crosstalk.

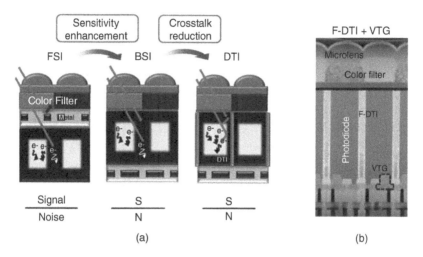

Figure 8. (a) Back-side illumination maximizes sensitivity by eliminating optical obstacles to incident light, whereas deep trench isolation (DTI) reduces crosstalk; (b) ISOCELL structure, where DTI is combined with a vertical transfer gate.

The sensitivity is decreased as the pixel size scales down because the number of photons is reduced. In order to maintain high sensitivity, the pixel illumination technology has migrated from front-side illumination (FSI) to back-side illumination (BSI).[34] Since there is no optical obstacle for incident light from microlens to photodiode in the BSI technology, as shown in Fig. 8(a), the sensitivity is maximized. Furthermore, BSI sensors can be manufactured using a bonding technology that allows for 3D integration. For example, stacked image sensors can be manufactured by bonding separately processed pixel and circuit wafers. Using this stacked sensor, chip size can be effectively shrunk and special processes can be used on the pixel wafer, thus enhancing the chip performance.

The undesired crosstalk becomes more severe as the pixel size shrinks due to the diffraction of incident light. The nondiagonal elements of color conversion matrix (CCM) become larger as the crosstalk increases, worsening the SNR of final image, as illustrated in Fig. 9. Therefore, it is very important to minimize the

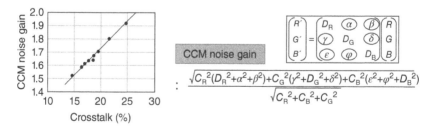

Figure 9. Crosstalk and color conversion matrix noise gain.

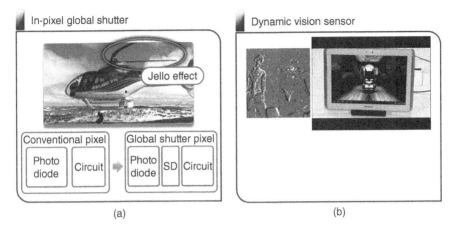

Figure 10. Advanced pixel technologies: In-pixel global shutter (a) and dynamic vision sensor (b).

crosstalk between the pixels to enhance SNR at both low and high illumination. In order to reduce the crosstalk, physically complete isolation between neighboring pixels can be introduced, for example, via deep trench isolation (DTI). Recently, a combination of DTI structure with a vertical transfer gate, known as the ISOCELL process, was introduced for better color and image quality, see Fig. 8(b). The ISOCELL technology will allow us to overcome the 1 μm pixel barrier of the CIS technology for further pixel downscaling.

Novel pixel technologies can add new value to CISs. Among them, one of most beneficial techniques is the global shutter sensor, whose main function is to eliminate the so-called "jello" effect of blurring in images of fast-moving objects or when a camera vibrates as shown in Fig. 10. This distorted image occurs because of time lag in sequential gathering of the information from the pixels. With the in-pixel global shutter, light is captured by all pixels simultaneously, eradicating the time lag and the jello effect. This is achieved by temporarily storing photoelectrons in each photodiode by adding a storage element to each pixel.

Another useful innovation is the DVS, which mimics the human photoreceptor cells that respond only changes in the scene. Since DVS has an event-based asynchronous operation rather than the traditional synchronized and frame-based one, we can detect relatively faster movements. This technology will be useful for mobile, surveillance, and obstacle detection for drones.

6.   Packaging technology

As recent IT paradigm shifts from PC to mobile applications, integration of devices becomes more important in electronic systems. In the previous PC era, semiconductor chips were massively produced in a certain standardized form.

Primary requirements for semiconductor packaging were the simple and discrete forms and low cost.[35] This trend has been completely changed by the advent of smartphones, in which portability and small form factor are very critical characteristics for packaging. Therefore, 3D packaging based on die and package stacking has become a key technology.[36] There are two major components in 3D packaging: TSV-based die stacking and package stacking.

Originally, TSV technology was adopted in server-specific DDR4 DRAM memory units. Cloud servers require high-speed, high-capacity, and low-power DRAM memory. It is difficult to satisfy these requirements with conventional wire bonding because of the restricted I/O density and long interconnect lines. Compared to conventional package stacks, TSVs increase the I/O density by a factor of 17 and shorten interconnect length by a factor of 40. Consequently, TSVs can have high performance and high capacity with low power consumption and small form factor.[37] One of the possible concerns in TSVs is excessive loading, but it is not an issue because only the master chip is required to communicate with the controller regardless of the number of stacks, as shown in Fig. 11. Evaluation of an RDIMM unit with TSV technology showed a 24% reduction in power consumption compared to a wire-bonded LRDIMM unit as shown in Fig. 12.[9]

The other emerging memory technology using TSVs is a high bandwidth memory (HBM), which aims at high bandwidth DRAM applications such as high-performance computing, networks, and graphics. It is possible to have high bandwidth in HBM through wide I/O interconnects, which can be achieved by fine pitch TSVs. TSVs allow up to 1024 I/O's per chip, while the number of I/O's per chip is limited to 32 in conventional wire bonding technology. As a result,

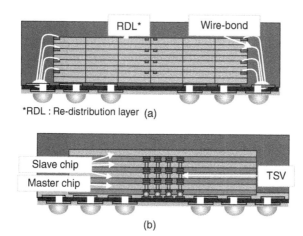

Figure 11. Comparison between a wire-bonding package and a TSV package.

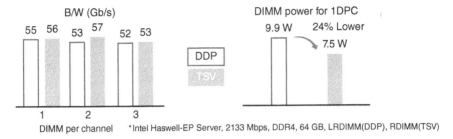

Figure 12. Improvements in performance and power consumption with TSV.

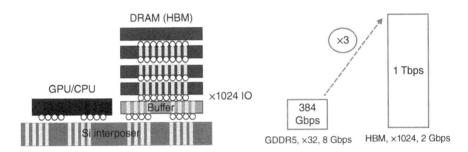

Figure 13. HBM system configuration and increase in bandwidth due to TSV.

bandwidth of over 1 Tb/s can be achieved with relatively low I/O speed as shown in Fig. 13.[38]

Package stacking, also known as package-on-package (PoP), can be used to achieve higher system-level integration. Previously, devices were packaged individually and mounted onto a board. However, Internet of things (IoT) and wearable devices are encouraging the convergence of IT technologies, requiring the integration of various devices and passive elements onto a single small package platform.[39] Early PoP versions involved integrated application processors (APs) and mobile DRAM stacks. Embedded package-on-packages (ePoPs) were an extension of this, where the surrounding embedded memory and passive elements were included into the integrated system in response to demand for smaller form factors, as shown in Fig. 14. The DRAM, NAND, buffer chip, and passive elements were stacked onto the largest component of the AP. This resulted in an overall package footprint reduction of over 60%.

In the near future, we will need to integrate numerous semiconductor devices in a single package – such as AP, memory, CIS, MEMS, and so on. This system integration in a package is expected to provide a huge size reduction and great performance enhancement for future IoT and wearable devices. The major challenges will be focused on improving thermal performance, increasing I/O density,

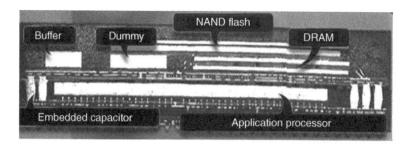

Figure 14. Embedded PoP structure. Stacking of DRAM, NAND, and buffer chips onto an application processor reduces the footprint by 60%.

and designing for power noise reduction. There are current research efforts in all of these areas.

7. Silicon photonics technology

Data center traffic has been and will be continuously increasing as people use more connected devices. There are forecasts of Internet-connected devices increasing from 4.9 billion in 2014 to 25 billion in 2020. The global data center Internet protocol (IP) traffic forecasts estimate a threefold increase from 2013 to 2018.[40] In order to meet the bandwidth demand, data centers will expand the deployment of optical interconnects. Infonetics Research expects that 100G optical modules will make up 50% of data center transmission capacity by 2019.[41]

Research on Si photonics is ongoing to reduce the cost of optical interconnects and accelerate the replacement of copper wire. Silicon photonics will allow optical devices to be fabricated at low cost using standard semiconductor technologies and integrated with microelectronic devices.[42, 43] Figure 15 summarizes the recent research activities on Si photonics at Samsung.[44] Photonic devices are fabricated using locally crystallized silicon on a bulk Si substrate that allows lower material cost and process compatibility with CMOS transistors. Lasers for light sources are formed with III/V gain layers that are wafer-bonded to Si-photonic die, shown in Fig. 15(b). Bonded lasers are expected to provide light sources at lower cost compared to external laser diode modules. Integrated microlenses on the top surface enable passive alignment and correspondingly cheaper package cost, shown in Fig. 15(c). Figure 15(d) presents major device building blocks made on a bulk Si platform, including modulators, photodetectors, waveguides, and vertical grating couplers.

It is expected that optical interconnection technology using Si photonic devices will be widely used in data-driven world. High bandwidth capability of optical interconnection technology will be essential in data centers, high-bandwidth consumer electronics, and so on.

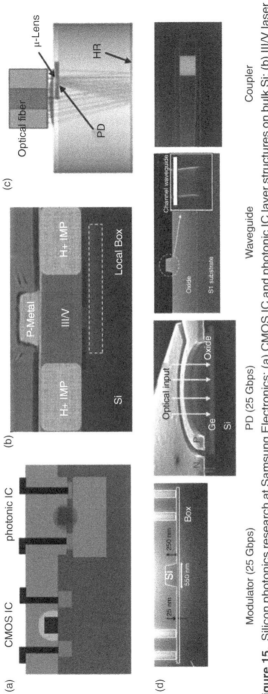

Figure 15. Silicon photonics research at Samsung Electronics: (a) CMOS IC and photonic IC layer structures on bulk Si; (b) III/V laser bonded to Si photonic IC; (c) light coupling from fiber to photodetector using a microlens; (d) major Si photonic building blocks.

8. Concluding remarks

In this chapter, the evolution and prospects for the future Si technologies have been reviewed. The Si-based memory and logic technologies have been success-fully scaled down to 1X nm node. From the device point of view, all of these Si devices face no fundamental physical limitations down to sub-10 nm nodes. Practically, fabrication cost and manufacturability are of increasing concern. Patterning difficulties, as well as tight overlay and uniformity tolerances, will increase fabrication costs. However, these difficulties will be eventually overcome by innovative integration schemes, new materials, and so on.

Along with the individual technology evolution, the convergence of vari-ous technologies will generate new areas of functional diversification. Packaging technologies, including TSV, will not only improve the performance but also merge various functions of many different devices. Integrated CISs will collect more useful data that will be processed, analyzed, and stored by advanced Si technologies. Huge amounts of data will be transferred between the devices, gen-erating heavy IP traffic – an issue that can be resolved by novel interconnection technology such as optical interconnects. It is expected that all of these advanced Si technologies will open the era of a truly data-driven world.

Acknowledgments

We would like to thank Duckhyung Lee, Jungchak Ahn, Sayoon Kang, Tae-Je Cho, Sujin Ahn, Kyungho Ha, and Daewon Ha for the helpful technical discus-sions.

References

1. IDC, "The digital universe of opportunities: rich data and the increasing value of the Internet of things," 2014.
2. D. Evans, *The Internet of Things: How the Next Evolution of the Internet is Changing Everything*, Cisco Systems, 2011.
3. A. McAfee and E. Brynjolfsson, "What makes company good at IT?," *Wall St. J.* (2011).
4. M. Guild and T. Danaher, "Big data comes to the farm," *Financial Sense* (2014), www.financialsense.com/contributors/guild/big-data-farm.
5. Editorial, "Data driven healthcare," *MIT Technol. Rev.* (2014).
6. M. Lorenz, "Data driven journalism: What is there to learn?" Presented at *Innovation Journalism Conf.*, Stanford (2010), www.technologyreview.com/business-report/data-driven-health-care.
7. K. Kim, U.-I. Chung, Y. Park, *et al.*, "Extending the DRAM and flash mem-ory technologies to 10 nm and beyond," *Proc. SPIE* **8326**, 832605 (2012).

8. K. Kim, "From the future Si technology perspective: Challenges and opportunities," *Tech. Dig. IEDM* (2010), pp. 1–9.
9. K. Kim, "Silicon technologies and solutions for the data-driven world," *Proc. ISSCC* (2015), pp. 1–7.
10. Y. Park and J. Lee, "Device considerations of planar NAND flash memory for extending towards sub-20 nm regime," *Proc. Intern. Memory Workshop* (2013) pp. 1–4.
11. Y. Park, J. Lee, S. S. Cho, G. Jin, and E. Jung, "Scaling and reliability of NAND flash devices," *Proc. Intern. Reliability Phys. Symp.* (2014), no. 6860599.
12. J. Seo, K. Han, T. Youn, *et al.*, "Highly reliable M1X MLC NAND flash memory cell with novel active air-gap and *p*+ poly process integration technologies," *Tech. Dig. IEDM* (2013), no. 6724554.
13. K.-T. Park, M. Kang, D. Kim, *et al.*, "A zeroing cell-to-cell interference page architecture with temporary LSB storing and parallel MSB program scheme for MLC NAND flash memories," *IEEE J. Solid State Circ.* **43**, 919–927 (2008).
14. H. Tanaka, M. Kido, K. Yahashi, *et al.*, "Bit cost scalable technology with punch and plug process for ultra high density flash memory," *Proc. VLSI Symp.* (2007), pp. 14–15.
15. J. Jang, H.-S. Kim, W. Cho, *et al.*, "Vertical cell array using TCAT (terabit cell array transistor) technology for ultra high density NAND flash memory," *Proc. VLSI Symp.* (2009), pp. 192–193.
16. J. Choi and K. S. Seol, "3D approaches for non-volatile memory," *Proc. VLSI Symp.* (2011), pp. 178–179.
17. K.-T. Park, J.-M. Han, D. Kim, *et al.*, "Three-dimensional 128 Gb MLC vertical NAND flash-memory with 24-WL stacked layers and 50 MB/s high-speed programming," *Tech. Dig. ISSCC* (2014), pp. 334–335.
18. J.-W. Im, W.-P. Jeong, D.-H. Kim, *et al.*, "A 128 Gb 3b/cell V-NAND flash memory with 1 Gb/s I/O rate," *Proc. ISSCC* (2015), article 7.2.
19. J. Elliott and E. S. Jung, "Ushering in the 3D memory era with V-NAND," Presented at *Flash Memory Summit* (2013).
20. T. Skotnicki, C. Fenouillet-Beranger, C. Gallon, *et al.*, "Innovative materials, devices, and CMOS technologies for low-power mobile multimedia," *IEEE Trans. Electron Devices* **55**, 96 (2008).
21. K. Mistry, C. Allen, C. Auth, *et al.*, "A 45 nm logic technology with high-κ+ metal gate transistors, strained silicon, 9 Cu interconnect layers, 193 nm dry patterning, and 100% Pb-free packaging," *Tech. Dig. IEDM* (2007), pp. 247–250.
22. T. Ghani, M. Armstrong, C. Auth, *et al.*, "A 90 nm high volume manufacturing logic technology featuring novel 45 nm gate length strained silicon CMOS transistors," *Tech. Dig. IEDM* (2003), pp. 978–980.
23. C. Auth, C. Allen, A. Blattner, *et al.*, "A 22 nm high performance and low-power CMOS technology featuring fully-depleted tri-gate transistors,

self-aligned contacts and high density MIM capacitors," *Proc. VLSI Symp.* (2012), pp. 131–132.

24. H. Kawasaki, V. S. Basker, T. Yamashita, *et al.*, "Challenges and solutions of finFET integration in an SRAM cell and a logic circuit for 22 nm node and beyond," *Tech. Dig. IEDM* (2009), no. 5424366.

25. J. P. Colinge, M. H. Gao, A. Romano-Rodriguez, H. Maes, and C. Claeys, "Silicon-on-insulator 'gate-all-around device'," *Tech. Dig. IEDM* (1990), pp. 595–598.

26. S. D. Suk, S.-Y. Lee, S.-M. Kim, *et al.*, "High performance 5 nm radius twin silicon nanowire MOSFET (TSNWFET): Fabrication on bulk Si wafer, characteristics, and reliability," *Tech. Dig. IEDM* (2005), pp. 717–720.

27. M. Radosavljevic, B. Chu-Kung, S. Corcoran, *et al.*, "Advanced high-κ gate dielectric for high-performance short-channel $In_{0.7}Ga_{0.3}As$ quantum well field effect transistors on silicon substrate for low power logic applications," *Tech. Dig. IEDM* (2009), no. 5424361.

28. O. Weber, Y. Bogumilowicz, T. Ernst, *et al.*, "Strained Si and Ge MOSFETs with high-κ/metal gate stack for high mobility dual channel CMOS," *Tech. Dig. IEDM* (2005), pp. 137–140.

29. K. Ikeda, M. Ono, D. Kosemura, *et al.*, "High-mobility and low-parasitic resistance characteristics in strained Ge nanowire pMOSFETs with metal source/drain structure formed by doping-free processes," *Proc. VLSI Symp.* (2012), pp. 165–166.

30. Y. Morita, T. Mori, S. Migita, *et al.*, "Synthetic electric field tunnel FETs: Drain current multiplication demonstrated by wrapped gate electrode around ultrathin epitaxial channel," *Proc. VLSI Symp.* (2013), pp. 236–237.

31. K. E. Moselund, H. Schmid, C. Bessire, M. T. Bjork, H. Ghoneim, and H. Riel, "InAs–Si nanowire heterojunction tunnel FETs,". *IEEE Electron Device Lett.* **33**, 1453 (2012).

32. J. C. Ahn, C.-R. Moon, B. Lee, *et al.*, "Advanced image sensor technology for pixel scaling down toward 1.0 μm," *Tech. Dig. IEDM* (2008), no. 4796671.

33. H. Wakabayashi, K. Yamaguchi, M. Okano, *et al.*, "A 1/2.3-inch 10.3 Mpixel 50 frame/s back-illuminated CMOS image sensor," *Tech. Dig. ISSCC* (2010), pp. 410–411.

34. K. Lee, J. C. Ahn, B. Kim, *et al.* "SNR performance comparison of 1.4 μm pixel: FSI, light-guide, and BSI," *Intern. Image Sensor Workshop* (2011), pp. 9–11.

35. R. R. Tummala, "Packaging: Past, present, and future," *Proc. 6th Intern. Conf. Electronic Packaging Technol.* (2005).

36. R. R. Tummala, *System on Package: Miniaturization of the Entire System*, New York: McGraw Hill, 2008.

37. J. Burns, "TSV-based 3D integration," chapter in: A. Papanikolaou, D. Soudris, R. Radojcic, eds., *Three-Dimensional System Integration*, New York: Springer, 2011, pp. 13–32.

38. H. Goto, "Memory evolution for next generation GPU," PC Watch (2015).

39. G. A. Riley, "How 3D is stacking-up," *Adv. Packag.* **17**, 16–18 (2008).
40. Cisco Systems, "Cisco global cloud index: Forecast and methodology, 2013–2018," 2014.
41. Infonetics Research, "10G/40G/100G data center optics biannual market size and forecast," April 2015.
42. Editorial, "Simply silicon," *Nature Photonics* **4**, 491 (2010).
43. M. Streshinsky, R. Ding, Y. Liu, *et al.*, "The road to affordable, large-scale silicon photonics," *Opt. Photonics News* **24**, 32–39 (2013).
44. H. Byun, J. Bok, K. Cho, *et al.*, "Bulk-Si photonics technology for DRAM interface," *Photonics Res.* **2**, A25–A33 (2014).

1.2

How Lithography Enables Moore's Law

J. P. H. Benschop
ASML, Veldhoven, The Netherlands

1. Introduction

Over the last 50 years, Moore's Law has set the pace for the electronics industry, delivering increasing computing capabilities at stable cost. This was driven by the steady pace of the increase of components in an integrated circuit (IC), which has to a large extent been enabled by optical lithography printing increasingly smaller electronic features on a silicon wafer.

We will quantify what the contribution of lithography to Moore's Law has been in the past and then discuss the future lithography options to extend Moore's Law into the future.

2. Moore's Law and the contribution of lithography

In 1965, Moore observed that the number of components in an IC or (micro)chip, doubled every year[1] and he predicted this would continue. Due to the high fixed costs to produce a single chip, increasing density would lead to lower cost per component. This empirical observation of economics became later known as "Moore's Law."

In a later paper, Moore quantified the contributions to a doubling of components per year[2]: die size reduction (~30%), dimension reduction (~30%), and "device and circuit cleverness" (~40%). Lithography, enabling the dimension reduction, thus contributed to a doubling of components per chip every 3.3 years. In that same paper Moore predicted that the components per chip would double every 2 years, compared with his previous prediction of 1 year. This revised "shrink" rate has since continued and has set the pace for the semiconductor industry over the past 40 years.

In a 1995 paper[3] by Moore, the contribution of lithography to the increased density was shown. From this graph, it can be concluded that lithography leads to a doubling of components per square centimeter every 2.8 years. Hence the contribution of lithography to Moore's Law has been fairly constant from the beginning, see Fig. 1.

Future Trends in Microelectronics: Journey into the Unknown, First Edition.
Edited by Serge Luryi, Jimmy Xu, and Alexander Zaslavsky.
© 2016 John Wiley & Sons, Inc. Published 2016 by John Wiley & Sons, Inc.

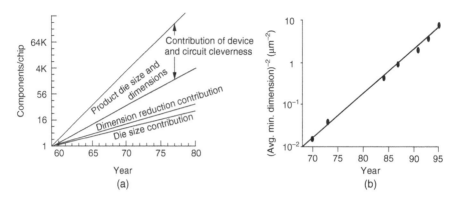

(a)   (b)

Figure 1. Contribution of lithography to Moore's Law: (a) 1960–1975 "dimensional reduction" doubles component density every 3.3 years; (b) 1970–1995 "minimum dimension" doubles component density every 2.8 years.

3. Lithography technology: past and present

Optical lithography has always been the workhorse for integrated circuit manufacturing. Since the late 1970s, optical projection has been used where a mask (containing the pattern to be printed) is demagnified and repeatedly imaged on a silicon wafer.

The resolution of the optical lithography tool is usually expressed by

$$R = k_1 \lambda_0 / \mathrm{NA}, \tag{1}$$

where R is the half pitch, k_1 is a process factor, λ_0 is the wavelength in vacuum, and NA is the numerical aperture, given by the product of the refractive index n and $\sin(\alpha)$ where α is the opening angle: $\mathrm{NA} = n \sin(\alpha)$. Note that the minimum k_1 for a single exposure equals 0.25. This corresponds to two plane waves interfering from opposite sites, as illustrated in Fig. 2.

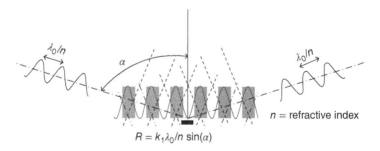

$$R = k_1 \lambda_0 / n \sin(\alpha)$$

Figure 2. Resolution is defined as the half pitch of a dense pattern of repeated lines and spaces. Minimum resolution R, resulting from two waves interfering from the opposite sides is given by $0.25 \lambda_0 / \mathrm{NA}$.

To improve the resolution, and thus enable Moore's Law, all three parameters in Eq. (1) have been improved over time. Thus, for the process factor k_1 the main developments have been as follows:

- Pushing the single exposure as close as possible to physical limit of 0.25 using image enhancement[4] (e.g., source mask optimization[5]) and feedback loops (metrology to measure printed patterns feeding back to many actuators in the lithography tool, such as focus, dose, and illuminator mode,[6] bring IC printing closer to the physical limit);

- Frequency doubling by multiple exposure[7] or spacer technologies[8] that enable a $k_1 < 0.25$.[9]

Progress in the wavelength and numerical aperture parameters is summarized in Table 1. We observe that over the years the wavelength λ_0 has decreased from 436 nm (g-line of mercury) to 365 nm (i-line of mercury) to 248 nm (KrF laser) to 193 nm (ArF laser). It also shows that the NA was increased from 0.28 to 1.35 (using a water-based "immersion"-extension of the lens to replace the air between the lens and the wafer). The most recent step to extreme ultraviolet (EUV) wavelength (EUV of 13.5 nm using a plasma source) is so significant (15× improvement) that the NA could be relaxed to 0.33 from 1.35.

Table 1 shows how decreases in wavelength and improvements in NA, in combination with a decrease in k_1, have led to an increase of the pixels per field by a factor of over 10^5 over the last decades.

	David Mann (GCA) 4800	ASML/ 40	ASML/ 300	ASML/ 1100	ASML 19×0i	ASML 3100	ASML 3300
Year of 1st prototype	1975	1987	1995	2000	2007	2009	2012
Wavelength (nm)	436	365	248	193	193	13.5	13.5
NA	0.28	0.4	0.57	0.75	1.35	0.25	0.33
k_1	0.90	0.77	0.57	0.39	0.27	0.50	0.44
CD (nm)	1400	700	250	100	38	27	18
Step/scan field (mm)	10×10	14×14	22×27	26×33	26×33	26×33	26×33
Pixels per field	50×10^6	400×10^6	10×10^9	86×10^9	600×10^9	1.2×10^{12}	2.6×10^{12}
Weight (kg)	2	20	250	400	1080	700	1600

Table 1. The evolution of wavelength, NA and k_1 over the years.

4.   Lithography technology: future

Here we review the leading lithography candidates to extend Moore's Law.

•   *Multiple patterning immersion lithography*

There is no fundamental limit to frequency multiplication. Octuple patterning has resulted in 5.5 nm lines,[10] 35× smaller than the wavelength of the light used to print the pattern. There are, however, practical limitations:[11] increased process complexity leads to increased cost and cycle time and there is the overlay/edge placement challenge.[12]

•   *Directed self-assembly*

Block copolymers are two different types of polymer chains (e.g., PS and PMMA) connected at one end. The polymer blocks phase-separate as a result of the chemical differences between the polymers and self-assemble in regular patterns at the molecular scale, see Fig. 3. The size (from a few to tens of nanometer) and shape (cylinders, lamellae, etc.) depend on the molecular weight and composition of the copolymer.

Lithography or any other top-down prepatterning technique is needed to induce a long-range ordering.[13–15] Directed self-assembly (DSA) should therefore be seen as a technology complimentary to optical lithography. It enables a smaller feature size[16] and/or better line edge roughness. It has been demonstrated that a 71–83 nm diameter contact hole printed with immersion lithography can

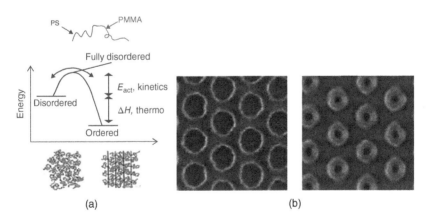

(a)   (b)

Figure 3. Principle and practical results for directed self-assembly (DSA). (a) Ordered state energetically more favorable; (b) SEM picture of contact holes created by ArFi before (left) and after (right) DSA.

be shrunk to a 29 nm diameter hole while reducing the three-sigma uniformity of the hole diameter from 3.5 nm (immersion) to 1.3 nm (DSA).[17]

The critical issues with DSA are defect density[18] (compared to immersion lithography standard of tens per square centimeter) and placement error[17] (which adds ~2 nm additional error).

- *Electron beam*

Electron beam lithography offers excellent resolution but at very low productivity. It has been used for many years to write the masks used in optical lithography.

Multiple direct write e-beam lithography concepts are being pursued. In order to increase productivity, various concepts, leading to many beams used in parallel, are being pursued – see Fig. 4 for progress in recent years. These include the following:

- MAPPER,[19] a 5 kV raster wafer writer with a single source, multiple lenses (target 13,000), and multiple spots per lens (49 realized);

- IMS,[20] 50 kV raster mask writer, single source, many spots (262,144 realized) in single lens field;

- multibeam[21] wafer writer, vector scan, multiple column, one beam per column.

These concepts each have their own pros and cons. However, a critical factor for any e-beam system is the extendibility. Next to the fact that shot noise will mandate an increased dose to reach smaller resolution (a factor e-beam has in

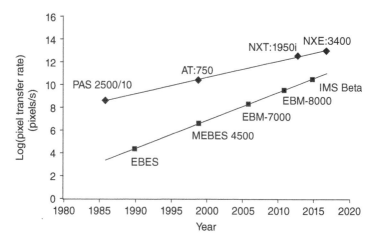

Figure 4. Pixel transfer rate of optical and e-beam lithography tools over the years.

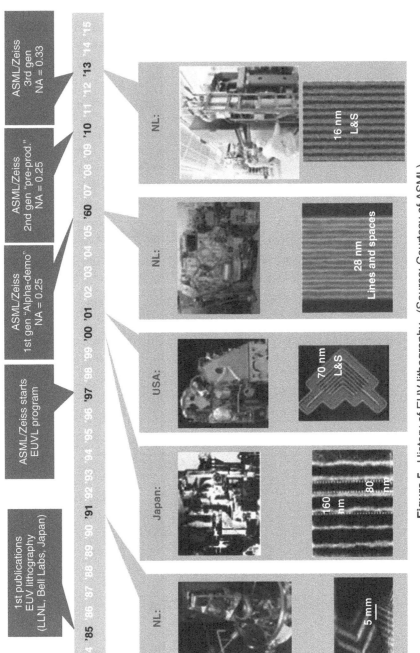

Figure 5. History of EUV lithography. (Source: Courtesy of ASML).

common with optical photon-based lithography), electron–electron interaction makes it extremely difficult, if not impossible, to maintain productivity for smaller resolutions. Without concept changes, the estimated productivity decrease can be as large as a factor eight per node.[22]

• *Imprint*

The "step and flash" imprint technology is being pursued for use in IC lithography.

A master template is written with e-beam, inspected, and repaired. Imprint is used to replicate this master; the replica templates are used to imprint a wafer.[23]

The 1X mask is inherently difficult to write, inspect, and repair. The fact that it is repeatedly used in contact with the wafer makes lifetime a critical issue.

Another critical issue is overlay. In particular, mix and match to optical systems is more difficult since deforming the mask has limited correction capability compared to the many parameters officered in an optical lens.

Despite significant progress over the recent years,[24] the achievable numbers for defect density, contamination, and overlay need significant improvement before this technology can be used in volume IC production (see Table 2 in Section 5).

• *Extreme ultraviolet (EUV)*

EUV lithography uses 13.5 nm light and multilayer reflective optics. It has been intensively pursued since the 1980s, as illustrated in Fig. 5.[28–30]

Over the last decade, several generations of full-field EUV scanners have been shipped.[31–33] The latest NA = 0.33 EUV scanner has demonstrated excellent imaging and overlay,[26] see Fig. 6. Throughput of 1000 wafers/day has been

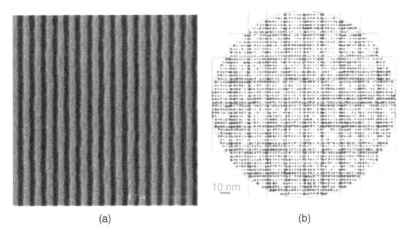

(a)   (b)

Figure 6. State-of-the-art imaging and overlay results: 13 nm dense lines and spaces (a); <2 nm overlay $\mu + 3\sigma$ in a single EUV tool (b).

	ArFi[25]	EUVL[26]	DSA[17,18]	e-Beam[19,27]	Imprint[23]
Resolution	38 nm L/S 19 nm SADP 10 nm SAQP	13 nm L/S at 0.33NA 9 nm SADP	14 nm L/S with PS-PMMA <10 nm L/S with high chi materials	16 nm L/S trade-offs in dose and throughput	Can reproduce e-beam resolution: <15 nm L/S
Overlay (mixed machine)	<2.3 nm Mixed machines	<2 nm Single machine	2 nm added to ArFi overlay (RSS)	10 nm	4.8 nm Single machine
Throughput	>250 wafers/hour	Current: ~55 wafers/hour at 80 W Target: 125 wafers/hour at 250 W	~150 wafers/hour track process with 5 min anneal time	1 wafers/hour 10 wafers/hour with 49 × 13 K beams	10 wafers/hour Per station
Defect density	<1 cm^{-2}	<1 cm^{-2} mask <1 cm^{-2} wafer	24 cm^{-2}	<1 cm^{-2} mask	9 cm^{-2}

Table 2. Comparison of state-of-the-art lithography results as of the time of writing.

reached using an 80 W source. Yet consistently high productivity remains a critical issue. Availability needs further improvement, followed by a 4× increase in source power.

Once EUV is inserted into high-volume manufacturing, it will extend Moore's Law by resolution-enhancing techniques[34] (lower k_1) and by introducing high NA.[35]

Another wavelength reduction to ~6 nm has been explored and proof-of-principle of source[36] and coatings[37] have been demonstrated. Any wavelength change comes with significant challenges to the tool, as well as to the mask and resist infrastructures, and therefore it is not expected to happen anytime soon.

5. Summary

Status of today's and potential future lithography solutions is summarized in Table 2.

6. Conclusion

Optical lithography has enabled Moore's Law to continue for the last 50 years by a combination of wavelength decrease, numerical aperture increase, and process improvements (k_1 factor decrease).

Optical lithography has been extended over the last decade by immersion lithography and multiple patterning. As such, it is today's workhorse for the semiconductor industry.

The next step for optical lithography is EUV, which will greatly simplify patterning and thus promises faster yield ramp and lower cost. It has far more industry support than alternative patterning techniques and is the leading candidate to succeed immersion lithography. Significant progress in EUV lithography has been made, yet sources need further improvement, as well as the tool availability.

As for the alternative patterning techniques, DSA still needs optical lithography to guide the patterns and should thus be seen as a complementary technology; e-beam is the technology of choice for mask writing but lacks the speed to compete in high-volume wafer patterning, whereas imprint is unlikely to overcome the challenges inherent to a 1X contact mask.

Acknowledgments

To all lithographers worldwide whose innovations enabled Moore's Law.

References

1. G. E. Moore, "Cramming more components onto integrated circuits," *Electronics* **38**(8), 114–117 (1965).
2. G. E. Moore, "Progress in digital integrated electronics," *Tech. Dig. IEDM* (1975), pp. 11–13.
3. G. E. Moore, "Lithography and the future of Moore's Law," *Proc. SPIE* **2438**, 2–17 (1995).
4. J. Finders, M. Eurlings, K. V. I. Schenau, M. Dusa, and P. Jenkins, "Low-k_1 imaging: How low can we go?," *Proc. SPIE* **4226**, 1–15 (2000).
5. R. Socha, X. Shi, and D. LeHoty, "Simultaneous source mask optimization (SMO)," *Proc. SPIE* **5853**, 180–193 (2005).
6. J. Mulkens, P. Hinnen, M. Kubis, A. Padiy, and J. Benschop, "Holistic optimization architecture enabling sub-14-nm projection lithography," *J. Micro/Nanolith. MEMS MOEMS* **13**, 011006 (2014).
7. M. Dusa, J. Quaedackers, O. F. A. Larsen, *et al.*, "Pitch doubling through dual patterning lithography challenges in integration and litho budgets," *Proc. SPIE* **6520**, 65200G (2007).
8. W.-Y. Jung, C.-D. Kim, J.-D. Eom, *et al.*, "Patterning with spacer for expanding the resolution limit of current lithography tool," *Proc. SPIE* **6156**, 61561J (2006).
9. G. Capetti, P. Cantu, E. Galassini, *et al.*, "Sub $k_1 = 0.25$ lithography with double patterning technique for 45 nm technology node flash memory devices at $\lambda = 193$ nm," *Proc. SPIE* **6520**, article no. 011006 (2007).
10. K. Oyama, S. Yamauchi, S. Natori, A. Hara, M. Yamato, and H. Yaegashi, "Robust complementary technique with multiple-patterning for sub-10 nm node device," *Proc. SPIE* **9051**, 90510V (2014).
11. W. Arnold, M. Dusa, and J. Finders, "Manufacturing challenges in double patterning lithography," *IEEE Intern. Symp. Semicond. Manufacturing (ISSM-2006)*, paper MC-233.
12. J. Mulkens, M. Hanna, H. Wei, V. Vaenkatesan, H. Megens, and D. Slotboom, "Overlay and edge placement control strategies for the 7-nm node using EUV and ArF lithography," *Proc. SPIE* **9422**, 94221Q (2015).
13. M. P. Stoykovich and P. F. Nealey, "Block copolymers and conventional lithography," *Mater. Today* **9**, 20–29 (2006).
14. C. T. Black, "Polymer self-assembly as a novel extension to optical lithography," *ACS Nano* **1**, 147–150 (2007).
15. I. Bita, J. K. W. Yang, S. J. Yeon, C. A. Ross, E. L. Thomas, and K. K. Berggren, "Graphoepitaxy of self-assembled block copolymers on 2D periodic patterned templates," *Science* **321**, 939–943 (2008).
16. R. Ruiz, H. Kang, F. A. Detcheverry, *et al.*, "Density multiplication and improved lithography by directed block copolymer assembly," *Science* **321**, 936–939 (2008).

17. S. F. Wuister, D. Ambesi, T. S. Druzhinina, *et al.*, "Fundamental study of placement errors in directed self-assembly," *J. Micro/Nanolith. MEMS MOEMS* **13**, 033005 (2014).

18. H. Pathangi, B. T. Chan, H. Bayana, *et al.*, "Defect mitigation and root cause studies in 14 nm half-pitch chemoepitaxy directed self-assembly LiNe flow," *J. Micro/Nanolith. MEMS MOEMS* **14**, 031204 (2015).

19. G. de Boer, M. P. Dansberg, R. Jager, *et al.*, "MAPPER: Progress towards a high volume manufacturing system," *Proc. SPIE* **8680**, 86800O (2013).

20. E. Platzgummer, E. Klein, and H. Loeschner, "Electron multi-beam technology for mask and wafer writing at 0.1 nm address grid," *Proc. SPIE* **8680**, 868001 (2013).

21. D. Lam, D. Liu, and T. Prescop, "E-beam direct write (EBDW) as complementary lithography," *Proc. SPIE* **7823**, 78231C (2010).

22. M. Wieland, "Progress towards a manufacturing system," Presented at *Lithography Workshop* (2015), available at www.lithoworkshop.org/2015-Lithography-Workshop-Program-Book-v1-2.pdf.

23. H. Takeishi and S. V. Sreenivasan, "Nanoimprint system development and status for high volume semiconductor manufacturing," *Proc. SPIE* **9423**, 94230C (2015).

24. K. Ichimura, K. Yoshida, S. Harada, T. Nagai, M. Kurihara, and N. Hayashi, "HVM readiness of nanoimprint lithography templates: Defects, CD, and overlay," *Proc. SPIE* **9423**, 94230D (2015).

25. S. Weichselbaum, F. Bornebroek, T. De Kort, *et al.*, "Immersion and dry scanner extensions for sub-10 nm production nodes," *Proc. SPIE* **9426**, 942616 (2015).

26. A. Pirati, R. Peeters, D. Smith, *et al.*, "Performance overview and outlook of EUV lithography systems," *Proc. SPIE* **9422**, 94221P (2015).

27. I. Servin, N. A. Thiam, P. Pimenta-Barros, *et al.*, "Ready for multi-beam exposure at 5 kV on MAPPER tool: Lithographic and process integration performances of advanced resists/stack," *Proc. SPIE* **9423**, 94231C (2015).

28. H. Kinoshita, T. Kaneko, H. Takei, N. Takeuchi, and S. Ishihara, "Study on X-ray reduction projection lithography," Presented at *47th Autumn Meeting Japan Soc. Appl. Phys.* (1986), paper no. 28-ZF-15.

29. W. T. Silfvast and O. R. Wood II, "Tenth micron lithography with a 10 Hz 37.2 nm sodium laser," *Microelectron. Eng.* **8**, 3–11 (1988).

30. A. M. Hawryluk and L. G. Seppala, "Soft X-ray projection lithography using an X-ray reduction camera," *J. Vac. Sci. Technol. B* **6**, 2162–2166 (1988).

31. H. Meiling, H. Meijer, V. Y. Banine, *et al.*, "First performance results of the ASML alpha demo tool," *Proc. SPIE* **6151**, 651508 (2006).

32. C. Wagner, J. Bacelar, N. Harned, *et al.*, "EUV lithography at chipmakers has started: Performance validation of ASML's NXE:3100," *Proc. SPIE* **7969**, 79691F (2011).

33. R. Peeters, S. Lok, J. Mallman, *et al.*, "EUV lithography: NXE platform performance overview," *Proc. SPIE* **9048**, 90481J (2014).

34. S. Hsu, R. Howell, J. Jia, *et al.*, "EUV resolution enhancement techniques (RETs) for $k_1 = 0.4$ and below," *Proc. SPIE* **9422**, 94221I (2015).

35. B. Kneer, S. Migura, W. Kaiser, J. T. Neumann, and J. Van Schoot, "EUV lithography optics for sub 9 nm resolution," *Proc. SPIE* **9422**, 94221G (2015).

36. K. Koshelev, V. Krivtsun, R. Gayasov, *et al.*, "Experimental study of laser produced gadolinium plasma emitting at 6.7 nm," Presented at *Intern. Workshop EUV Sources*, Dublin, Ireland (2010).

37. T. Tsarfati, R. W. E. van de Kruijs, E. Zoethout, E. Louis, and F. Bijkerk, "Reflective multilayer optics for 6.7 nm wavelength radiation sources and next generation lithography," *Thin Solid Films* **518**, 1365–1368 (2009).

1.3

What Happened to Post-CMOS?

P. M. Solomon

IBM, T. J. Watson Research Center, Yorktown Heights, NY 10598, USA

1. Introduction

At this date of writing, we should be approaching the end of CMOS as the premier IT logic technology with new "beyond CMOS" nanotechnologies waiting in the wings. However, CMOS is surging ahead, with noteworthy modifications, while its putative rivals have not materialized. The rivals with self-described revolutionary impact have barely progressed to the single device level and rarely to the level of simple circuits and never to CMOS-competitive switching speeds. Here we examine what went wrong and whether there is any path to a true post-CMOS logic technology.

A previous article,[1] part of this series of conferences dating back 20 years, reviewed a number of nanodevices that had the possibility of going beyond CMOS in their ability to achieve high performance at much lower switching energies. At the time, CMOS appeared to be rapidly approaching its scaling limits. Today, 6 years later, scaling of gate lengths has slowed down, yet the technology has continued aggressively to higher densities and improved functionality.[2, 3] Competing nanotechnologies have advanced in a different direction – not toward commercialization, in the main, but rather toward the application of novel physical principles to logic devices. The commercialization threshold is lower for memory[4] devices, where novel materials offer data retention capabilities at densities beyond CMOS, and efficient error correction can compensate for device deficiencies, but this is not the subject of this chapter.

We discuss these two streams in the context of what it takes for a new device technology to succeed in displacing an established technology and, as background, review some basic constraints these are subject to.

2. General constraints on speed and energy

Classical digital logic (as opposed to quantum computing,[5] which is beyond the scope of this chapter) is constrained by some rather inflexible rules of nature that

Future Trends in Microelectronics: Journey into the Unknown, First Edition.
Edited by Serge Luryi, Jimmy Xu, and Alexander Zaslavsky.

make replacement of a successful technology difficult since this technology has evolved to best accommodate these rules.

First, logic signals are propagated by electromagnetic voltage signals on wires. Other modes of transport are possible, but this is by far the fastest (see Table 1).

Logic may thus propagate distances of ~1 mm without incurring a large penalty. Nonvoltage (or noncharge)-based logic schemes have been proposed to circumvent some of the energy costs of charge-based logic.[1] These rely on the transmission of noncharge-based tokens that could be spins, phonons, or electronic or magnetic polarization. These tokens propagate at much lower speeds but still fast enough to cover intradevice distances (nm) in picoseconds. For longer distance communication, the signals need to be transduced to voltages and then back again at the receiving end, placing an extra burden on such schemes.

Optical communication is another option, which is rapidly replacing high-speed interconnects at distances greater than the current frontier of ~1 cm. Signal propagation is much less lossy and less dispersive than propagation down wires; however, it is still impractical for shorter distances because of the optical to electrical energy costs of transduction.

Since integrated processors, barring some unforeseen change in architecture, will always involve high-speed communication over distance, the need to propagate logic signals on wires will be common to all future logic technologies.

A large part of the energy cost of computation derives from the dissipation of the stored electrostatic energy of the wires $\sim 0.5 C_{W0} L_W V^2$, where C_{W0} is the wiring capacitance per unit length, L_W the wire length, and V the logic voltage. This relationship is fairly inflexible since C_{W0} is practically technology independent[6] and L_W is common to all technologies sharing the same wiring rules. The parameter setting different technologies apart is V, the logic voltage. Theis and Solomon[7] determined that the minimum usable voltage, for any

Medium	Velocity	Time of flight	
	(cm/s)	(mm^{-1})	(nm^{-1})
Electromagnetic (light)	2×10^{10}	5 ps	–
Transverse EM mode (metal wire)	2×10^{10}	5 ps	–
Electrons (crystal maximum)	$\sim 10^8$	1 ns	1 fs
Electrons (Si)	10^7	10 ns	10 fs
Spin wave	10^6	100 ns	0.1 ps
Sound	5×10^5	200 ns	0.2 ps
Domain wall	10^4	10 ms	10 ps

Table 1. Propagation speed of various physical mechanisms.

technology, is at least the capacitor reset noise $(kT/C)^{1/2}$, with a suitable error margin. From Fig. 1(a), we see that $V > 10$ mV is typically needed. For a specific technology the limit can be higher, for example, CMOS needs $V > 300$ mV to ensure a large enough on/off ratio.

The device capacitance itself may be swamped in the dense environment of an integrated chip – see Fig. 1(b). The capacitance of the surrounding wires and vias can be larger than the intrinsic device capacitance. This typically affects competing non-CMOS technologies, which tout lower intrinsic capacitances as part of their competitive advantage, more than CMOS itself.

Around 1960, Feynman[9] famously said, "There is plenty of room at the bottom." Well, that room has been shrinking exponentially ever since. Measuring progress in terms of feature size we have already covered 63% of the distance (on a log scale), and about the same amount in terms of energy – see Table 2. It will be an increasingly difficult engineering challenge to exploit the remaining 1/3.

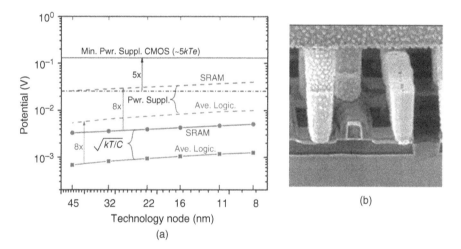

Figure 1. (a) Minimum voltage for logic derived from capacitor reset noise. (Source: Reproduced with permission from Ref. 7); (b) CMOS transistor showing surrounding wires and vias. (Source: Haensch.[8] Reproduced with permission of IEEE).

	1970	2015	Limit
Feature size	10 μm	14 nm	3 Å
Voltage	10 V	1 V	10 mV
Speed	50 ns	5 ps	Power-limited
Energy (kT)	2.4×10^{10}	10^5	40

Table 2. Room at the bottom: limits on voltage set by capacitor reset noise from Fig. 1; energy limit derived for a bit-error rate $< 10^{-19}$.

3. Guidelines for success

As a guideline for a future technology we should look to the past. There have been several technological transformations in the history of information technology, all leading to orders of magnitude improvement over their predecessors. This is shown in Table 3 (the numbers are approximate, meant to convey order of magnitude).

As shown in Table 3, vacuum tubes displaced mechanical relays with a 1000-fold speed improvement. Then, bipolar transistors, in turn, displaced vacuum tubes with a 100-fold volume and power reduction and a huge improvement in reliability. While bipolar transistors were initially discrete devices, the invention of ICs enabled huge increases in their speed and performance over the years, and again greatly improved their reliability. In turn, CMOS replaced bipolar transistors due to its negligible standby power, which allowed scaling to continue. Today power, both standby and dynamic, has become an issue for CMOS. Even though power per circuit has decreased, the number of circuits has greatly increased.

Importantly, CMOS was successful in spite of the fact that device performance was initially inferior to existing bipolar technology. This is due to the huge advantage of complementary switching in that power need only be dissipated during the switching transient. The advantage lies both in the fact that power is dissipated over a small fraction of the clock cycle, and that only a fraction of the circuits are switching in any clock cycle. The combination of these two factors results in a ~1000-fold power reduction compared to a noncomplementary technology. The power efficiency of CMOS makes it very difficult for a noncomplementary technology to compete.

The criteria for a successful technology are listed in Table 4. The only device requirements that are absolutely essential for the performance of logic are

Logic device	Volume (m^3)	Circuit delay (s)	Average circuit power (W)	Reliability	Epoch of dominance
Mechanical relay	– 10^{-4}	– 10^{-2}	– 10^{-1}	Poor	1930s 1940s
Vacuum tube	10^{-4} 10^{-5}	10^{-5} 10^{-6}	1	Very poor	1950s 1960s
Bipolar transistor	10^{-7} 10^{-14}	10^{-6} 10^{-11}	10^{-2} 10^{-3}	Good Very good	1960s 1980s
CMOS	10^{-15} 10^{-18}	10^{-9} 10^{-11}	10^{-6} 10^{-8}	Excellent ?	1980s 2016–

Table 3. Evolution of logic devices. The CMOS power estimates assume an activity factor of 1% and a 10% switching delay/clock period.

Property	Need	Explanation
Concatenability	B	Communication with minimum penalty
Nonlinearity	A	Needed to make logic threshold decisions
Level restoration	B	Robust logic propagation
Isolation	B	Stable concatenation of many stages
Gain	B, C	Sufficient voltage gain for level restoration and current gain for fan out
Universal logic	A	Easy to achieve
Error margin	D	Requirement of very low error rate
Speed	E	Competitiveness: can be traded for energy
Low energy	E	Competitiveness: can be traded for speed
Low standby power	E	Essential for very large systems
Nonvolatility	F	Useful additional property
Four-terminal isolation	F	Useful for power distribution in extremely large systems and signal distribution in noisy systems

Table 4. Requirements for competitive logic with the following classification: A = absolutely necessary; B = interstage buffer can provide at extra cost in area, power, and performance; C = may be propagated by a clock, even without voltage gain; D = error correction possible at extra cost; E = necessary for performance but not functionality; F = benefit for future systems.

nonlinearity and the possession of a complete logic set (the *and* operation combined with inversion). However, there are some other desirable characteristics listed in Table 4, with CMOS possessing all but the last two: four-terminal isolation and nonvolatility. Any technology that lacks some of these characteristics can still do functional logic but is at a competitive disadvantage. For instance, a technology that lacks voltage gain and isolation, such as a majority gate formed by nonlinear resistors, can do logic locally, but the logic signals have to be amplified and isolated by a buffer amplifier in order to propagate further, and often the buffer amplifier is larger and more power hungry than the logic itself. Another example is cellular logic[10, 11] where a polarization state, using majority logic, is propagated forward using a clock, in the absence of voltage gain. While the energy dissipated in logic devices might be small, the power dissipation of the clock circuitry could be much larger.

Apart from these criteria, technologies have succeeded because they have some built-in natural advantages that enable them to get a foothold. As the technology gathers momentum and accumulates invention and processing expertise, it becomes less dependent on this initial advantage. For instance,

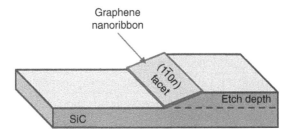

Figure 2. Graphene nanoribbon grown on a crystal facet. (Source: After Sprinkle *et al.*[13]).

silicon MOSFET technology based its early success on the superb properties of SiO_2 as a gate dielectric with a low interface state density and low leakage, but now SiO_2 has been replaced by the higher performance HfO_2. Similarly, the III–V heterojunction field-effect transistors (FETs)[12] first took advantage of the nearly perfect $GaAs/Al_xGa_{1-x}As$ lattice match, but today higher-performance material pairs can be used, such as InGaAs/GaAs, despite the built-in stress due to lattice mismatch. This learning has, in turn, been exploited by the mainstream silicon technology, which now uses strained Si_xGe_{1-x} to improve performance. A possible example for the future is the case of the graphene nanoribbon FET. Graphene in extended sheets does not have a band gap and is therefore unsuitable for logic devices. However, if it can be patterned into nanoribbons a few nanometer wide with perfectly defined edges, then a band gap will open up, along with logic possibilities. Doing this by brute force to the required precision is infeasible; however, researchers at Georgia Tech[13] have succeeded in growing graphene nanoribbons on etched facets defined by step height, as shown in Fig. 2. Such advances could open up possibilities that seemed beyond reach.

On the other hand, when Hisamoto *et al.*[14] proposed the delta-FET in 1990, shown in Fig. 3(a), the community, including the present author, thought this could never be done reproducibly, since the smallest dimension of the FET, the channel thickness, needed to be precisely controlled. Now the fin-FET – new name, same idea, see Fig. 3(b) – is the backbone of high-performance CMOS technology. The secret is that the FET characteristics, in this channel thickness range and for an undoped fin, are not very sensitive to the channel thickness.

4.   Benchmarking and examples

The practice of benchmarking a new and as yet untried technology has become an important way of estimating its potential *in lieu* of actual performance. Here we consider technologies investigated by Nikonov and Young,[15, 16] as part of the Nanoelectronics Research Initiative (NRI) and StarNet programs.[17] While not

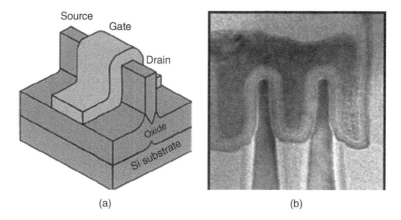

Source
Gate
Drain
Oxide
Si substrate

(a)   (b)

Figure 3. (a) Original proposal of delta-FET from 1990[14] compared to (b) modern 14 nm fin-FET, with a fin pitch of 42 nm. (Source: Natarajan *et al.*[3] Reproduced with permission of IEEE).

comprehensive, this list represents a good cross section of current research. The technologies we consider, using terminology and references from Nikonov and Young[16] for this exercise, represent semiconductor switching devices: CMOS (as a reference) both low and high power, and vdWFETs using 2D graphenoid channels; five varieties of tunnel field-effect transistors (TFETs), homojunction, heterojunction and graphene nanoribbon, 2D–2D, and dichalcogenide (HetJFETS, HomJTFETs, GnrTFETs, ThinTFET, TMDTFET); directional switching from angled graphene (GpnJ); bilayer pseudospin field-effect transistor (BISFET) relying on a Bose–Einstein condensate; ferroelectric FET utilizing negative capacitance (NCFET) and memory (FEFET) effects; FETs utilizing strain (PiezoFET), the excitonic field-effect transistor (exFET); the FET utilizing a metal–insulator transition (MITFET); and magnetic (spintronics) devices: the spin field-effect transistor (spinFET) where spin mediates gate voltage modulation of drain current, the all spin logic (ASL) device and charge-spin logic (CSL) where signals are propagated as spin or charge currents, the spin transfer torque/domain wall (STT/DW) device, the spintronic majority gate (SMG), spin-dynamic devices such as the spin wave device (SWD) and spin torque oscillator (STO), and nanomagnetic dipole logic (NML).

This device list is long and represents research into many possible proposed physical mechanisms capable of being used for logic. While, CMOS, the vdWFET, and the TFETs are electronic devices based on standard electron-gas physics, the other electronic devices invoke collective effects: ferroelectricity, piezoelectricity, metal–insulator transitions, excitonic effects, and Bose–Einstein condensation (BISFET). Still other devices are magnetic or electric-field-mediated magnetic, based on such effects as electric-field modulation of spin lifetime plus magnetic tunnel junction (MTJ) spin-filtering (spinFET), electric current-driven domain-wall motion (STT/DW), MTJ spin injection (SMG and STO), spin torque flip of magnetization (ASL), spin diffusion (ASL), and interaction between magnetic dipoles (NML).

To these devices, we add the following: carbon nanotube (CNT) field-effect transistor,[18, 19] piezoelectronic transistor (PET),[20] and nanoelectromechanical systems (NEMS).[21] The first is an FET but with greatly improved transport parameters and scaling characteristics, the other two are nonelectronic "transduction" devices capable, in principle, of operating at lower voltages, hence lower power, than conventional FETs. In fact, NEMS circuits have operated at voltages as low as 10 mV.[22] The PET should be distinguished from the PiezoFET.[23] In spite of an incorrect reference, the PiezoFET of Ref. 15 is a device where surface charge-generated piezoelectricity directly modulates an FET channel, whereas in the PET a piezoelectric crystal exerts force on a piezoelectric element causing switching.

The performance benchmarks for these technologies are indicated in Fig. 4(a) with solid circles. They have included factors such as wiring capacitance based on proposed circuit layouts for a large enough logic block to provide a realistic estimate of the wiring capacitance. The author's own estimates are shown in Fig. 4(b) for the added CNT, PET, and NEMS relay technologies (in addition to CMOS). These latter estimates assume a 2-NAND logic gate with an average wiring load and parasitic capacitance based on closely packed electrodes characteristic of a future scaled technology. Estimates of untried technologies do involve large uncertainties, but benchmarking does help to focus on technologies of interest.

In Table 5, we look at a subset of these technologies and measure them against the requirements shown in Table 4. While all the technologies listed can, in principle, be made to work, there are factors that militate against them. Having a poor error margin is probably the most serious liability. A reliance on non-electromagnetic propagation, lack of concatenability, or lack of voltage gain and isolation mean complicated and power-consuming external circuitry is needed to sustain logic propagation. A poor on/off ratio is a liability since the alternative, clocking the power supply (power clocking) is itself power consuming. This table represents our present understanding of these exploratory technologies and explains why, barring a breakthrough, none of the magnetic technologies (last three columns of Table 5) is likely to supplant CMOS.

The situation becomes even clearer when we look at the maturity of the various device concepts laid out in Table 6 (which includes III–V FETs as an example of a mature technology). The feasibility of some devices is unclear even in their physical concept. The BISFET relies on never yet seen presence of a Bose–Einstein condensate at room temperature and a completely speculative mechanism for its abrupt quenching. The spinFET, after decades of pursuit, has yet to be demonstrated and it has been argued[24] that the physical basis of operation is not sound. The TFET, while deceptively simple has yet to achieve convincing subthermal turn-on slopes after many years of work, and this might

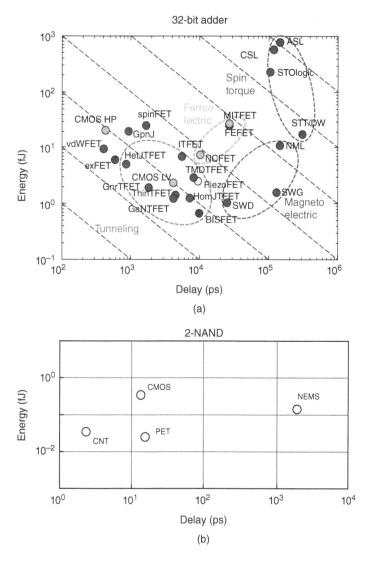

Figure 4. (a) Benchmarking of a 32-bit adder against various post-CMOS technologies. (Source: Nikonov and Young.[16] Reproduced with permission of IEEE); (b) Benchmarking of the 2-NAND gate against CMOS-like (CNT, CMOS) and four-terminal relay-type (PET, NEMS) technologies by the author.

be due to an as-yet unexamined physical limitation. This applies as well to the GpnJ.

A whole class of magnetic, and magnetoelectric devices has emerged[16, 25] (MTJ, STT/DW, SMG, ASLD, and NML) based on the successful application of spin-valve and spin-torque concepts to MRAM.[26] These are based on sound

Logic class	Three-terminal electronic	Four-terminal electronic	Four-terminal magnetic	Two-terminal dipole	Majority gate magnetic
Members	CMOS, TFET, CNT, other FETs	NEMS, PET	STT/DW	QCA, NML	SMG, ASL, other spin
Universal logic	Y	Y	Y	Y	Y
Nonlinearity	Y	Y	Y	Y	Y
Concatenability	Y	Y	Y	Y	N, Y
Level restoration	Y	Y	Y	N	N,Y
Isolation	Y	Y	Poor	N	Poor
Gain	Y	Y	Y	N	Poor
Propagation	e/m	e/m	e/m	Dipole	Diffusion
Error margin	Good	Good	Poor	Very poor	Poor
On/off ratio	Good	Good	Poor	NA	N
Nonvolatility	N	N	Y	Y	Y
Four-terminal DC isolation	N	Y	Y	N	N

Table 5. Logic requirements versus proposed technologies.

Device	Device principle	Device demo	Circuit demo	IC demo	Commercial IC
CMOS	Y	Y	Y	Y	Y
III–V FET	Y	Y	Y	Y	Y
vdWFET	Y	Y	–	–	–
CNT	Y	Y	Y	–	–
BISFET	–	–	–	–	–
TFET	Y?	Y?	–	–	–
GpnJ	Y?	–	–	–	–
spinFET	–	–	–	–	–
ASL	Y	–	–	–	–
NML	Y	Y	Y	–	–
STT/DW	Y	Y	Y	–	–
NEMS	Y	Y	Y	–	–
PET	Y	Y	–	–	–

Table 6. Maturity of selected technologies.

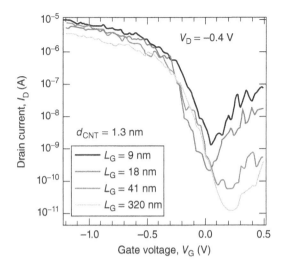

Figure 5. Carbon nanotube transistor transfer characteristics versus gate length L_G. (Source: Franklin *et al.*[19] Reproduced with permission of IEEE). The CNT diameter is 1.3 nm.

principles, yet for logic they are very slow and have on/off ratios scarcely viable for logic. A major advance in the last 2 years, under the NRI and StarNet programs, has been the elucidation of magnetoelectric switching that has the potential of greatly reduced switching energies, making spin-based options much more competitive (see Fig. 4). The property that makes the magnetic options very interesting for logic is nonvolatility, of which more is said later.

Of the devices that have reached the stage of circuit demonstration, the CNT, STT/DW, and NEMS stand out as having distinct possibilities for future applications. The CNT is the only device with speed and low-power potential well beyond CMOS, but its fabrication will require a revolution in bottom-up process technology, some already underway.[18] Its small cross section makes it highly vulnerable to stray charge and noise and this may prove to be an insoluble obstacle (see Fig. 5). There is also a whole class of FETs based on atomically thin 2D materials (vdWFETs). While devices have been demonstrated, the technology is nascent and not clearly superior to CMOS.

Two other technologies appear in Fig. 4(b), as well as Tables 5 and 6: the NEMS relay and the PET. Both are relay-type technologies based on transduction, with the output electrostatically isolated from the input. The transduction principle permits scaling to very low voltages. For instance, a NEMS circuit has demonstrated switching at 10 mV with on/off ratio of six orders of magnitude.[22] The PET[20] will not go that far, but should be much faster. Both devices have yet to demonstrate the endurance (number of switching cycles) suitable for logic.

5. Discussion

At this point in our discussion, it must be apparent why CMOS is such a pervasive technology and why it has been so difficult for new technologies to emerge. At their conception, semiconductor devices promised such large improvements over the extant technologies that within 5 years or so of the first demonstration of a working device, there were commercial applications, an outcome that has eluded putative nanotechnologies after ~5 years of intensive research. Moreover, CMOS has even swallowed up its erstwhile competitors. Witness the CMOS RF technology displacing III–V technology from its microwave niche. As long as Moore's Law scaling continued unabated, it was virtually impossible for a new technology to catch up. Now that the scaling is finally slowing down, there is precious little room left at the bottom to nurture a new technology.

While direct replacement of CMOS is highly unlikely in any foreseeable future there are specialized applications that could favor a new technology. This is also an era of heterogeneous integration, where it is economically viable to integrate several technologies on the same silicon carrier. Thus, openings exist for new technologies suited for special applications, and technologies that complement CMOS. Such applications include innovative memory technologies,[4] which are not discussed in this chapter, but where substantial progress had been made during the last decade. Also optics and electro-optic interfaces may provide opportunities for III–V and graphene technologies,[27] as well as electro-optic amplifiers such as the exFET[28]. Beyond this, our survey has

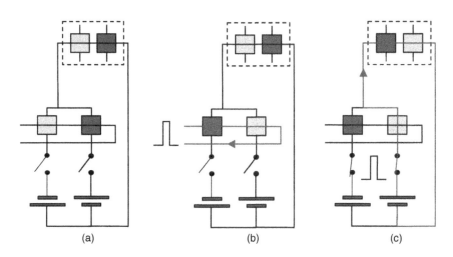

Figure 6. Operation of nonvolatile logic: (a) powered-down logic with devices holding their states; (b) programming current from preceding logic block changes state of the devices; (c) logic block is momentarily powered up, evaluating the logic state of the block and providing current to program the succeeding block (enclosed in the dashed rectangle).

revealed two properties that have potential to change how logic is implemented. The first is nonvolatility of magnetic devices (see also Ref. 15) and the second is the four-terminal property of transduction devices.

Nonvolatility presents a different way of doing logic, illustrated in Fig. 6. The logic states of devices in a logic block may be set by signals from outside the block, while the block is powered off, as in Fig. 6(b). When the power is pulsed on, the state of the logic is evaluated and the result propagated to the next block – see Fig. 6(c). This example applies to a technology such as STT/DW where current flow in a control line of a magnetic device changes its state. This has the obvious application to collection of data in a powered-down block and evaluating a logical function of that data at a later time.

Four-terminal isolation is inherent in transduction devices, such as the NEMS relay, the PET and the STT/DW. In these devices, the input terminals are isolated electrostatically from the output terminals, allowing blocks of logic to be run "off ground" at different common-mode voltages. This can drastically simplify the power distribution in future computers operating at low power supply voltage but very high currents, since with the four-terminal option logic blocks may be placed in series with respect to a higher voltage power supply.

6. Conclusion

To date, the quest for an alternate switch to replace CMOS has come up empty, with no contenders for the foreseeable future. This has to do with the power, simplicity, and adaptability of the CMOS concept, which has taken it so far down the scaling path that there is little room left at the bottom. An opportunity does exist for technologies that complement CMOS by providing additional functionality. The new devices under study provide three such opportunities: electro-optic interfacing, nonvolatility, and four-terminal isolation.

Acknowledgments

I wish to acknowledge the insight and help of Tom Theis.

References

1. P. M. Solomon, "Device proposals beyond silicon CMOS," chapter in: S. Luryi, J. M. Xu, and A. Zaslavsky, eds., *Future Trends in Microelectronics: From Nanophotonics to Sensors and Energy*, Hoboken, NJ: Wiley/IEEE Press, 2010, pp. 127–140.
2. C.-H. Lin, B. Greene, S. Narasimha, *et al.*, "High performance 14 nm SOI FinFET CMOS Technology with $0.0174 \, \mu m^2$ embedded DRAM and 15 levels of Cu metallization," *Tech. Dig. IEDM* (2014), pp. 74–76.

3. S. Natarajan, M. Agostinelli, S. Akbar, *et al.*, "A 14 nm logic technology featuring 2nd-generation fin-FETs, air-gapped interconnects, self-aligned double patterning and a $0.0588 \, \mu m^2$ SRAM cell size," *Tech. Dig. IEDM* (2014), pp. 71–73.

4. J. Singh Meena, S. M. Sze, U. Chand, and T.-Y. Tseng, "Overview of emerging nonvolatile memory technologies," *Nano. Res. Lett.* **9**, 526–558 (2014).

5. T. D. Ladd, F. Jelezko, R. Laflamme, Y. Nakamura, C. Monroe, and J. L. O'Brien, "Quantum computers," *Nature* **464**, 45–53 (2010).

6. P. M. Solomon, "A comparison of semiconductor devices for high speed logic," *Proc. IEEE* **70**, 489–509 (1982).

7. T. N. Theis and P. M. Solomon, "In quest of the 'next switch': Prospects for greatly reduced power dissipation in a successor to the silicon field-effect transistor," *Proc. IEEE* **98**, 2005–2014 (2010).

8. W. Haensch, "High performance computing beyond 14 nm node – is there anything other than Si?," *Proc. 13th Intern Conf. Ultimate Integr. Silicon (ULIS)* (2012), pp. 33–36.

9. R. P. Feynman, "There's plenty of room at the bottom," Presented at 1959 APS Meeting, text available in *Caltech Eng. Sci.* **23**, 22 (1960).

10. C. S. Lent and B. Isaksen, "Clocked molecular quantum-dot cellular automata," *IEEE Trans. Electron Dev.* **50**, 1890–1896 (2003).

11. R. P. Cowburn and M. E. Welland, "Room temperature magnetic quantum cellular automata," *Science* **287**, 1466–1468 (2000).

12. P. M. Solomon and H. Morkoc, "Modulation-doped GaAs/AlGaAs heterojunction field effect transistors (MODFETs), ultrahigh-speed devices for supercomputers," *IEEE Trans. Electron Dev.* **31**, 1015–1027 (1984).

13. M. Sprinkle, M. Ruan, Y. Hu, *et al.*, "Scalable templated growth of graphene nanoribbons on SiC," *Nature Nanotechnol.* **5**, 727–731 (2010).

14. D. Hisamoto, T. Kaga, Y. Kawamoto, and E. Takeda, "A fully depleted lean-channel transistor (DELTA) – A novel vertical ultra-thin SOI MOS-FET," *Tech. Dig. IEDM* (1989), pp. 833–836.

15. D. M. Nikonov and I. A. Young, "Overview of beyond-CMOS devices and a uniform methodology for their benchmarking," *Proc. IEEE* **101**, 2498–2533 (2013).

16. D. M. Nikonov and I. A. Young, "Benchmarking of beyond-CMOS exploratory devices for logic integrated circuits," *IEEE J. Expl. Sol.-State Comp. Dev. Circ.*, **1**, 3–11 (2015).

17. See www.src.org/program/nri/ and www.src.org/program/starnet/ (accessed on April 9, 2016).

18. P. M. Solomon, "Carbon-nanotube solutions for the post-CMOS-scaling world," chapter in: S. Luryi, J. M. Xu, and A. Zaslavsky, eds., *Future Trends in Microelectronics: Up the Nano Creek*, New York: Wiley, 2007, pp. 212–223.

19. A. D. Franklin, S.-J. Han, G. S. Tulevski, *et al.*, "Sub-10 nm carbon nanotube transistor," *Tech. Dig. IEDM* (2011), pp. 525–527.

20. D. Newns, B. Elmegreen, X. H. Liu, and G. Martyna, "A low-voltage high-speed electronic switch based on piezoelectric transduction," *J. Appl. Phys.* **111**, article no. 084509 (2012).

21. T.-J. King Liu, E. Alon, V. Stojanovic, and D. Markovic, "The relay reborn," *IEEE Spectrum* **49**, 40–43 (2012).

22. U. Zaghloul and G. Piazza, "Sub-1-volt piezoelectric nanoelectromechanical relays with millivolt switching capability,". *IEEE Electron Device Lett.* **35**, 669–671 (2014).

23. S. Agarwal and E. Yablonovitch, "A nanoscale piezoelectric transformer for low-voltage transistors," *Nano Lett.* **14**, 6263–6268 (2014).

24. S. Bandyopadhyay, "Spintronic devices – Friendly critique," Presented at 2014 ITRS Emerging Research in Devices Meeting, Albuquerque, NM, August 2014.

25. S. Bandyopadhyay, "Information processing with electron spins," *ISRN Mater. Sci.* 2012, 697056 (**2012**).

26. W. J. Gallagher and S. S. P. Parkin, "Development of the magnetic tunnel junction MRAM at IBM: From first junctions to a 16-Mb MRAM demonstrator chip," *IBM J. Res. Develop.* **50**, 5–23 (2006).

27. Y. T. Hu, M. Pantouvaki, S. Brems, *et al.*, "Broadband 10 Gb/s graphene electro-absorption modulator on silicon for chip-level optical interconnects," *Tech. Dig. IEDM* (2014), pp. 128–130.

28. P. Andreakou, S. V. Poltavtsev, J. R. Leonard, *et al.*, "Optically controlled excitonic transistor," *Appl. Phys. Lett.* **104**, 091101 (2014).

1.4

Three-Dimensional Integration of Ge and Two-Dimensional Materials for One-Dimensional Devices

M. Östling, E. Dentoni Litta, and P.-E. Hellström
School of Information and Communication Technology, KTH Royal Institute of Technology, 16440 Kista, Sweden

1. Introduction

The outstanding progress experienced by the semiconductor industry, and the electronics and information technology industries as a consequence, in the past decades is aptly represented by the well-known Moore's Law,[1] which states that the number of components that can be manufactured on a chip at the same cost has increased exponentially over time. Traditionally, Moore's Law has been supported by two-dimensional (2D) scaling, initially captured by Dennard's scaling rules[2] but recently complemented by the continuous introduction of novel technologies and device structures, such as Cu/low-κ back end of line (BEOL),[3] strain engineering,[4] high-κ/metal gate,[5] channel engineering,[6] fully depleted silicon-on-insulator (FD-SOI),[7] and fin-FETs,[8] as well as by the introduction of increasingly complex multiple patterning techniques[9] – see Fig. 1. However, continued 2D scaling of complementary metal-oxide-semiconductor (CMOS) technology past the 7 nm node looks extremely complex for technical and economic reasons, as well as fundamental physical limits. Complex technical issues make it extremely challenging to achieve both area scaling and the expected performance improvements, due to the increased impact of quantum effects, self-heating, and performance bottlenecks in contacts and BEOL.[10] Even if such issues are resolved, fundamental scaling showstoppers are within sight, since critical device dimensions are already in the nanometer range in current nodes and there are no known technologies that can work at subatomic or even near-atomic length scales.[10] Finally, the foundation of Moore's Law has always been economic, and the economics of 2D scaling have been put into question recently, with claims that foundry nodes past the 28 nm node do not provide the expected benefits in terms of cost per transistor, and that manufacturing costs are expected to increase even further due to the need for multiple patterning.[11]

Future Trends in Microelectronics: Journey into the Unknown, First Edition.
Edited by Serge Luryi, Jimmy Xu, and Alexander Zaslavsky.
© 2016 John Wiley & Sons, Inc. Published 2016 by John Wiley & Sons, Inc.

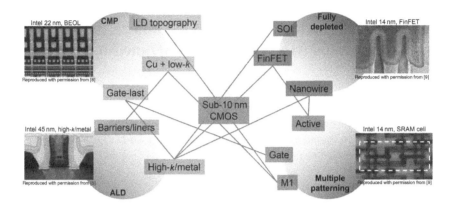

Figure 1. Modern CMOS technology nodes rely on the integration of novel technologies and device structures to complement and support the conventional 2D device scaling. Breakthrough technologies enabling sub-10 nm CMOS nodes include the use of chemical-mechanical polishing (CMP) for planarization and etch-back in FEOL and BEOL, atomic layer deposition (ALD) of high-κ dielectrics and conformal metal films, fully depleted device structures for enhanced scalability, and multiple patterning for subresolution lithography. (Source: Mistry *et al.*[5], Auth *et al.*[8], and Natarajan *et al.*[9] Reproduced with permission of IEEE).

Even if the introduction of extreme ultraviolet (EUV) lithography can partly alleviate the cost issues, it is still expected that future technology nodes will be very expensive both from the manufacturing and the design points of view. Accordingly, they may be suitable only for large-scale high-cost products, such as microprocessors and high-end mobile systems-on-chips (SoCs). A different scaling path, which is less reliant on 2D shrinking, is therefore desired to cater to both the majority of the mobile market and the growing Internet of Things (IoT) market.[12]

An alternative scaling path has indeed been identified in leveraging the third dimension, that is, in stacking additional functionality in the vertical direction. This approach has already resulted in commercial products, chiefly in the memory market,[13, 14] and holds the potential to enable higher performance and functional integration in the CMOS platform. The potential for economic benefit lies in the possibility of simplifying and optimizing different parts of the fabrication sequence. For example, moving the SRAM to a layer separate from CMOS logic can result in simplified processing and better performance compared to realizing both elements in parallel on the same die. This concept can be extended to the idea that different layers may enable different functionalities and that some of the layers may be fabricated in older (and much less expensive) process nodes – see Fig. 2.

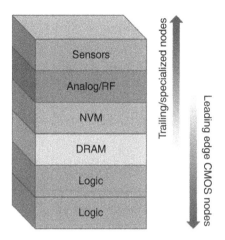

Figure 2. 3D integration concept: several layers realizing different functionalities can be fabricated in different process nodes, providing better cost/performance benefits, and stacked on top of each other using vertical interconnections.

Stacking of additional functionality can be obtained via a number of techniques, which can be broadly divided into three classes:[15]

- Three-dimensional (3D) packaging. Chip stacking is achieved by wire bonding or package-on-package techniques, resulting in minimal modifications to the chip fabrication process but also in the lowest interconnect density.

- Parallel 3D. Chips are processed separately and subsequently stacked on top of each other, using through-silicon vias (TSVs) to interconnect them. Parallel 3D introduces some modifications to the fabrication sequence, mostly in the BEOL but also in the front-end of line (FEOL), and achieves an intermediate interconnect density, limited by the TSV size (1–10 μm in current technology).

- Sequential or monolithic 3D. In this case, 3D chips are built sequentially in layers on top of previously processed layers, as shown in Fig. 3. The upper device layers can be formed in a number of ways (e.g., by epitaxial growth via seed windows), but wafer bonding is at present the technique achieving the highest integration flexibility and crystalline quality.[16] Sequential 3D requires the greatest modifications to the fabrication flow, with a strong impact on both FEOL and BEOL, but achieves the highest possible interconnect density, comparable to conventional vias.

Although all three classes of 3D integration hold significant technical and commercial potential, this chapter focuses primarily on sequential 3D, since only

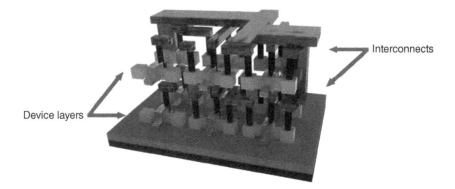

Figure 3. Sequential 3D integration can provide an alternative scaling path. Two or more CMOS device layers can be stacked vertically and interconnected using intralayer BEOL metallization and interlayer vias.

the high interconnection density achieved in this case is capable of supporting a scaling path in the vertical dimension.

The successful implementation of a sequential 3D integration scheme requires a strong research effort in developing low-thermal-budget FEOL processing. Specifically, novel materials and technological solutions are needed to ensure that the additional steps introduced by the formation of additional device layers can be performed without degrading lower-layer devices and without sacrificing the performance of upper-layer devices. The status of current research in FEOL processing compatible with sequential 3D integration, as well as a view of the main outstanding challenges, is summarized in Section 2. Besides enabling scaling of "conventional" CMOS circuits in the vertical direction, 3D integration also opens the possibility of cointegrating additional "more than Moore" functionality, such as analog/RF, sensors, emerging memories, or energy sources, as illustrated in Fig. 4. The possibility of integrating novel concepts based on 2D materials is explored in Section 3. In addition, sequential 3D integration has a strong impact not only at the device level but also at the system level, requiring the development of a 3D design ecosystem. Section 4 describes the different technological options available in the context of 3D integration and their impact on design and manufacturing. Finally, we present some conclusions that can be drawn from the current status of research and the expected future challenges.

2. FEOL technology and materials for 3D integration

Sequential 3D integration involves strong changes to the conventional silicon processing, both in the FEOL and in the BEOL, since it is based on bonding unprocessed semiconductor layers and processing them on top of fully processed

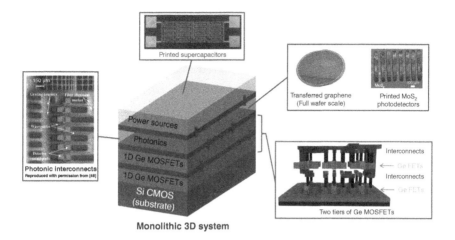

Figure 4. Conceptual illustration of a sequential 3D system, integrating two Ge-channel CMOS logic layers, a photonic interconnect layer, and graphene supercapacitors. (Source: Naiini et al.[17] Reproduced with permission of IEEE).

substrates. This results in the need to develop a FEOL process flow for the upper layers that does not degrade the FEOL and BEOL in the lower layers.

Since upper-layer devices would naturally be fabricated in a thin semiconductor layer on top of a dielectric film, the natural choice for transistor architecture is one based on FDSOI technology, be it planar, fin-FET, or nanowire (NW) – see Fig. 5. Although several options can be considered in the context of sequential 3D integration, only advanced device structures would render the scheme compatible with sub-10 nm CMOS nodes. In particular, the use of 1D devices such as gate-all-around nanowire (GAA NW) FETs appears promising for integration in the 7 or 5 nm node, in the form of stacked horizontal or vertical NWs, which would allow high I_{ON} with small footprint while benefiting from the improved electrostatics of a GAA architecture to minimize I_{OFF}.[18]

Different choices of channel materials are also being considered for integration in CMOS. Even though (strained) Si has traditionally been the only viable choice for the channel, extremely scaled technology can derive significant advantages from the use of alternative semiconductors, mainly group IV elemental and alloy semiconductors (Ge, SiGe, GeSn) and III–V alloy semiconductors (e.g., GaAs, InGaAs). A substantial body of literature has explored the integration of alternative channel materials on the CMOS platform, showing high-performance scaled Ge-channel FETs with gate length down to 25 nm and equivalent oxide thickness below 1 nm,[19, 20] high mobility in Ge devices achieved via strain engineering and NW patterning (record-high hole mobility of 1490 cm^2/V·s),[21, 22] improved reliability in SiGe pFETs,[23] cointegration of SiGe and III–V semiconductors on SOI,[24] and promising results for conventional and unconventional device architectures based on III–V materials.[25–27]

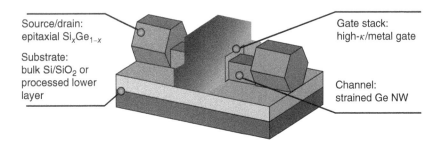

Figure 5. Fully depleted SOI devices are the natural device structures for sub-10 nm CMOS nodes in the context of sequential 3D integration. Even though planar, fin-FET, and nanowire SOI FETs would all be compatible with sequential 3D integration, fin-FETs and gate-all-around structures would provide the best performance in terms of I_{ON} and I_{OFF} currents.

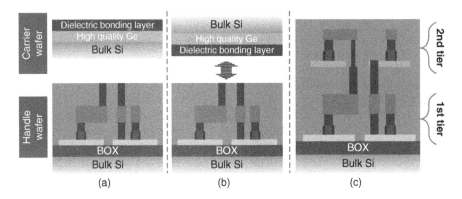

Figure 6. Transfer of device layer on fully processed substrate. (a) Processed substrate is completed with both FEOL and BEOL of the first layer. A device layer is grown on a carrier wafer and capped with a dielectric bonding layer; (b) After wafer bonding, the carrier wafer and buffer layers are removed; (c) The second layer is processed at low thermal budget. (Source: Garidis et al.[28] Reproduced with permission of IEEE).

Sequential 3D integration schemes can be developed using most channel materials, as long as the layers can be grown with high quality on 300 mm wafers and a proper wafer bonding technology can transfer the device layer onto the processed substrate, as shown in Fig. 6.[28] Depending on the type of 3D integration scheme, pFETs and nFETs may be fabricated on separate layers or on the same layer. While the latter approach provides more flexibility for circuit design, the former has strong advantages in terms of manufacturability and yield. As a matter of fact, fabricating both types of transistors on each layer would result in duplication of the complete FEOL process for each additional layer, with detrimental effects on costs and yield. Allocating nFETs and pFETs to different layers, on the other hand, would strongly simplify the process flow and allow a simple separate optimization of the two transistor types, that is, using different channel materials, orientations, device architectures, and gate stacks.

This is especially advantageous since alternative channel materials are not perfectly suitable for both transistor types. In fact, Ge presents the highest hole mobility and is the most favorable choice for pFETs,[29-31] whereas Ge nFETs have been demonstrated,[32, 33] but without strong advantages over Si. Conversely, the highest electron mobility is obtained in III–V semiconductors, but it is not matched by a similarly high hole mobility.[34-36] An advanced 3D integration scheme may therefore use III–V nFETs and Ge pFETs on separate layers, significantly simplifying the integration flow as compared to their cointegration on the same layer. On the other hand, in applications where III–V semiconductors are not suitable (e.g., due to their limited inversion thickness scalability) or high integration costs are not justified, a (strained) Si nFET layer combined with a SiGe or Ge pFET layer may be preferable. A Si/SiGe option would particularly benefit low-cost applications, such as IoT devices, since the relevant technologies are already highly developed and the solutions for further scaling are either known or in advanced research stages.[37-44]

The thermal budget in the FEOL processing of the upper layers needs to be severely limited (compared to a conventional budget of ~1000 °C), since high temperatures would degrade the performance and reliability of the bottom layers, especially regarding the stability of the source/drain silicide and of the BEOL. Depending on the materials used in the bottom layer, the maximum thermal budget is typically in the 400–550 °C range. While this range is compatible with many steps in the FEOL, at least three steps need higher temperatures, namely, high-κ annealing, dopant activation, and epitaxy.[45] Solutions are currently being researched for these steps, mainly by using plasma or reactive annealing for high-κ dielectrics, laser annealing, or solid-phase epitaxial regrowth for dopant activation, and optimized precursors and processes for epitaxial growth. Even though the applicability of low-thermal-budget processing has been demonstrated for Si-based sequential 3D,[46] the use of advanced channel materials like Ge and III–Vs holds the potential of a better compatibility with thermal budgets, which may allow Cu metal/low-κ BEOL in the bottom layers.

3. Integration of "more than Moore" functionality

Besides providing an alternative path for traditional CMOS scaling, 3D integration is also extremely appealing for its potential to integrate completely different functionality on top of the CMOS platform. Specialized functions, such as sensors, optoelectronics, analog and RF circuits, which today are mostly realized on different chips, could be integrated directly on top of a CMOS chip, with huge savings in terms of area, delay, and power.[12] Major system performance improvements may also be achieved by directly stacking the logic and memory chips, where the latter could be fabricated in a standard SRAM, DRAM, or NAND process[47] or realized via emerging concepts such as RRAM.[48] In addition, completely new nonCMOS ("more than Moore") functionality can be integrated if the 3D integration approach is combined with novel materials such as Ge and graphene.

Beyond their applicability to CMOS circuits, the integration of Ge and III–V materials has the promise to unleash on-chip photonic interconnects. As the delay in long BEOL Cu lines increasingly dominates the overall chip performance, the replacement of long-distance metallic connections with photonic links has been proposed as a way to drastically improve both performance and power efficiency. The availability of alternative channel materials would then allow the fabrication of all the required components for optical links, that is, lasers, modulators, and photodetectors, as well as the passive components – see Fig. 7.[17] For example, the availability of Ge in a CMOS process flow is advantageous for high-bandwidth photonics, since the direct energy band gap in Ge is only slightly larger than the indirect band gap, permitting the fabrication of efficient photodetectors and laser diodes at a wavelength of 1.55 µm.[50] Recently, photonic components based on 2D materials have shown great promise, on the basis of the excellent optical properties observed in monolayer graphene and MoS_2.[49, 51, 52]

Integration of graphene and other 2D materials in the CMOS platform also holds the promise of integrating completely new functionalities, since the very large surface to volume ratio of these materials has enormous potential in such applications as energy storage and sensing. An example of a CMOS-compatible

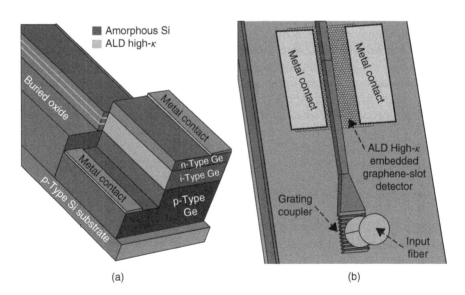

(a)   (b)

Figure 7. Schematics of the integration of (a) Ge and (b) graphene photodetectors with dielectric waveguides. (Source: Naiini *et al.*[17, 49] Reproduced with permission of IEEE).

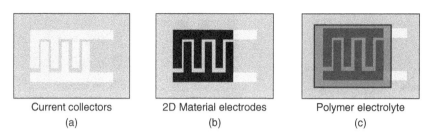

| Current collectors | 2D Material electrodes | Polymer electrolyte |
| (a) | (b) | (c) |

Figure 8. Back end of the line fabrication of an all-solid-state planar supercapacitor with 2D material electrodes: (a) metal current collectors by lift-off, (b) inkjet-printed 2D material electrodes, and (c) printed polymer electrolytes.

supercapacitor energy storage device is illustrated in Fig. 8. The integration of 2D materials in a sequential 3D integration flow can be obtained by transferring them on top of the BEOL of a fully processed chip, using techniques such as wafer-scale graphene transfer[53] and inkjet printing.[54] Integration of graphene on a CMOS chip can provide added functionality through the use of hot-electron transistors,[55, 56] pressure sensors,[57] and supercapacitors[54], leading to the possibility of extending SoC designs to include high-speed analog circuits, sensors, and energy sources.

4. Implications of 3D integration at the system level

Besides having a strong impact on FEOL and BEOL processing and on material choices and integration schemes, sequential 3D integration also requires innovations on many levels of the VLSI design flow, from layout generation to place-and-route to system partitioning and IP blocks.[58] The specific impact on these three areas depends strongly on how circuits are allocated to different layers, whether at the transistor level, gate level, or block level.[59]

In transistor-level allocation, the individual transistors within each circuit can be allocated to separate layers, with the aim of achieving area savings within each cell. This methodology is in principle the most flexible, especially if the designer is free to allocate the transistors to the same or different layers independently of the device polarity. However, this would, on the one hand, lead to an explosion in the number of cells in the standard cell library and, on the other hand, negate the positive impacts on yield and cost of dedicating each layer to one transistor type. In a practical scenario, all cells should be designed so that nFETs and pFETs are on separate, noninterchangeable layers, as shown in Fig. 9. While some individual cells have been shown to benefit from such an allocation, with area savings in the 30–45% range,[60, 61] to date there has been no clear demonstration that an entire cell library can be designed with consistent area savings and meeting all specifications, especially considering that transistor

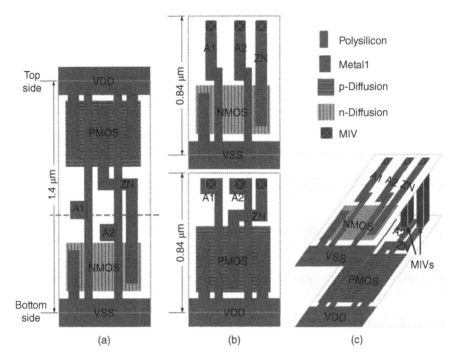

Figure 9. Example of a 3D standard cell. The conventional inverter can be redesigned in 3D by localizing the nFET and pFET in different layers with appropriate intracell and external connections. (Source: Lee et al.[60] Reproduced with permission of IEEE).

matching will necessarily be worse compared to a traditional design (where all transistors in a cell are fabricated at the same time in close proximity, significantly reducing intertransistor variability). On the other hand, if a fully 3D cell library with good performance and high density can be designed, then the impact on place-and-route and system-level partitioning would be minimal, although some IP blocks would still need to be redesigned.

In cell-level allocation, the standard cell design is kept two-dimensional, even though different versions of the cells would likely be necessary to take into account the fact that transistors fabricated on different layers will, to some degree, exhibit different characteristics. This type of allocation requires a full redesign of IP blocks and place-and-route algorithms, placing a heavy burden on the design community. Some demonstrations have been achieved in this area, showing that 3D place-and-route can be achieved via specialized algorithms or by applying relatively simple modifications to the standard algorithms implemented in commercial EDA tools.[12, 62] The main impact of cell-level allocation on the underlying technology is that full CMOS FEOL is required on each layer. This limits the cost and yield benefits of 3D integration to possible die area savings or reduced manufacturing costs compared to manufacturing two separate chips.

In block-level allocation, both standard cells and IP blocks are kept two-dimensional, with the same caveat of possibly different transistor performance on different layers as in cell-level allocation. Also in this case, the underlying technology needs to support fabrication of both nFETs and pFETs on each layer. The advantages of block-level allocation lie in the reduced impact on the design phase, since most conventional blocks can be reused and since some degree of block-level 3D partitioning is already supported in commercial EDA tools, given its applicability to parallel 3D integration.[15]

At the time of this writing, there is no evidence indicating that one type of allocation is clearly superior to the others. The best allocation strategy will probably be application dependent, striking a balance between technology constraints (possibility of CMOS FEOL on each layer, degree of transistor matching between different layers, size and tolerance of interlayer vias, etc.), and design considerations (type of standard cell library, availability of place-and-route strategies, possibility of adapting existing IP blocks, etc.). Finally, there is still much research to be done at the system level to identify the applications (e.g., memory on logic, logic on logic, FPGA, analog on digital), where 3D integration can provide the greatest advantages in terms of performance, area, power, and cost.[63]

5.   Conclusion

Three-dimensional integration is widely considered today as a plausible alternative and complement to traditional 2D scaling, eliciting strong interest and support from both the academic community and the industry. As of today, different types of 3D integration are at very different stages of maturity: 3D packaging has reached a reasonably mature stage, parallel 3D is in the initial commercial stage, and sequential 3D is at an advanced research stage. This can be easily comprehended by considering that the three approaches have completely different impacts on both the VLSI design flow and on manufacturing, with sequential 3D requiring the greatest development effort.

Besides the obvious need to fully develop the processing technology required for sequential 3D integration and the necessary supporting design methodologies, there are other issues that need to be addressed before sequential 3D integration can be seriously considered in commercial products.

The first issue to address is the scalability of the approach. While linearly adding layers can be naively considered a scaling path, there is no indication so far that area and performance advantages would scale with the number of layers. On the other hand, cost and yield would likely degrade linearly with the number of layers, leading to the intuitive conclusion that there should be a maximum number of layers for the benefits to still outweigh the costs. This places a clear scaling limit on the approach. However, this question is not limited to sequential 3D but can be applied to all 3D techniques, requiring a serious analysis of the costs and benefits of 3D integration.

The second issue is specific to sequential 3D integration, as compared to parallel 3D and 3D packaging. Given that a sequential 3D chip is fabricated monolithically on wafer scale, there is no way to limit processing to only known good dies, with the consequence that the yield of the final 3D chip is the direct product of the yield in each layer. On the other hand, in 3D packaging and in most forms of parallel 3D integration, it is possible to discard failed dies and only produce 3D chips from known good dies. This results in improvements of the total yield, even though the yield loss in the chip-stacking phase has to be considered. There is still no clear demonstration, at the time of writing, that sequential 3D integration can compete with parallel 3D and with 3D packaging on yield and cost.

The third issue is related to the formation of a 3D ecosystem. All 3D approaches require a new design ecosystem, but sequential 3D has the greatest impact on the VLSI design flow and needs a completely different approach to standard cell design, place-and-route algorithms, and/or IP block design.

In conclusion, one can observe that the technological foundation for sequential 3D integration is almost in place, since great progress has been made to develop low-thermal-budget processing and integration of novel materials. In particular, encouraging results have been reported on the integration of alternative channel materials in 1D device structures suitable for future CMOS nodes. Moreover, much effort has been devoted to developing low-temperature processing both for Si and alternative channel material devices in the context of sequential 3D integration, as well as to address the scientific and technical questions for a potential CMOS integration of disruptive functionalities based on 2D materials. However, clear demonstrations of advantages at the system level and strong support from the EDA community are necessary to transform sequential 3D integration from promising lab results to a successful commercial solution.

Acknowledgments

This work was supported by the Swedish Foundation for Strategic Research, the Swedish Research Council, the European Union ERC Advanced Grant 228229 OSIRIS, the EUFP7 GRADE, and the ERC PoC iPublic. The authors would like to thank Jiantong Li, Saul Rodriguez, and Anderson Smith for helpful discussions.

References

1. G. E. Moore, "Cramming more components onto integrated circuits," *Electronics* **38**, 114–117 (1965).
2. R. H. Dennard, F. H. Gaensslen, H.-N. Yu, V. L. Rideout, E. Bassous, and A. R. LeBlanc, "Design of ion-implanted MOSFET's with very small physical dimensions," *IEEE J. Solid-State Circuits* **9**, 256–268 (1974).
3. P. Smeys, V. Mcgahay, I. Yang, *et al.*, "A high performance 0.13 μm SOI CMOS technology with Cu interconnects and low-κ BEOL dielectric," *Symp. VLSI Technol.* (2000), pp. 184–185.
4. M. Wiatr, T. Feudel, A. Wei, *et al.*, "Review on process-induced strain techniques for advanced logic technologies," *15th Int. Conf. Adv. Therm. Process. Semicond.* (2007), pp. 19–29.
5. K. Mistry, C. Allen, C. Auth, *et al.*, "A 45 nm logic technology with high-κ + metal gate transistors, strained silicon, 9 Cu interconnect layers, 193 nm dry patterning, and 100% Pb-free packaging," *Tech. Dig. IEDM* (2007), pp. 247–250.
6. K. Cheng, A. Khakifirooz, N. Loubet, *et al.*, "High performance extremely thin SOI (ETSOI) hybrid CMOS with Si channel NFET and strained SiGe channel PFET," *Tech. Dig. IEDM* (2012), pp. 18.1.1–18.1.4.
7. Q. Liu, M. Vinet, J. Gimbert, *et al.*, "High performance UTBB FDSOI devices featuring 20 nm gate length for 14 nm node and beyond," *Tech. Dig. IEDM* (2013), pp. 9.2.1–9.2.4.
8. C. Auth, C. Allen, A. Blattner, *et al.*, "A 22 nm high performance and low-power CMOS technology featuring fully-depleted tri-gate transistors, self-aligned contacts and high density MIM capacitors," *Symp. VLSI Technol.* (2012), pp. 131–132.
9. S. Natarajan, M. Agostinelli, S. Akbar, *et al.*, "A 14 nm logic technology featuring 2nd-generation FinFET transistors, air-gapped interconnects, self-aligned double patterning and a 0.0588 μm² SRAM cell size," *Tech. Dig. IEDM* (2014), pp. 3.7.1–3.7.3.
10. I. L. Markov, "Limits on fundamental limits to computation," *Nature* **512**, 147–154 (2014).
11. G. Yeap, "Smart mobile SoCs driving the semiconductor industry: Technology trend, challenges and opportunities," *Tech. Dig. IEDM* (2013), pp. 16–23.
12. K. Arabi, K. Samadi, and Y. Du, "3D VLSI: A scalable integration beyond 2D," *Int. Symp. Phys. Des.* (2015), pp. 1–7.
13. H. Tanaka, M. Kido, K. Yahashi, *et al.*, "Bit cost scalable technology with punch and plug process for ultra high density flash memory," *Symp. VLSI Technol.* (2007), pp. 14–15.
14. N. Chandrasekaran, "Challenges in 3D memory manufacturing and process integration," *Tech. Dig. IEDM* (2013), pp. 344–348.
15. J. Lau, "Evolution, challenge, and outlook of TSV, 3D IC integration and 3D silicon integration," *Int. Symp. Mater.* (2011), pp. 462–488.

16. P. Batude, M. Vinet, B. Previtali, *et al.*, "Advances, challenges and opportunities in 3D CMOS sequential integration," *Tech. Dig. IEDM* (2011), pp. 151–154.

17. M. M. Naiini, H. H. Radamson, G. Malm, and M. Östling, "Integrating 3D PIN germanium detectors with high-κ ALD fabricated slot waveguides," *15th Intern. Conf. Ultim. Integr. Silicon (ULIS)* (2014), pp. 45–48.

18. C. Dupré, A. Hubert, S. Bécu, *et al.*, "15 nm-diameter 3D stacked nanowires with independent gates operation: ΦFET," *Tech. Dig. IEDM* (2008), pp. 1–4.

19. R. Pillarisetty, B. Chu-Kung, S. Corcoran, *et al.*, "High mobility strained germanium quantum well field effect transistor as the p-channel device option for low power ($V_{CC} = 0.5$ V) III–V CMOS architecture," *Tech. Dig. IEDM* (2010), pp. 6.7.1–6.7.4.

20. B. Duriez, G. Vellianitis, M. J. H. van Dal, *et al.*, "Scaled p-channel Ge FinFET with optimized gate stack and record performance integrated on 300 mm Si wafers," *Tech. Dig. IEDM* (2013), pp. 522–525.

21. S. Gupta, V. Moroz, L. Smith, Q. Lu, and K. C. Saraswat, "A group IV solution for 7 nm FinFET CMOS: Stress engineering using Si, Ge and Sn," *Tech. Dig. IEDM* (2013), pp. 641–644.

22. W. Chern, P. Hashemi, J. T. Teherani, *et al.*, "High mobility high-κ-all-around asymmetrically-strained Germanium nanowire trigate p-MOSFETs," *Tech. Dig. IEDM* (2012), pp. 387–390.

23. J. Franco, B. Kaczer, M. Toledano-Luque, *et al.*, "Superior reliability of high mobility (Si)Ge channel pMOSFETs," *Microelectronics Eng.* **109**, 250 (2013).

24. L. Czornomaz, N. Daix, K. Cheng, *et al.*, "Cointegration of InGaAs n- and SiGe p-MOSFETs into digital CMOS circuits using hybrid dual-channel ETXOI substrates," *Tech. Dig. IEDM* (2013), pp. 2.8.1–2.8.4.

25. U. E. Avci and I. A. Young, "Heterojunction TFET scaling and resonant-TFET for steep subthreshold slope at sub-9 nm gate-length," *Tech. Dig. IEDM* (2013), pp. 96–99.

26. S. W. Chang, X. Li, R. Oxland, *et al.*, "InAs N-MOSFETs with record performance of $I_{ON} = 600$ µA/µm at $I_{OFF} = 100$ nA/µm ($V_D = 0.5$ V)," *Tech. Dig. IEDM* (2013), pp. 417–420.

27. T.-W. Kim, D.-H. Kim, D. H. Koh, *et al.*, "Sub-100 nm InGaAs quantum-well (QW) tri-gate MOSFETs with Al_2O_3/HfO_2 (EOT < 1 nm) for low-power logic applications," *Tech. Dig. IEDM* (2013), pp. 425–428.

28. K. Garidis, G. Jayakumar, A. Asadollahi, E. Dentoni Litta, P. Hellström, and M. Östling, "Characterization of bonding surface and electrical insulation properties of inter layer dielectrics for 3D monolithic integration," *Joint EUROSOI–ULIS Workshop* (2015), pp. 165–168.

29. C. Claeys, J. Mitard, G. Hellings, *et al.*, "Status and trends in Ge CMOS technology," *ECS Trans.* **54**, 25–37 (2013).

30. P. Hashemi and J. L. Hoyt, "High hole-mobility strained-Ge/$Si_{0.6}Ge_{0.4}$ P-MOSFETs with high-κ/metal gate: Role of strained-Si cap thickness," *IEEE Electron Device Lett.* **33**, 173–175 (2012).

31. S. Takagi, R. Zhang, S.-H. Kim, *et al.*, "MOS interface and channel engineering for high-mobility Ge/III–V CMOS," *Tech. Dig. IEDM* (2012), pp. 505–508.

32. A. Toriumi, C. Lee, C. Lu, and T. Nishimura, "High electron mobility n-channel Ge MOSFETs with sub-nm EOT," *ECS Trans.* **64**, 55–59 (2014).

33. R. Zhang, N. Taoka, P.-C. Huang, M. Takenaka, and S. Takagi, "1-nm-thick EOT high mobility Ge n- and p-MOSFETs with ultrathin GeO_x/Ge MOS interfaces fabricated by plasma post-oxidation," *Tech. Dig. IEDM* (2011), pp. 642–645.

34. J. A. del Alamo, D. Antoniadis, A. Guo, *et al.*, "InGaAs MOSFETs for CMOS: Recent advances in process technology," *Tech. Dig. IEDM* (2013), pp. 24–27.

35. R. T. P. Lee, R. J. W. Hill, W.-Y. Loh, *et al.*, "VLSI processed InGaAs on Si MOSFETs with thermally stable, self-aligned Ni-InGaAs contacts achieving enhanced drive current and pathway towards a unified contact module," *Tech. Dig. IEDM* (2013), pp. 44–47.

36. N. Waldron, C. Merckling, W. Guo, *et al.*, "An InGaAs/InP quantum well FinFet using the replacement fin process integrated in an RMG flow on 300 mm Si substrates," *Symp. VLSI Technol.* (2014), pp. 26–27.

37. L.-Å. Ragnarsson, S. A. Chew, H. Dekkers, *et al.*, "Highly scalable bulk Fin-FET devices with multi-V_T options by conductive metal gate stack tuning for the 10-nm node and beyond," *Tech. Dig. VLSI Symp.* (2014), no. 6894359.

38. A. Veloso, L.-Å. Ragnarsson, T. Schram, *et al.*, "Integration challenges and options of replacement high-κ/metal gate technology for (sub-)22 nm technology nodes," *ECS Trans.* **52**, 385–390 (2013).

39. K. Choi, T. Ando, E. A. Cartier, *et al.*, "The past, present and future of high-κ/metal gates," *ECS Trans.* **53**, 17–26 (2013).

40. M. M. Frank, Y. Zhu, S. W. Bedell, T. Ando, and V. Narayanan, "Gate stacks for silicon, silicon germanium, and III–V channel MOSFETs," *ECS Trans.* **61**, 213–223 (2014).

41. E. Dentoni Litta, P.-E. Hellström, and M. Östling, "Integration of $TmSiO/HfO_2$ dielectric stack in sub-nm EOT high-κ/metal gate CMOS technology," *IEEE Trans. Electron Devices* **62**, 934–939 (2015).

42. E. Dentoni Litta, P.-E. Hellström, and M. Östling, "Enhanced channel mobility at sub-nm EOT by integration of a TmSiO interfacial layer in HfO_2/TiN high-κ/metal gate MOSFETs," *IEEE J. Electron Devices Soc.* **3**, 397–404 (2015).

43. M. Östling, E. Dentoni Litta, and P.-E. Hellström, "Recent advances in high-κ dielectrics and inter layer engineering," *12th IEEE Int. Conf. Solid-State Integr. Circuit Technol.* (2014), pp. 1–6.

44. P.-E. Hellström, E. Dentoni Litta, and M. Östling, "Interfacial layer engineering using thulium silicate/germanate for high-κ/metal gate MOSFETs," *ECS Trans.* **64**, 249–260 (2014).

45. M. Vinet, P. Batude, C. Fenouillet-Beranger, *et al.*, "Monolithic 3D integration: A powerful alternative to classical 2D scaling," *IEEE SOI-3D-Subthreshold Microelectron. Technol. Unified Conf.* (2014), pp. 1–3.

46. P. Batude, C. Fenouillet-Beranger, L. Pasini, *et al.*, "3DVLSI with Cool-Cube process: An alternative path to scaling," *Symp. VLSI Technol.* (2015), pp. 48–49.

47. C. H. Shen, J. M. Shieh, T. T. Wu, *et al.*, "Monolithic 3D chip integrated with 500 ns NVM, 3 ps logic circuits and SRAM," *Tech. Dig. IEDM* (2013), pp. 232–235.

48. M. M. Shulaker, T. F. Wu, A. Pal, *et al.*, "Monolithic 3D integration of logic and memory: Carbon nanotube FETs, resistive RAM, and silicon FETs," *Tech. Dig. IEDM* (2014), pp. 638–641.

49. M. M. Naiini, S. Vaziri, A. D. Smith, M. C. Lemme, and M. Östling, "Embedded graphene photodetectors for silicon photonics," *Device Res. Conf. (DRC)* (2014), p. 43.

50. F. K. Hopkins, K. M. Walsh, A. Benken, *et al.*, "Germanium on silicon to enable integrated photonic circuits," *Proc. SPIE* **8876**, 88760X (2013).

51. T. J. Echtermeyer, L. Britnell, P. K. Jasnos, *et al.*, "Strong plasmonic enhancement of photovoltage in graphene," *Nature Commun.* **2**, article no. 458 (2011).

52. O. Lopez-Sanchez, D. Lembke, M. Kayci, A. Radenovic, and A. Kis, "Ultrasensitive photodetectors based on monolayer MoS_2," *Nature Nanotechnol.* **8**, 497–501 (2013).

53. A. D. Smith, S. Vaziri, S. Rodriguez, M. Östling, and M. C. Lemme, "Large scale integration of graphene transistors for potential applications in the back end of the line," *Solid State Electron.* **108**, 61–66 (2015).

54. J. Li, F. Ye, S. Vaziri, M. Muhammed, M. C. Lemme, and M. Östling, "Efficient inkjet printing of graphene," *Adv. Mater.* **25**, 3985–3992 (2013).

55. S. Vaziri, G. Lupina, C. Henkel, *et al.*, "A graphene-based hot electron transistor," *Nano Lett.* **13**, 1435–1439 (2013).

56. S. Vaziri, M. Belete, E. Dentoni Litta, *et al.*, "Bilayer insulator tunnel barriers for graphene-based vertical hot-electron transistors," *Nanoscale* **7**, 13096–13104 (2015).

57. A. D. Smith, F. Niklaus, A. Paussa, *et al.*, "Electromechanical piezoresistive sensing in suspended graphene membranes," *Nano Lett.* **13**, 3237–3242 (2013).

58. O. Billoint, H. Sarhan, I. Rayane, *et al.*, "A comprehensive study of monolithic 3D cell on cell design using commercial 2D tool," *Design Autom. Test Europe (DATA) Conf.* (2015), pp. 1192–1196.

59. S. Panth, K. Samadi, Y. Du, and S. K. Lim, "Design and CAD methodologies for low power gate-level monolithic 3D ICs," *Intern Symp. Low Power Electron. Design (ISLPED)* (2014), pp. 171–176.

60. Y.-J. Lee, P. Morrow, and S. K. Lim, "Ultra high density logic designs using transistor-level monolithic 3D integration," *Int. Conf. Comput. Design* (2012), pp. 539–546.

61. C. Liu and S. K. Lim, "Ultra-high density 3D SRAM cell designs for mono-lithic 3D integration," *IEEE Intern. Interconnect Technol. Conf.* (2012), pp. 1–3.

62. S. Bobba, A. Chakraborty, O. Thomas, *et al.*, "CELONCEL: Effective design technique for 3-D monolithic integration targeting high performance integrated circuits," *16th Asia South Pacific Design Autom. Conf.* (2011), pp. 336–343.

63. M. S. Ebrahimi, G. Hills, M. M. Sabry, *et al.*, "Monolithic 3D integration advances and challenges: From technology to system levels," *IEEE SOI-3D-Subthreshold Microelectron. Technol. Unified Conf.* (2014), pp. 1–2.

1.5

Challenges to Ultralow-Power Semiconductor Device Operation

Francis Balestra
IMEP-LAHC, Minatec, Grenoble-Alpes University, 38016 Grenoble Cedex 1, France

1. Introduction

The historic trend in micro-/nanoelectronics over the last 40 years has been to increase both speed and density by scaling down the electronic devices, together with reduced energy dissipation per binary transition, and to develop many novel functionalities for future electronic systems. We are facing today dramatic challenges for "more Moore" and "more than Moore" applications: substantial increase of energy consumption and heating that can jeopardize future IC integration and performance, reduced performance due to limitation in traditional high-conductivity metal/low-κ dielectric interconnects, limit of optical lithography, heterogeneous integration of new functionalities for future nanosystems, and so on.

Therefore, many breakthroughs, disruptive technologies, novel materials, and innovative devices are needed in the next two decades.

With respect to the substantial reduction of the static and dynamic power of future high-performance/ultralow-power terascale integration logic and autonomous nanosystems, new materials and novel device architectures are mandatory for ultimate complementary metal-oxide-semiconductor (CMOS) and beyond-CMOS eras, as well as new circuit design techniques, architectures, and embedded software.

Alternative memories, especially PCRAM, RRAM, or MRAM will be useful for pushing the limit of integration and performance beyond those afforded by present nonvolatile, DRAM and SRAM memories.

In the interconnect area, optical and RF interconnects, carbon or other 2D materials can overcome the present limitations of copper interconnects for power consumption and performance.

Concerning ultimate processing technologies, extreme ultraviolet (EUV) lithography, immersion multiple patterning, multi-e-beam maskless or imprint

Future Trends in Microelectronics: Journey into the Unknown, First Edition.
Edited by Serge Luryi, Jimmy Xu, and Alexander Zaslavsky.
© 2016 John Wiley & Sons, Inc. Published 2016 by John Wiley & Sons, Inc.

lithography, as well as self-assembly of nanodevices could be used for different applications.

Future autonomous nanosystems will also need the development of nanoscale high-performance novel functionalities, which could be integrated on CMOS platforms. In this respect, nanostructures and nanodevices, especially nanowires (NWs), are of great interest, for instance, for improving the sensitivity, resolution, selectivity, energy consumption, and response time of nanosensors, and for solar, mechanical, and thermal energy harvesting.

Parallel or sequential processes could be used for the integration of these future high-performance and low-power sustainable, secure, ubiquitous and pervasive systems, which will be of high added value for many applications in the field of (autonomous) detection and communication of health problems, environment quality and security, secure transport, building and industrial monitoring, entertainment, education, and so on.[1–14]

This chapter focuses on the main trends, challenges, limits, and possible solutions for ultralow-power nanoscale devices in the CMOS and beyond-CMOS arena, including many novel materials, ultrathin films, multigates, nanowires, and small slope switches.

2. Ultimate MOS transistors

The slowdown of V_{DD} scaling and the substantial increase of the subthreshold leakage lead to a dramatic enhancement of the dynamic and static power consumption. This power challenge is due to the subthreshold slope limit S, which is 60 mV/dec at room temperature for MOSFETs. A lower limit in energy per operation can be reached with minimum $V_{DD} \sim S$ and minimum $E \sim CS^2$, where C is the device capacitance, as illustrated schematically in Fig. 1.[15]

There are several ways of reducing the swing SS, given by

$$\text{SS} = \frac{\partial V_G}{\partial(\log I_D)} = \frac{\partial V_G}{\partial \psi_S}\frac{\partial \psi_S}{\partial(\log I_D)} = 2.3\left(1 + \frac{C_S}{C_{ins}}\right)\frac{k_B T}{q} = 2.3m\frac{k_B T}{q}. \quad (1)$$

One can decrease of the transistor body factor m, using ultrathin body silicon-on-insulator (SOI), multigate or nanowire MOSFETs, carbon nanotube or 2D channels (leading to $m \sim 1$), and negative capacitance field-effect transistors (FETs), or MEMS/NEMS structures (leading to $m < 1$). Alternatively, one can resort to low-temperature operation, which cannot be applied to traditional applications. Finally, one can modify the carrier injection mechanisms by turning to impact ionization or band-to-band tunneling (BTBT).

In the field of MOSFETs, several very interesting advances have been recently reported. Ultrathin semiconductor films and 3D-FETs can improve sub-20 nm CMOS node performance and substantially reduce supply voltage and short channel effects.

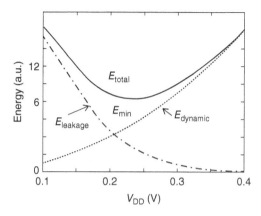

Figure 1. Lower limit in energy per operation obtained for optimal drain bias V_{DD}.

The best MOSFET devices leading to SS close to its minimum value are using fully depleted channels[16–19] (e.g., FD-SOI with very thin buried oxide[19]) or fully inverted ones with volume inversion[20] that is even more effective into optimizing electrostatic control[21] of the structure (e.g., double-gate (DG), bulk, or SOI tri-gate/fin-FET, gate-all-around (GAA) MOSFET or nanowire FET).

The first fully depleted SOI MOSFETs have been fabricated on SOS materials[16] but with Si films of poor quality. The first FD-SOI MOSFET with an ideal swing of about 60 mV/dec was demonstrated by numerical simulation in 1985 (see Fig. 2).[17] The first experimental demonstration of an FD-SOI MOSFET with SS ~ 60 mV/dec followed in 1986.[18]

Short channel effects (DIBL and charge sharing) in multigate MOSFETs are also reduced with decreasing the silicon film thickness t_{Si} (down to 5 nm) and doping, regardless of whether the architecture is single- or DG (see Fig. 3). Furthermore, a reduced sensitivity to t_{Si} and doping is observed for the DG devices, which is promising for the optimization of their electrical properties.[21]

DG devices with gate underlap have also shown to lead to a very good I_D, a reduction in the drain-induced barrier lowering (DIBL) and drain-to-gate leakage current, and a decrease in propagation delay and power.[22]

Small diameter (3 nm) nanowire FETs have demonstrated experimentally very good I_{ON}/I_{OFF} performance,[23] with the best SS ~ 75 mV/dec obtained by numerical simulation down to 5 nm gate length for both strained Si and InGaAs channel materials, as shown in Fig. 4.[24]

In addition to III–V materials, there are other high-mobility materials that may replace Si in future VLSI applications. In this respect, heterogeneous 2D atomic crystals, especially transition-metal dichalcogenides (TMDs), have atomically smooth surface without dangling bounds or thickness fluctuations, as well as good mobility in chemical vapor deposition (CVD)-deposited films. They are thus very attractive enablers of ultimately scaled transistors and 3D ICs. The first

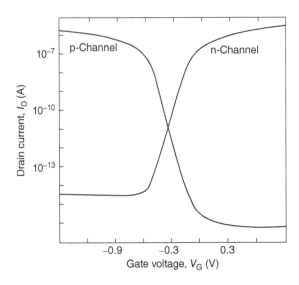

Figure 2. Transfer $I_D(V_G)$ characteristics obtained by numerical simulation for fully depleted n- and p-channel SOI MOSFETs showing a 60 mV/dec subthreshold swing (10^{-14} Si doping, 600 nm Si film).[17]

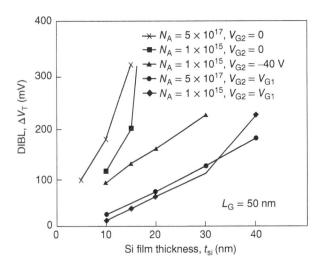

Figure 3. DIBL effect versus silicon film thickness t_{Si} for $L_G = 50$ nm devices: single-gate SOI MOSFETs ($t_{OX1} = 380$ nm; $t_{OX2} = 3$ nm) with high doping ($N_A = 5 \times 10^{17}$ cm^{-3}, $V_{G2} = 0$), low doping ($N_A = 10^{15}$ cm^{-3}, $V_{G2} = 0$), back channel accumulation ($N_A = 10^{15}$ cm^{-3}, $V_{G2} = -40$ V); and for double-gate SOI MOSFETs ($t_{OX1} = t_{OX2} = 3$ nm, $N_A = 10^{15}$ cm^{-3} and $N_A = 5 \times 10^{17}$ cm^{-3} with $V_{G1} = V_{G2}$).

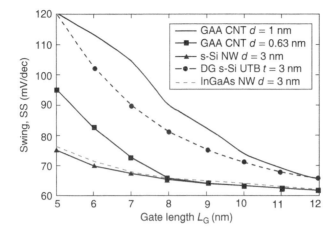

Figure 4. Subthreshold swing SS obtained by quantum transport numerical simulation as a function of device architecture for gate length L_G down to 5 nm: double-gate strained-Si ultrathin-body MOSFET ($t_{Si} = 3$ nm), Ω-gate strained-Si and InGaAs nanowires with 3 nm diameter, gate-all-around carbon nanotubes (GAA CNTs) with 1 and 0.63 nm diameters.

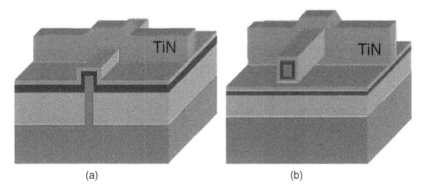

Figure 5. Si/MoS$_2$/HfO$_2$/TiN hybrid fin-FET (a) and nanowire transistor (b).

CMOS-compatible TMD 3D transistor technology using novel hybrid Si/MoS$_2$ channel fin-FET and NW FET was demonstrated recently (see Fig. 5).

These 3D TMD transistors exhibit improved electron mobility compared to Si-based 3D FETs as well as higher drive current I_{ON}, as illustrated in Figs 6 and 7, respectively.[25] It has also been shown that MoS$_2$ FETs can meet high-performance (HP) requirement up to 6.6 nm node by employing bilayer MoS$_2$ as the channel material, while low-standby-power (LSTP) requirements present significant challenges for sub-10 nm nodes. On the other hand, the high

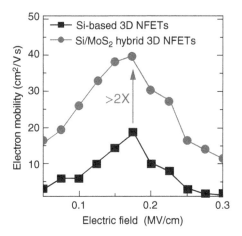

Figure 6. Field effect mobility of n-type Si/MoS$_2$ fin-FET showing a substantial increase compared with Si fin-FET.

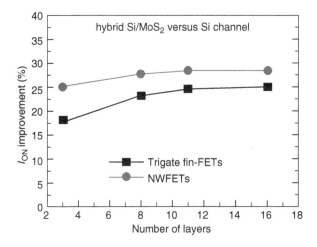

Figure 7. Improvement of I_{ON} current drive for hybrid Si/MoS$_2$ devices compared with Si fin-FET and NW FETs as a function of the number of MoS$_2$ layers.

mobility and the low effective mass of tungsten diselenide (WSe$_2$) may enable 2D FETs for both HP and LSTP applications down to the 5 nm node.

The subthreshold swing and DIBL were evaluated for both DG and SOI structures from monolayer (1L) to three-layer (3L) MoS$_2$, and compared to Si-based ultrathin-body (UTB) DG FETs. It was shown that MoS$_2$ FETs generally have better electrostatics than Si devices. Single-gate SOI topology can only sustain good electrostatics for 1L TMDs, while DG topology can present interesting performance for up to 3L TMDs, as shown in Fig. 8.[26]

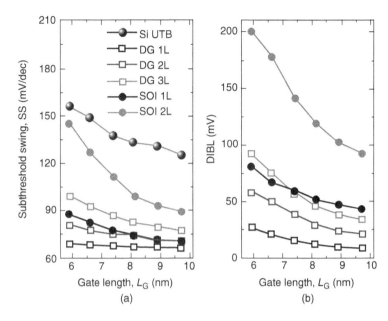

Figure 8. (a) Subthreshold swing and (b) DIBL versus L_G scaling for MoS$_2$ FETs realized with 1, 2, or 3 monolayers, compared to Si-based UTB DG FETs with a symmetrical top and bottom gate dielectric (in SOI the bottom dielectric is 50 nm thick SiO$_2$ and the bottom gate is grounded).

Carbon nanotube FETs, with a very good control of short channel effects and with the highest current drive to-date ($>100\,\mu A/\mu m$ at 400 nm channel length and 1 V drain bias) have been demonstrated while simultaneously achieving reasonable $I_{ON}/I_{OFF} > 5000$.[27]

On the other hand, silicon fin-FETs with silicided source/drain, exploiting biased Schottky barriers, have been demonstrated. By combining a positive feedback induced by weak impact ionization with a dynamic modulation of the Schottky barriers, the device achieves a minimal subthreshold slope of 3 mV/dec and an average subthreshold slope of 6.0 mV/dec over 5 dec of current at room temperature. However, a drain bias of several volts is needed.[28]

A comparative analysis of ring oscillators based on different channel materials, different spacer materials, and different transistor architectures suggests that the largest benefits of speed gain and power consumption reduction for MOSFET architectures is achieved by switching from 7 nm Si baseline fin-FET process to 5 nm vertical Si NWs.[29]

Another interesting possibility at the end of the roadmap could be to use semimetals for channel and source/drain. Indeed, for 1D devices, the band gap of semiconductors and semimetals increases with decreasing NW diameter. Therefore, a semimetal could become a semiconductor in the sub-10 nm range, as shown in Fig. 9. In this case, 1D structures could be used for the channel

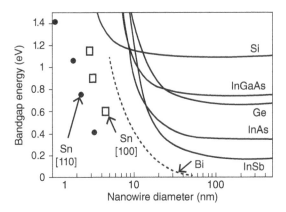

Figure 9. Band gap versus NW diameter for various semiconductors and semimetals.

and thicker metallic source/drain regions could be made from the same material (e.g., Sn, Bi) without any extrinsic doping. Very good electrical properties have been shown by *ab-initio* simulations for 1 nm GAA Sn NWs with a 2.3 nm gate length, leading to a 72 mV/dec swing and $I_{ON} \sim 3$ mA/µm at $V_D = 250$ mV.[30]

3. Small slope switches

In the domain of small slope switches, the tunneling field-effect transistors (TFETs) seem very promising. However, so far, no one has demonstrated experimentally a TFET that has simultaneously both an I_{ON} comparable to a CMOS FET and a subthreshold swing of less than 60 mV/dec for several decades. The highest current for which a subthreshold swing of 60 mV/dec is observed is approximately 10 nA/µm. While the experimentally observed I_{ON} are in reasonable agreement with theoretical predictions, there is disparity in the subthreshold characteristics due to the fact that simulations usually neglect nonidealities and defects and that these advanced technologies using novel materials, technologies, and device architectures have to be optimized.[31]

The originally studied TFET configurations also suffered from low I_{ON} current levels usually in the order of few µA/µm caused by the misalignment of the tunneling direction and the gate-induced electric field. A new class of TFETs was introduced in which both BTBT direction and gate electric field are kept parallel, which allows I_{ON} to be proportional to the gate length. Among such transistors, the electron–hole bilayer tunneling field-effect transistor (EHBTFET) was proposed to exploit BTBT between a 2D electron gas and a 2D hole gas. Tunneling between 2D regions was demonstrated to be the most advisable scenario to optimize switching behavior. Semiclassical simulations for Ge EHBTFET showed impressive results that later on needed to be corrected

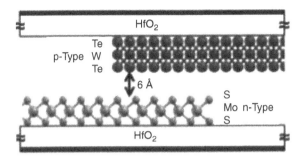

Figure 10. Proposed MoS_2–WTe_2 double-gate TFET structure.

for quantum mechanical effects and parasitic lateral tunneling and its undesired implications for the device performance.

It has been demonstrated that the inclusion of quantum confinement in Ge EHBTFET gives rise to the appearance of tunneling paths misaligned with the gate electric field that were not present in the semiclassical framework. This unwanted leakage between the overlap and underlap regions of the device reduced the steepness of the I–V curves and degraded SS. To overcome this problem, a heterogate EHBTFET may provide a potential solution to suppress the lateral tunneling in the subthreshold region while maintaining the high I_{ON} due to vertical BTBT in the overlap region.[32]

Simulated DG TFETs with 2D layers using MoS_2–WTe_2 vertical tunneling, shown in Fig. 10, also showed very good subthreshold swing.[33]

Finally, TFETs realized with III–V materials, for example, InAs, strained InAs, GaSb–InAs heterostructures, graded AlGaSb source H-TFET and quantum well (QW) InAs–GaSb source with InAs channel have been compared by quantum simulations. The best results were obtained for the QW-TFET, with graded source TFET coming second, and both significantly outperforming the reference MOSFET – see Fig. 11.[34]

4. Conclusion

We are facing dramatic challenges dealing with future nanoscale devices, including performance, power consumption, new materials, device integration, interconnects, ultimate technological processes, and novel functionalities, needing disruptive approaches and inducing fundamental trade-offs for future ICs and nanosystems. This chapter reviewed recent advances and promising solutions for future ultralow power devices combining novel materials and innovative device architectures, including 2D layers, TMDs, heterostructures and quantum wells using strained Si, Ge, and III–V thin films, as well as multigate, nanowire, CNT transistors, and tunnel FETs.

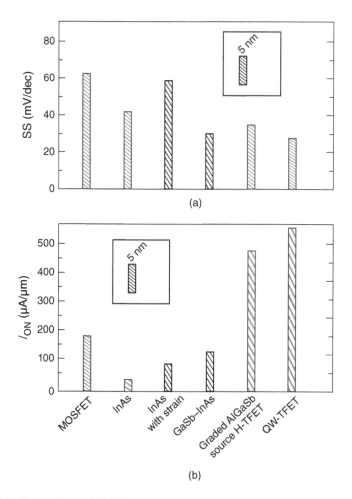

(a)

(b)

Figure 11. Comparison of MOSFETs performance (current drive I_{ON} and subthreshold swing SS) with InAs TFET, strained InAs TFET, GaSb–InAs heterostructure TFET, graded source AlGaSb–InAs HTFET, and QW InAs–GaSb–InAs TFET. In all cases, I_{OFF} is set at 5 nA/μm.

Acknowledgments

The author would like to thank the Sinano Institute Members and Partners of Nanosil and Nanofunction FP7 European Networks of Excellence.

References

1. F. Balestra, ed., *Nanoscale CMOS: Innovative Materials, Modeling and Characterization*, London: ISTE-Wiley, 2010.

2. F. Balestra, ed., *Beyond CMOS Nanodevices*, vols. 1 and 2, London: ISTE-Wiley, 2014.

3. J. Brini, M. Benachir, G. Ghibaudo, and F. Balestra, "Subthreshold slope and threshold voltage of the volume inversion MOS transistor," *Proc. IEE Part G: Circ. Devices Syst.* **138**, 133–136 (1991).

4. F. Balestra, M. Benachir, J. Brini, and G. Ghibaudo, "Analytical models of subthreshold swing and threshold voltage for thin and ultra-thin film SOI MOSFETs," *IEEE Trans. Electron Devices* **37**, 2303–2311 (1990).

5. F. Conzatti, M. G. Pala, D. Esseni, E. Bano, and L. Selmi, "A simulation study of strain induced performance enhancements in InAs nanowire tunnel-FETs," *Tech. Dig. IEDM* (2011), p. 95.

6. G. Dewey, B. Chu-Kung, J. Boardman, and J. M. Fastenau, "Fabrication, characterization and physics of III–V heterojunction tunneling FET for steep subthreshold swing," *Tech. Dig. IEDM* (2011), p. 785.

7. J. Appenzeller, Y.-M. Lin, J. Knoch, and Ph. Avouris, "Band-to-band tunneling in carbon nanotube field-effect transistors," *Phys. Rev. Lett.* **93**, 193805 (2004).

8. A. Padilla, C. W. Yeung, C. Shin, C. Hu, and T.-J. King Liu, "Feedback FET: A novel transistor exhibiting steep switching behavior at low bias voltages," *Tech. Dig. IEDM* (2008), p. 171.

9. K. Boucart, A. M. Ionescu, and W. Riess, "Asymmetrically strained all-silicon tunnel FETs featuring 1 V operation," *Proc. ESSDERC* (1989), p. 452.

10. F. Mayer, C. Le Royer, J.-F. Damlencourt, *et al.*, "Impact of SOI, $Si_{1-x}Ge_xOI$ and GeOI substrates on CMOS-compatible tunnel FET performance," *Tech. Dig. IEDM* (2008), pp. 1–5.

11. D. Esseni and M. G. Pala, "Interface traps in InAs nanowire tunnel FETs and MOSFETs – Part II: Comparative analysis and trap-induced variability," *IEEE Trans. Electron Devices* **60**, 2802–2807 (2013).

12. S. Brocard, M. Pala, and D. Esseni, "Design options for hetero-junction tunnel FETs with high on current and steep sub-threshold voltage slope," *Tech. Dig. IEDM* (2013), p. 5.4.1.

13. M. H. Lee, J.-C. Lin, Y.-T. Wei, *et al.*, "Ferroelectric negative capacitance hetero-tunnel field-effect-transistors with internal voltage amplification," *Tech. Dig. IEDM* (2013), p. 104.

14. S. Brocard, M. Pala, and D. Esseni, "Large on-current enhancement in hetero-junction tunnel-FETs via molar fraction grading," *IEEE Trans. Electron Devices* **35**, 184–186 (2014).

15. S. Hanson, M. Seok, D. Sylvester, and D. Blaauw, "Nanometer device scaling in subthreshold logic and SRAM," *IEEE Trans. Electron Devices* **55**, 175–185 (2008).

16. F. Balestra, J. Brini, and P. Gentil, "Deep depleted SOI MOSFETs with back potential control: A numerical simulation," *Solid State Electron.* **28**, 1031–1037 (1985).

17. F. Balestra, Characterization and Simulation of SOI MOSFETs with Back Potential Control, PhD thesis, INP-Grenoble, 1985.

18. J. P. Colinge, "Subthreshold slope of thin-film SOI MOSFETs," *IEEE Electron Device Lett.* **7**, 244–246 (1986).

19. F. Andrieu, O. Weber, J. Mazurier, O. Thomas, J.-P. Noel, and C. Fenouillet-Béranger, "Low leakage and low variability ultra-thin body and buried oxide (UT2B) SOI technology for 20 nm low-power CMOS and beyond," *Proc. VLSI Symp.* (2010), pp. 57–58.

20. F. Balestra, S. Cristoloveanu, M. Benachir, J. Brini, and T. Elewa, "Double-gate silicon-on-insulator transistor with volume inversion: a new device with greatly enhanced performance," *IEEE Electron Device Lett.* **8**, 410–412 (1987).

21. E. Rauly, O. Potavin, F. Balestra, and C. Raynaud, "On the subthreshold swing and short channel effects in single and double gate deep submicron SOI MOSFETs," *Solid State Electron.* **43**, 2033–2037 (1999).

22. A. Bansal, B. C. Paul, and K. Roy, "Impact of gate underlap on gate capacitance and gate tunneling current in 16 nm DGMOS devices," *Proc. IEEE SOI Conf.* (2004), pp. 94–95.

23. C. Dupré, A. Hubert, S. Becu, *et al.*, "15 nm-diameter 3D stacked nanowires with independent gates operation: ΦFET," *Tech. Dig. IEDM* (2008), p. 549.

24. M. Luisier, M. Lundstrom, D. A. Antoniadis, and J. Bokor, "Ultimate device scaling: Intrinsic performance comparisons of carbon-based, InGaAs, and Si FETs for 5 nm gate length," *Tech. Dig. IEDM* (2011), p. 251.

25. M.-C. Chen, C.-Y. Lin, K.-H. Li, *et al.*, "Hybrid Si/TMD 2D electronic double channels fabricated using solid CVD few-layer-MoS_2 stacking for Vth matching and CMOS-compatible 3DFETs," *Tech. Dig. IEDM* (2014), p. 808.

26. W. Cao, J. Kang, D. Sarkar, W. Liu, and K. Banerjee, "Performance evaluation and design considerations of 2D semiconductor based FETs for sub-10 nm VLSI," *Tech. Dig. IEDM* (2014), p. 729.

27. M. M. Shulaker, T. F. Wu, A. Pal, *et al.*, "Monolithic 3D integration of logic and memory: Carbon nanotube FETs, resistive RAM, and silicon FETs," *Tech. Dig. IEDM* (2014), p. 812.

28. J. Zhang, M. De Marchi, P.-E. Gaillardon, and G. De Micheli, "A Schottky-barrier silicon FinFET with 6.0 mV/dec subthreshold slope over 5 decades of current," *Tech. Dig. IEDM* (2014), p. 339.

29. V. Moroz, L. Smith, J. Huang, *et al.*, "Modeling and optimization of group IV and III–V fin-FETs and nanowires," *Tech. Dig. IEDM* (2014), p. 180.

30. L. Ansari, G. Fagas, J.-P. Colinge, and J. C. Greer, "A proposed confinement modulated gap nanowire transistor based on a metal (tin)," *Nano Lett.* **12**, 2222–2227 (2012).

31. H. Lu and A. C. Seabaugh, "Tunnel field-effect transistors: State-of-the-art," *J. Electron Devices Soc.* **2**, 44–49 (2014).

32. J. L. Padilla, C. Alper, F. Gámiz, and A. M. Ionescu, "Assessment of pseudo-bilayer structures in the heterogate germanium electron-hole bilayer tunnel field-effect transistor," *Appl. Phys. Lett.* **106**, 262102 (2015).

33. K.-T. Lam, G. Seol, and J. Guo, "Performance evaluation of MoS_2–WTe_2 vertical tunneling transistor using real-space quantum simulator," *Tech. Dig. IEDM* (2014), p. 721.

34. M. Pala and S. Brocard, "Exploiting heterojunctions to improve the performance of III–V nanowire tunnel-FETs," *IEEE J. Electron Devices Soc.* **3**, 115–121 (2015).

1.6

A Universal Nonvolatile Processing Environment

T. Windbacher, A. Makarov, V. Sverdlov, and S. Selberherr
Institute for Microelectronics, TU Wien, 1040 Vienna, Austria

1. Introduction

After many decades of stunning progress in the shrinking of complementary metal-oxide-semiconductor (CMOS) devices, the steadily increasing difficulty in handling physical limitations as well as the rapidly increasing production and investment costs for each new technology generation will stop CMOS scaling in the not-too-distant future. Among the most challenging problems for further performance gains today are the static power dissipation as well as the interconnection delay and the associated energy for information transport.[1, 2] A very efficient solution to the static leakage power problem is to simply turn off unused parts of a circuit. However, this causes the previously stored information to vanish and requires energy- and time-wasting recovery cycles, when the dormant circuit parts are powered up. Thus, in order to avoid information loss during shutdown, nonvolatile elements must be incorporated.

Due to its CMOS compatibility, nonvolatility, high endurance, and fast operation, spintronics is a promising avenue for adding nonvolatility to circuits.[3] The term spintronics is very general and covers a vast number of devices with an extreme variety in operating principles and practical feasibility for commercial applications.[3, 4] In this chapter, we concentrate on what, in our opinion, appears to be the most feasible technology for large-scale integration in the next few years: the combination of CMOS with nonvolatile magnetoresistive random-access memory (MRAM). Indeed, the integration of CMOS and magnetic tunnel junctions (MTJs) is not only likely but already available in the form of nonvolatile stand-alone MRAM arrays and embedded DRAM,[5] and the introduction of further commercial products will surely follow.[6-9]

Importantly, the all-electrical magnetization manipulation in modern MRAM by spin transfer torque (STT) renders the wires for separate magnetic field generation superfluous and also significantly reduces the MTJ switching energy. Technological advances, such as the exploitation of free-layer perpendicular magnetic anisotropy and use of MgO tunnel barriers, have led to a further reduction in switching energy, as well as improved scalability.[10] Promising

Future Trends in Microelectronics: Journey into the Unknown, First Edition.
Edited by Serge Luryi, Jimmy Xu, and Alexander Zaslavsky.
© 2016 John Wiley & Sons, Inc. Published 2016 by John Wiley & Sons, Inc.

spintronic solutions with respect to speed and power consumption have been able to compete with CMOS-only solutions.[9, 11]

However, the usual CMOS/MTJ hybrid structures use MTJs only for storage, while the actual computation is still carried out via CMOS logic. The addition of MTJs also increases circuit complexity and footprint, as extra transistors are needed to read and write the MTJs to access the stored memory data.

2. Universal nonvolatile processing environment

Shifting the actual computation into the magnetic domain would help not only to simplify the layout but also increase the integration density. We propose a universal nonvolatile processing environment, schematically illustrated in Fig. 1, consisting of spin torque majority gates (STMGs)[12] and nonvolatile magnetic flip-flops,[13] illustrated in Figs 2 and 3, respectively. By arranging the STMGs and flip-flops in an array, the flip-flops can be exploited as shared buffers available for neighboring STMGs. In order to elaborate the concept further, a possible realization of an easily extendable 1-bit full adder with the aid of just a single STMG and three nonvolatile flip-flops is explained in the following section.

To fully understand the operation of the nonvolatile processing environment, one has to address first the operation principle of the devices as well as how the information is transferred between them. As illustrated in Figs 2 and 3, the devices are basically spin valves (nonmagnetic metal interconnection layers) or MTJs (oxide interconnection layers) with a common free layer. Thus, these devices are operated via current pulses and the polarities of the applied pulse represent the two logic states "0" and "1," respectively. The flip-flop has two

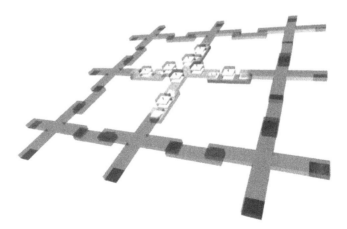

Figure 1. The spin torque majority gates (crosses) perform the computation, while the nonvolatile flip-flops (rectangles) act as shared buffers. A single STMG and three flip-flops are sufficient to realize a concatenated 1-bit full adder.

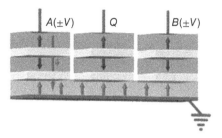

Figure 2. The nonvolatile magnetic flip-flop comprises three synthetic antiferro-magnetic polarizer stacks with perpendicular magnetization orientation. Two polar-izer stacks, A and B, are for inputs and one polarizer stack Q is for readout. They are connected via nonmagnetic interconnection layers to a common free magnetic layer featuring a uniaxial perpendicular magnetic anisotropy.

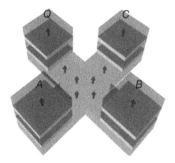

Figure 3. The STMG shares the basic structural features with the nonvolatile flip-flop but exhibits a cross-like common free layer and four polarizer stacks (A, B, C for input and Q for readout). The device is operated via polarity-encoded current pulses and the result of an operation is stored by the magnetization orientation of the free layer. The stored information can be accessed either by using the TMR or GMR effect or by employing the common free layer as a polarizer for a neighboring device.

inputs A and B. Four input combinations are possible. Depending on the applied pulse polarity in relation to the free layer orientation, the acting torque either drives growing precessions, which may lead to a change in the magnetization orientation, or it damps the magnetization movement and tries to keep it in its current position. In the case of two synchronous input pulses with the same polarity, two spin-polarized currents enter the free layer and cause two STTs acting in the same fashion. They add up and cause either the *set* or the *reset* of the common free layer orientation. On the other hand, if two synchronous input pulses with opposing polarities are applied, the two torques compensate each other and the initial magnetization orientation is preserved (*hold* operation). The STMG has the same operation principle, but instead of two input pulses there are always three synchronous pulses. Thus, for any input combination there is always at least one uncompensated STT contribution that prevails and determines the final state of the common free layer. This behavior replicates the *majority* function[12] and can facilitate combinational logic. One has to note that the *majority* function requires additionally the logic negation to form a computationally complete basis. The simplest solution for this is to assume that the *not* operation is carried out by inverting the polarity of the corresponding input pulse.

Finally, it is necessary to elucidate how information is transferred between these devices. As already pointed out above, the key to the operation is the orientation and interaction of the applied torques. Therefore, by applying a current pulse to one of the overlapping regions between neighboring devices, the electrons entering the first common free layer are polarized along the first layer's magnetization orientation – see Fig. 4. When they cross over into the second free layer, they relax to the magnetization orientation in the second free layer and by that create an STT encoded with the information stored in the first free layer. Since it takes much longer for a free layer to switch with one active input compared to the case when all inputs are active, there is a safe time window for copying information from a free layer without switching its magnetization.

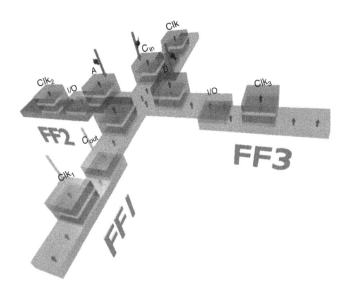

Figure 4. By performing a well-defined set of *majority*, *not*, and *copy* operations, an easily concatenable 1-bit full adder is realized. The overlapping region between the common free layers of the flip-flops and the STMG can be employed to first polarize electrons entering a free layer and then, when crossing over to an adjacent free layer of a neighboring device, to create a spin transfer torque orientation encoded with the information stored in the previous layer.

The assumed 1-bit full adder exhibits three inputs A, B, and C_{IN} and two outputs Sum and C_{OUT}. Furthermore, the logic function for the sum is given by[14]

$$Sum = A \; xor \; B \; xor \; C_{IN} \tag{1}$$

and the carry-out bit C_{OUT} is defined as

$$C_{OUT} = majority \left[A, B, C_{IN} \right]. \tag{2}$$

Due to the computational completeness of the proposed processing environment, any logic function can be transformed into a sequence of well-defined *majority* and *not* operations. For instance, in a first step one calculates *majority*[A, B, C_{IN}], with the resulting carry-out bit C_{OUT} subsequently copied into one of the adjacent flip-flops (e.g., FF1). In the next step, *majority*[A, B, *not*(C_{IN})] is performed and stored in another flip-flop (FF2). Finally, the *sum* bit is obtained by executing *majority*[*not*(FF1), FF2, C_{IN}] and moving the result into FF3. Thus, C_{OUT} and Sum are stored and accessible via FF1 and FF3, respectively. Since we chose to perform the calculation of C_{OUT} in the very first step, it can be already used by FF1's neighboring STMG as C_{IN} even before the calculation of sum is finished. In this way, the calculation is parallelizable and the exploitation of the flip-flops as shared local buffers reduces significantly information transport over the global bus.

3. Bias-field-free spin-torque oscillator

A further essential building block in modern electronics is the oscillator.[14] Unfortunately, spin-torque oscillators often require an external magnetic field or operate only at relatively low frequencies, which limits their practical implementation. Previously, we demonstrated that the nonvolatile magnetic flip-flop device intrinsically provides a spin-torque oscillator.[15] It operates without an external magnetic field at high frequencies and complements perfectly the proposed nonvolatile processing environment, thereby boosting the achievable integration density. In direct comparison to a CMOS ring oscillator (see Fig. 5) at an assumed half pitch of 15 nm, our proposed structure is approximately 30 times smaller.[3, 15]

In order to boost the output power as well as the operation frequency, a structure comprising two three-layer MTJ stacks with a shared free layer was proposed – see Fig. 6.[16] The oscillation frequency of the structure can be tuned by varying the amplitude of the applied current density of one of the MTJs, while the other one is kept fixed, as shown in Fig. 7. In this way, frequencies of up to 30 GHz can be excited.[17]

Furthermore, the oscillation spectrum contains a primary mode at frequency f and a secondary mode at frequency $f/2$. By increasing the applied current density j_B through MTJ_B, the amplitude of the secondary mode starts to increase and

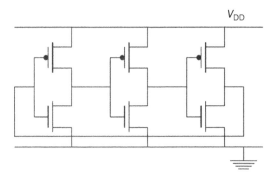

Figure 5. Circuit diagram for a minimal ring oscillator. The ring oscillator comprises an odd number of inverter stages (in this case three), where each output is fed to the input of the following stage. Above a certain supply voltage V_{DD} the circuit starts to oscillate.

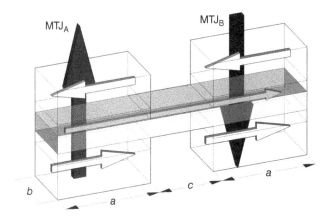

Figure 6. Schematic illustration of a spin-torque oscillator based on two MTJs. The arrows indicate the direction of the current for each of the MTJs.

eventually reaches more than half the maximum amplitude of the primary mode, whereas the output power of the primary mode decreases. There is also a significant dependence of the excited modes on the geometry of the free layer, shown in Fig. 8. Structures with free layer lengths ranging from 40 to 60 nm and current density combinations between 10^7 and 2.05×10^8 A/cm^2 have been investigated.

Increasing the length of the free layer and thereby the distance between the MgO-MTJs shifts the region where the additional large-amplitude oscillation mode appears in the direction of larger j_A/j_B ratios. Thus, the largest variation in the current density at which the additional mode does not reach large amplitudes is observed in the structure with a 40 nm free layer length.

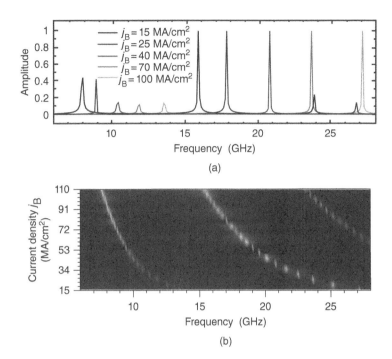

Figure 7. Illustration of the signal spectral density with normalized peak values. The length of the shared free layer has been set to 40 nm and the current density through MTJ_B was changed between 1.5×10^7 and 1.1×10^8 A/cm^2, while keeping the current density for MTJ_A fixed at $j_A = 2.05 \times 10^8$ A/cm^2.

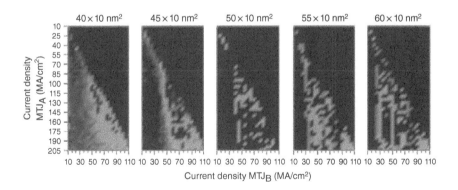

Figure 8. The geometrical dependence of the oscillation regime versus applied current density j_A through MTJ_A ($j_B \leq j_A/2$) for free layers of 40, 45, 50, 55, and 60 nm. Current density combinations for which the secondary oscillation mode $f/2$ reaches more than 50% of the primary mode f are shown in light gray.

4. Summary

The resulting nonvolatile processing environment features a highly regular structure, is computationally complete, and reduces the information transport due to its shared buffers. Thus, it is viable for a universal post-CMOS logic technology. The flip-flop is very versatile. The same device can be used stand alone or stacked for even higher integration density. It also offers the possibility of being used as a bias-field-free oscillator while preserving an extremely small footprint. For these reasons, we believe the presented processing environment and its components are very promising candidates for pushing the achievable integration density beyond the state-of-the-art CMOS-MTJ hybrids, while keeping the dissipated power and interconnection delay in check.

Acknowledgments

This research is supported by the European Research Council through the Grant #247056 MOSILSPIN.

References

1. International Technology Roadmap for Semiconductors, chapter PIDS (2013) available at http://www.itrs.net (accessed on Apr. 11, 2016).
2. R. Marculescu, U. Ogras, L.-S. Peh, N. Jerger, and Y. Hoskote, "Outstanding research problems in NoC design: System, microarchitecture, and circuit perspectives," *IEEE Trans. CAD Integr. Circ. Syst.* **28**, 3–21 (2009).
3. D. Nikonov and I. Young, "Overview of beyond-CMOS devices and a uniform methodology for their benchmarking," *Proc. IEEE* **101**, 2498–2533 (2013).
4. W. Zhao and G. Prenat, eds., *Spintronics-Based Computing*, New York: Springer, 2015.
5. See Everspin website (2015) available at www.everspin.com/mram-replaces-dram (accessed on Apr. 11, 2016).
6. Y. Zhang, W. Zhao, W. Kang, E. Deng, J.-O. Klein, and D. Revelosona, "Current-induced magnetic switching for high performance computing," chapter in: W. Zhao and G. Prenat, eds., *Spintronics-Based Computing*, New York: Springer, 2015, pp. 1–51.
7. E. Deng, Y. Zhang, W. Kang, B. Dieny, J.-O. Klein, G. Prenat, and W. Zhao, "Synchronous 8-bit non-volatile full-adder based on spin transfer torque magnetic tunnel junction," *IEEE Trans. Circ. Syst. I* **62**, 1757–1765 (2015).
8. E. Deng, W. Kang, Y. Zhang, J.-O. Klein, C. Chappert, and W. Zhao, "Design optimization and analysis of multicontext STT-MTJ/CMOS logic circuits," *IEEE Trans. Nanotechnol.* **14**, 169–177 (2015).

9. W. Zhao, L. Torres, Y. Guillemenet, *et al.*, "Design of MRAM based logic circuits and its applications," *ACM Great Lakes Symp. VLSI* (2011), pp. 431–436.
10. S. Ikeda, J. Hayakawa, Y. Ashizawa, *et al.*, "Tunnel magnetoresistance of 604% at 300 K by suppression of Ta diffusion in CoFeB/MgO/CoFeB pseudo-spin-valves annealed at high *T*," *Appl. Phys. Lett.* **93**, 082508 (2008).
11. D. Chabi, W. Zhao, E. Deng, Y. Zhang, N. B. Romdhane, J.-O. Klein, and C. Chappert, "Ultra low power magnetic flip-flop based on checkpointing/power gating and self-enable mechanisms," *IEEE Trans. Circ. Syst. I* **61**, 1755–1765 (2014).
12. D. E. Nikonov, G. I. Bourianoff, and T. Ghani, "Proposal of a spin torque majority gate logic," *IEEE Electron Device Lett.* **32**, 1128–1130 (2011).
13. T. Windbacher, H. Mahmoudi, V. Sverdlov, and S. Selberherr, "Spin torque magnetic integrated circuit," WIPO WO 2014/154997 A1 (published Oct. 2, 2014).
14. U. Tietze and C. Schenk, *Electronic Circuits – Handbook for Design and Applications*, 2nd ed, New York: Springer, 2008.
15. T. Windbacher, A. Makarov, H. Mahmoudi, V. Sverdlov, and S. Selberherr, "Novel bias field-free spin transfer oscillator," *J. Appl. Phys.* **115**, 17C901 (2014).
16. A. Makarov, V. Sverdlov, and S. Selberherr, "Concept of a bias-field-free spin-torque oscillator based on two MgO-MTJs," *Extended Abstracts SSDM* (2013), pp. 796–797.
17. A. Makarov, T. Windbacher, V. Sverdlov, and S. Selberherr, "Efficient high-frequency spin-torque oscillators composed of two three-layer MgO-MTJs with a common free layer," *Proc. 21st Iberchip Workshop* (2015).

1.7

Can MRAM (Finally) Be a Factor?

Jean-Pierre Nozières
*eVaderis, Minatec Entreprises BHT, 7 Parvis Louis Néel, 38040 Grenoble,
France; Spintec, Bat. 1005, 17 rue des Martyrs, 38054 Grenoble, France*

1. Introduction

Despite all the fuss about the "more-than-Moore" approaches, introduction of
new technologies in semiconductor fabs is always a challenge. Not all the blame
can be placed on conservative operations executives, though, and technologists
also carry their share of responsibilities. Despite never-ending improvements from
researchers, who always favor the "next-big-thing-that-will-make-it-work-better,"
displacing existing mature technologies, even scaled beyond reason (for example,
state-of-the-art DRAM), is a difficult decision to make. At the same time, an
early adopter may dominate the market by taking a timely risk and becomes a
leader, while competition is playing catch-up. It is really a game of cat and mouse
between the risks of being too early with a new technology or too late in the
marketplace ... and this may be exactly where magnetoresistive random-access
memory (MRAM) stands today.

Since its inception in the late 1990s and despite numerous promising
announcements, MRAM has thus far failed to live up to the expectations. The
recent advent of spin transfer torque (STT), however, has shed a new light
on MRAM with the promises of much improved performances and greater
scalability. Semiconductor giants, IDMs, fabless companies, and now pure-play
foundries are entering the game with promises of available technologies and
products ... in a year or so (but with no specific dates, though).

2. What is MRAM?

In an MRAM cell, unlike semiconductor memories which rely on electrical
charges, the information is stored in the magnetization direction of a small
nano-magnet.

The basic element of the MRAM is a magnetic tunnel junction (MTJ)
consisting in its simplest form in two magnetic layers separated by a very thin

Future Trends in Microelectronics: Journey into the Unknown, First Edition.
Edited by Serge Luryi, Jimmy Xu, and Alexander Zaslavsky.
© 2016 John Wiley & Sons, Inc. Published 2016 by John Wiley & Sons, Inc.

dielectric layer (around 1 nm thick). The nonvolatile binary information is stored in the direction of the magnetization of one of the layers, known as the free layer. The information is read using the second magnetic layer, known as the reference layer. Different resistance states of the MTJ are achieved depending upon the relative orientation of the two magnetic layers (parallel or antiparallel). The so-called tunnel magnetoresistance (TMR) is the metric of the read window, expressed in percent as $TMR = (R_{AP} - R_P)/R_P$, where R_P (R_{AP}) is the parallel (antiparallel) resistance value of the MTJ.

TMR values higher than 200% can commonly be obtained using MgO barriers with cells utilizing in-plane magnetic anisotropy. For perpendicular magnetic anisotropy cells, TMR values are somewhat lower, closer to 100%, for example, a factor of 2 in. resistance level between the "1" and "0" states – see Fig. 1. This is due, in part, to the reduced thickness of perpendicular anisotropy electrodes and in part to the incompatibility between the necessary bcc (001) MgO crystallization (for large TMR) and the fcc (111) orientation of magnetic layers (for large perpendicular anisotropy).

Each MRAM cell consists of an individual MTJ connected to a selection transistor (1T/1J), which acts as a switch. In conventional MRAM, the

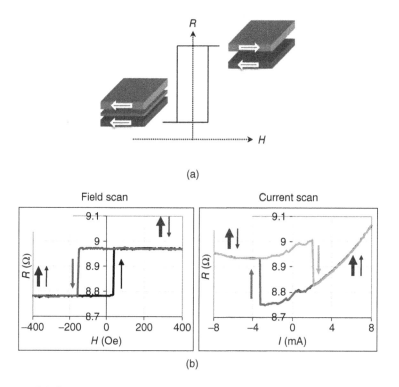

(a)

(b)

Figure 1. (a) Schematic magnetic tunnel junction R versus H field curve; (b) typical magnetoresistance in perpendicular CoFeB/MgO/CoFeB MTJ. (Source: After Ikeda et al.[1])

magnetization of the free layer is set by an external magnetic field created by the combined currents flowing in bit and word lines running above and below the cells. In 2006, the first commercial MRAM product was launched by Everspin. It has gained market acceptance and the company reports to have shipped more than 40 million chips, in applications as diverse as automotive, aerospace, storage, ... This approach, however, has two important shortcomings: a high write current (tens of mA/bit!) and a poor scalability.

Today, the focus is on STT technology, whereupon writing is achieved by a current injected perpendicular to the MTJ interfaces.[2, 3] The current is spin-polarized by the reference layer and transfers its angular momentum to the storage layer, which, under appropriate conditions, can lead to a direct reversal of said storage layer without any external magnetic field. The advantages of STT are numerous: lower cell footprint, since the MTJ and the transistor can now be stacked; reduced write current by one to two orders of magnitude; and full scalability, as what counts here is a current density.

Most groups worldwide are focusing on a "perpendicular STT" (p-STT) layout, in which the magnetizations of the MTJ layers are perpendicular to the wafer plane, as shown in Fig. 2. The p-STT geometry allows for greater scalability and better write efficiency, paving the way toward sub-20 nm technology, at the expense of a lower TMR. Perpendicular MTJ materials typically consist of Co/Pt or Co/Pd multilayers, but those have large Gilbert damping resulting in low TMR signals and larger critical currents. The most successful approaches have used the CoFeB/MgO electrodes, in which the perpendicular anisotropy comes from the metal/oxide interface, as first observed in 2002.[4]

The first STT products have been recently introduced by Everspin as a 64 Mb stand-alone chip, using a planar STT technology. Perpendicular STT demos have been reported by multiple players down to sub-20 nm MTJ dimensions, as shown in Fig. 3, with bit error rates (BERs) below 10^{-9}, and current pulse duration of a few nanoseconds, although practical programming speeds are closer to 10–20 ns.[5, 6]

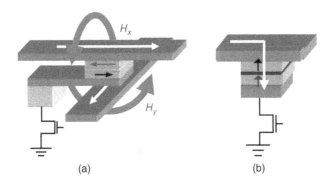

Figure 2. (a) Magnetic field-driven MRAM cell; (b) perpendicular STT-MRAM cell. White arrows indicate to current flow direction during the write operation.

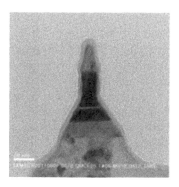

Figure 3. TEM image of a 17 nm STT-MRAM cell. (Source: Kim *et al.*[5] Reproduced with permission of IEEE).

3. Current limitations for stand-alone memories

Major advances have been achieved in the recent years, in particular, concerning reliability and BER. However, scaling STT RAM to sub-20 nm dimensions, which is the target for mass production market entry, remains a challenge due in part to manufacturing issues, in part to the usual dilemma encountered in all magnetics-based devices, such as the competing requirements of thermal stability (data retention) and write current (operating power and reliability). In addition for STT devices, write speed is also directly proportional to current, while read speed is linked to TMR amplitude. All this results in conflicting requirements and compromises. We describe below the major hurdles to scaling down to high-density, DRAM-compatible chips that still remain.

- *Data retention at sub-20 nm dimensions and elevated temperature*

One has to ensure that the thermal stability factor $\Delta = KV/k_B T$ that describes data retention of a bit cell of volume V and magnetic anisotropy K remains sufficiently large at operating temperature T. The requirement on Δ depends upon the required BER and memory capacity, with typical values falling between 60 and 80. The out-of-plane magnetic anisotropy K in p-MTJ decreases relatively quickly with T above room temperature both in Co/Pt(Pd)-based systems and at Co(Fe)/MgO interfaces. For the reference layer, one can design materials that are quite stable up to 250 °C. The problem is more critical in the storage layer, for which the retention must be large enough but whose magnetization must still be switchable by STT. For this reason, the brute-force approach of simply raising the room-temperature anisotropy cannot work. Careful materials engineering is needed to limit the decrease of magnetic anisotropy at higher T, in particular through control of Fe/Co ratio in CoFeB layers. This may add complexity and variability to the processing.

• *Stochastic switching within each cell*

In the conventional STT-MRAM embodiment, the equilibrium state has the storage layer and reference layer magnetizations parallel to one another. While this allows the maximum TMR amplitude, it implies a zero net torque, since STT is proportional to the cross product of the storage layer and reference layer magnetizations. Switching may not occur until triggered by a thermally activated magnetization fluctuation inducing a nonzero initial angle between storage layer magnetization and spin current polarization (e.g., reference layer magnetization). As a consequence, the switching dynamics is highly stochastic, as first observed in in-plane magnetized MTJs by Devolder *et al.*[7] For fast memory applications, this random incubation time must be minimized, with multiple approaches still under consideration. One of the most promising ideas consists in using two orthogonal spin-polarizing layers,[8, 9] as promoted by the Spin-Transfer Technologies start-up. In this way, deterministic (fast) switching can be achieved in an in-plane MTJ with an additional perpendicular polarizer, as shown in Fig. 4. The opposite configuration of an out-of-plane MTJ with an in-plane magnetized polarizer has been only studied so far in spin-valves.[10]

• *Cell-to-cell variability: impact of fabrication process*

Memory chip fabrication is a complex and extremely formatted process with numerous processing steps following a well-defined sequence. While the issue of materials contamination is now under control, due to the combined trends of simplified MTJ stack for p-STT (in terms of the number of materials involved) and a constant broadening of the materials implemented in the semiconductor industry (Pt, Co, Ni, Ta, Ru, ... all now widely used), several other challenges remain. One of the most critical is the etching of the MTJ cell at small feature size/cell pitch, which can have a severe impact on the cells magnetic properties.

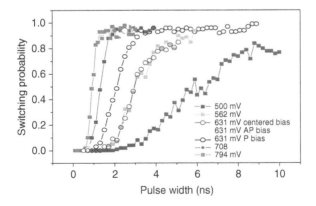

Figure 4. Deterministic switching in STT cells with a perpendicular polarizer. (Source: Marins de Castro *et al.*[9] Reproduced with permission of AIP Publishing LLC).

Taming the stochastic nature, and hence variability, of magnetization reversal requires thorough control of the etch process, both in terms of size distributions and structural and/or chemical defects at the edges of the pillar. Fluctuations in shape may change the magnetostatic/inter-layer coupling between the reference layer and the storage layer; local amorphization or chemical damage of the MgO tunnel barrier may affect the Co(Fe)/MgO interfacial anisotropy; chemical damages at the edges of the storage layer may facilitate the nucleation of reversed domains. All will have a direct impact on the switching current, thermal stability, TMR amplitude, and reliability, leading to variability in a large array. The jury is still out on the best cell etch process, with major equipment suppliers now involved so that rapid progress may be expected in the coming years.

4.   Immediate opportunities: embedded memories

For the past few years, the focus has been on cutting-edge technology, with players shooting for sub-20 nm feature sizes in densely packed, multi-Gb arrays, with DRAM replacement in sight. While significant strides have been made, it is still an uphill battle, against a constantly moving target, with the focus now on sub-10 nm technology. All the major DRAM players (Samsung, Micron, and Hynix) have ongoing STT-MRAM development programs, yet the jury is still out whether real products will ever be launched.

At the same time, the same core technology at more relaxed technology nodes may be the ideal embedded memory solution, in particular, when power consumption (or battery lifetime) is key. In today's SoC designs, memory power accounts for a large fraction of the dissipated power, be it static (leakage in standby mode) or active (dynamic power during read/write cycles), not to mention RC losses in the hundreds of kilometers of wires linking memory and logic. The best solution to circumvent this problem is to embed memory within the logic chip itself, in a so-called "logic-in-memory" architecture. However, while SRAM is ultrafast, it is by nature volatile (e.g., large standby power) and expensive in terms of footprint at ~$100F^2$ cell size; DRAM is dense but also volatile and difficult to embed; whereas Flash is indeed nonvolatile but sluggish at best and hampered by limited endurance and very large active power dissipation.

As exciting a technology as it is, STT-MRAM is by no means a champion: it is slower than SRAM, bigger and more expensive than Flash, and does not reach the truly infinite endurance of DRAM and SRAM. What makes STT-MRAM so unique in this landscape is its unique combination of nonvolatility, speed, and low active power, with endurance and the capacity to be implemented in any form factor as the icing on the cake. This unique combination allows embedding large(r) amounts of memory, therein simultaneously increasing the on-chip processing/storage capability and decreasing the overall chip power consumption.

As a first step, existing SRAM/NOR Flash memory blocks can be replaced, resulting in an immediate gain through zero leakage in sleep mode, instant power-up, reduced active power, and reduced send power (less data sent less often). As a second step, STT-MRAM can also be embedded within the logic blocks of the chips, for a further reduction of active/standby power, coupled to improved hardware/software flexibility. Finally, the entire system architecture may be redesigned ("full system approach"), resulting in a further improvement of the system efficiency.

All this of course has a price. Beyond the added processing costs, it requires a total rebuild of the circuit and system level architectures, which necessarily implies custom add-ons to the electronics design automation (EDA) tools that enable design and simulation. These must describe the behavior of MRAM cells through models close enough to the physical models (micromagnetics, transport phenomena, etc.), yet fast enough for allowing complex chip simulation in a reasonable runtime. The response of a p-STT MRAM cell is shown in Fig. 5 using the Callisto® design tool and the SPITT physical model.[11]

Thanks to the nonvolatile, fast-access/fast-write, and low-power STT-MRAM, it is now possible to combine high performance and intelligence levels, together with long lifetime operation, which is not possible in conventional SRAM-based and/or NOR-Flash-based MCUs. The major gains arise from on-the-fly, on-chip, compression and data logging which result in wireless transmission energy gain and fast, low-power on-chip storage enabling in-network, distributed data storage.

An additional gain stems for the energy management, with reduced sleep power thanks to the ability to save, without leakage, local individual states, and fast boot/power-up thanks to the ability to run at core speed with low energy. Note that boot and power-up are roughly equivalent and pretty much independent of code and configuration size (i.e., chip complexity), with the boot becoming massively parallel instead of the classical sequential initialization phase.

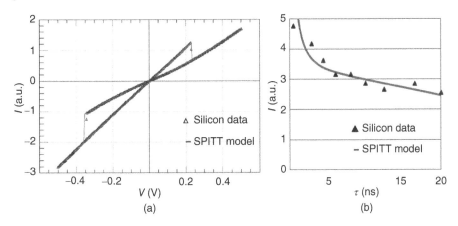

Figure 5. Simulated versus measured p-STT cell response: dots – silicon data, lines: SPITT model. (Source: From Bernard-Granger et al.[11])

Of course, the performance metrics are very much application-related and depend upon the data type (ECG, image, ...), data profile, whether one can tolerate to lose information and/or to delay its sending, etc. with the gain over conventional MCU getting higher with system complexity. Figure 6 shows a conventional microcontroller (MCU) based on a 200 MHz/Cortex-M4-like controller embedding 20 Mb of flexible memory (19 Mb of STT-MRAM and 1 Mb of SRAM), while Fig. 7 shows the expected energy gain compared to the equivalent

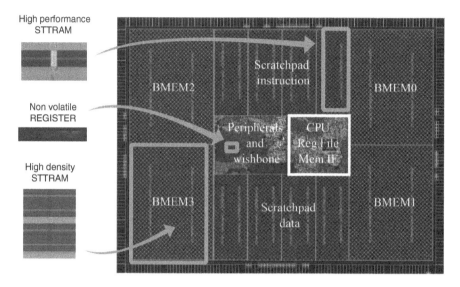

Figure 6. Microcontroller embedding 20 Mb of STT-MRAM.

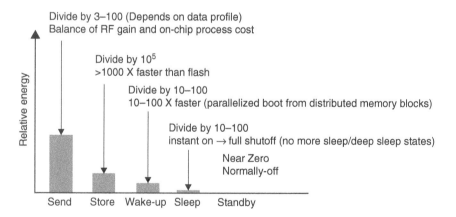

Figure 7. Approximate energy gains in an STT-MRAM-based MCU compared to the SRAM-/Flash-based equivalent.

SRAM/Flash-based MCU. Such a system can save the equivalent of 10^6 scalar measures, good enough for secure, intelligent microdata logging and/or advanced smart metering.

On top of these advantages, embedded STT-MRAM also offers a highly flexible platform to software developers through, for instance, quasi instant-on capabilities, full memory data/code partitioning (changing the code/data memory space ratio, enabling reliable firmware updates as code size increases), over-the-air, low-power, bit-level hardware programmability and/or repair (for hardware and/or data update), and instant-on snapshot-restore from involuntary shutdown (e.g., power loss).

5. Conclusion

Today, MRAM technology is taking multiple routes toward technological insertion. Not all may lead to the Promised Land, but this may just be the right time, with STT-MRAM technology nearing maturity and foundries finally entering the game. While stand-alone, high-capacity memory chips may still require significant developments before they hit the market, new applications in the Internet of things (IoT) that require storing and computing ever-increasing amounts of data in battery-powered systems may be the low-hanging fruit application that STT-MRAM has been awaiting for.

At the same time, because we are scientists, we cannot resist the excitement of "the next one," such as the burgeoning spin orbit torque technology,[12, 13] which holds promise of SRAM-like, ultrafast switching for L1 cache. This may pave the way to the Holy Grail of an all-nonvolatile data path in microprocessors and storage systems. Stay tuned!

References

1. S. Ikeda, K. Miura, H. Yamamoto, *et al.*, "A perpendicular-anisotropy CoFeB–MgO magnetic tunnel junction," *Nature Mater.* **9**, 721–724 (2010).
2. J. Slonczewski, "Current-driven excitations of magnetic multilayers," *J. Magn. Magn. Mater.* **159**, L1–L7 (1996).
3. L. Berger, "Emission of spin waves by a magnetic multilayer traversed by a current," *Phys. Rev. B* **54**, 9359–9358 (1996).
4. S. Monso, B. Rodmacq, S. Auffret, *et al.*, "Crossover from in-plane to perpendicular anisotropy in Pt/CoFe/AlO$_x$ sandwiches as a function of Al oxidation: A very accurate control of the oxidation of tunnel barriers," *Appl. Phys. Lett.* **80**, 4157–4159 (2002).
5. W. Kim, J. H. Jeong, Y. Kim, *et al.* "Extended scalability of perpendicular STT MRAM towards sub-20nm MTJ node," *Tech. Dig. IEDM* (2011), p. 24.1.1.

6. L. Thomas, J. Guenole, I. Zhu, *et al.*, "Perpendicular STT MRAM with high spin torque efficiency and thermal stability for embedded memory applications," *J. Appl. Phys.* **115**, article no. 172615 (2014).

7. T. Devolder, J. Hayakawa, K. Ito, *et al.*, "Single-shot time-resolved measurements of nanosecond-scale spin-transfer induced switching: Stochastic versus deterministic aspects," *Phys. Rev. Lett.* **100**, 057206 (2008).

8. H. Liu, D. Bedau, D. Backes, J. A. Katine, J. Langer, and A. D. Kent, "Ultra-fast switching in magnetic tunnel junctions based orthogonal spin transfer devices," *Appl. Phys. Lett.* **97**, 242510 (2010).

9. M. Marins de Castro, R. C. Sousa, S. Bandiera, *et al.*, "Precessional spin transfer switching in a magnetic tunnel junction with a synthetic anti-ferromagnetic perpendicular polarizer," *J. Appl. Phys.* **111**, 07C912 (2012).

10. R. Law, E. L. Tan, R. Sbiaa, T. Y. F. Liew, and T. C. Chong, "Reduction in critical current for spin transfer switching in perpendicular anisotropy spin valves using an in-plane spin polarizer," *Appl. Phys. Lett.* **94**, 062516 (2009).

11. F. Bernard-Granger, B. Dieny, R. Fascio, and K. Jabeur, "SPITT: A magnetic tunnel junction SPICE compact model for STT-MRAM," Presented at *MOS-AK Workshop, DATE* (2015).

12. I. M. Miron, K. Garello, G. Gaudin, *et al.*, "Perpendicular switching of a single ferromagnetic layer induced by in-plane current injection," *Nature* **476**, 189–193 (2011).

13. K. Jabeur, G. Di Pendina, F. Bernard-Granger, and G. Prenat, "Spin orbit torque nonvolatile flip-flop for high speed and low energy applications," *IEEE Electron Device Lett.* **35**, 408–410 (2014).

1.8

Nanomanufacturing for Electronics or Optoelectronics

M. J. Kelly

Department of Engineering, Centre for Advanced Photonics and Electronics, University of Cambridge, 9 JJ Thomson Avenue, Cambridge CB3 0FA, United Kingdom; MacDiarmid Institute for Advanced Materials and Nanotechnology, Victoria University of Wellington, Wellington 6140, New Zealand

1. Introduction

Artefacts in widespread use in electronic and optoelectronic systems have been manufactured by low-cost, high-volume processes. To get to that stage, the understanding and design of the artefact and the choice of materials are such that several key attributes apply[1]:

- The artefacts exhibit performance that is superior to what went before, or are otherwise much cheaper for the same performance.

- A high yield to an acceptable tolerance is achieved, with in-batch uniformity and interbatch reproducibility.

- A functional simulator is available for reverse engineering during development and right-first-time design in production.

- The artefacts are reliable and have an adequate service lifetime.

While we recognize these features, for example, in products for mainstream applications, we also note their absence in many, but not all, nanoscale artefacts made with comparable fabrication technologies such as deposition, etching, and lithography.

In practice, one can apply also this argument to complex molecules as nanoscale artefacts. If one translates six-sigma quality into four or five-nines purity, only a few relatively simple molecules are available at >99.99% purity, the level one would require if one were to make integrated circuits from such molecules in some bottom-up process, such as laying out a memory array based on such molecules as storage nodes. Here we are considering just the starting ingredients, before any considerations of a defect-free array of such molecules.

Future Trends in Microelectronics: Journey into the Unknown, First Edition.
Edited by Serge Luryi, Jimmy Xu, and Alexander Zaslavsky.
© 2016 John Wiley & Sons, Inc. Published 2016 by John Wiley & Sons, Inc.

While the move from three-dimensional electronic systems to two-dimensional (2D) systems (the quantum well, the 2D electron gas, etc.) has produced new physics and a wide range of new devices in widespread applications, the same is not true for one- or zero-dimensional electronic systems, and this insignificant part is because the level of reproducibility is inadequate for low-cost, high-volume production. In this chapter, we review recent progress and lack of progress on manufacturability of several nanoscale systems using the above criteria: tunnel devices, split-gate transistors/quantum point contacts, and other nanoscale systems.

2.   Nano-LEGO®

A good test example[2] is to consider a conventional LEGO® brick, typically a few tens of mm on each side, when scaled down by a factor of 10^N for $N < 7$. In Fig. 1, we show typical brick sizes as a function of the center-to-center pitch, P, of the pillars on the topside. We will assume a simplified underside where just the hole to receive the cylinders is retained, leaving the brick otherwise solid, rather than hollow as in conventional LEGO. A brick with a skin of only a few nanometers thickness would be too fragile. The manufacturing tolerance for conventional bricks is 10 µm, or typically of order 1/1000 of the brick size. This is used to allow tight mating of adjacent ricks and the possibility of building a tall vertical stack. Relaxing this tolerance at smaller scales is essential and compromising the ability to use the resulting bricks for assembling structures. At each scale we can consider the methods for (i) one-off fabrication, (ii) volume production, and (iii) the assembly and disassembly of structures, comprising on the order of 10^3 bricks, such as a scale model of the Eiffel Tower. Table 1 shows how first the practical use is constrained by the decreased size, then the reliability of manufacture and ultimately the reliability of even one-off fabrication.

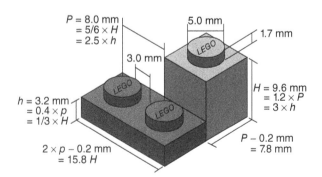

Figure 1. Dimensions of standard Lego® bricks, (Source: see http://www. lucasbrouwers.nl/ blog/2011/05/why-life-is-like-lego/ CC BY-SA 3.0).

Scale	Fabricate	Manufacture	Can use/user tool	Comment on use
1 (mm)	IM(p)	IM(p)	Yes/hands	Easy
10^{-1}	IM(p)	IM(p)	Yes/tweezers	Straightforward
10^{-2}	IS(p)	IS(p)	Yes/microtweezers	Tricky
10^{-3} (μm)	IS(p), DLE(s)	IS(p), DLE(s)	Yes/microtweezers, AFM	Demanding
10^{-4}	IS(p), DLE(s)	IS(p), DLE(s)	Yes/AFM	Only at relaxed tolerances
10^{-5}	IS(p), DLE(s)	None	None	Impossible
10^{-6} (nm)	AFM	None	None	Impossible

Table 1. Milli-, micro-, and nanobricks. The scale is set in comparison with real LEGO® bricks of Fig. 1; fabrication and manufacture techniques include IM(p) as injection molding of plastic, IS(p) as imprint stamping of plastic, and DLE(s) as deposition, lithography, and etching of semiconductors.

As the downscaling proceeds, the ability to use fails first, to manufacture fails second, and to fabricate fails last. The use fails for several reasons, one being the tolerance at 1/1000 is less easily scaled, another being the inability to construct a stable tower 100 bricks high at the 1 μm scale. Note that there are at least two technologies – (i) imprint stamping and (ii) deposition, lithography, and etching –that persist as fabrication technologies scale down to the nanoscale. There are many further aspects that can be considered instructively at each scale: yield, reproducibility, wear and tear, etc.

This exercise is one that all candidate artefacts for nanoscale manufacture must undertake successfully.

3.   Tunnel devices

Over the 40 years since tunneling and resonant tunneling was observed in thin barriers in semiconductor multilayers, there have been many device prototypes exploiting tunneling that exhibit superior figures of merit for switching and amplifying in both electronics and optoelectronics. There are functional simulators that can be used to model tunneling in devices and to reverse-engineer experimental results. The only problem has been the lack of in-wafer uniformity and wafer-to-wafer reproducibility, within and between growth runs, of the basic device characteristics.

Over the last 23 years, I have worked with various collaborators on this issue and we have recently made progress that achieves the required uniformity

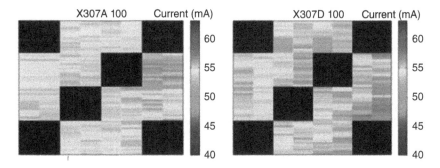

Figure 2. Wafer maps showing the current measured on (~320) 100 μm square diodes at a forward bias of 0.5 V. Note the extreme uniformity compared with previous reports.[5, 6] For both wafers, the standard deviation of the currents across the wafer is ~1% of the average current, and the average currents from the two wafers agree within 1%. Note: seven wafers were grown on a platen at a time in a configuration of a central wafer (D) and six around it (of which A is one): hence the claim of uniformity equivalent to a single 8 in. wafer. The black spaces are fields protected for further use. The mask has 10 subfields each with 32 100-μm square diodes, all of which have been measured: there is one mesa measured per pixel in the diagrams.

and reproducibility for a single AlAs barrier in GaAs designed for use as a millimeter-wave detector. Tunnel currents are very sensitive to the height and thickness of the tunnel barrier, with a 270% variation for a change in thickness of a single monolayer. It is exceedingly difficult to reach a prespecified target current–voltage characteristics with a <5% spread and with wafer-to-wafer reproducibility within 5%. Figure 2 presents maps[3] of the current for a fixed forward bias across 100 μm square diodes from two wafers grown that the same time in a RIBER V100HU MBE production growth machine that have a standard deviation in the electrical properties of 1% or less and a wafer average that differs by less than 1%. This result was achieved by taking a growth process that includes very slow growth of integer monolayer barriers proven on a research machine, where the interwafer and intergrowth uniformity for devices on the same part of a wafer were the same within a few percent,[4] but with a remaining cross-wafer linear trend of 20% in the current–voltage characteristics that remains from the off-axis geometry of the MBE aluminum source in the growth chamber. The process is now being examined in the context of double-barrier structures, as together with the detector diode, we have the elements of a high-efficiency but low-power communication system for operation up to 0.5 THz.

4. Split-gate transistors

Since the invention of the split-gate transistor (or quantum point contact) in 1986 and its subsequent use to discover the quantization of the one-dimensional ballistic resistance, most studies have used single gates or a few gates to investigate physics in 1D wires and laterally confined quantum dots. In terms of manufacturability, only one heroic study of 540 split-gate devices made in batches of 36 of different wafers gave any insight into the reproducibility of the quantization phenomenon in and between devices.[7] A part of the problem is the limited number of wires that can be taken from the cryostat for external measurements.

We have developed a first-generation multiplexing geometry that allows an array of 256 devices to be addressed with less than 20 wires – see Fig. 3(a).[8] We can distinguish between fabrication yield (94%), which can be improved to 100% with greater experience and attention to detail, and quantum yield (76% going to

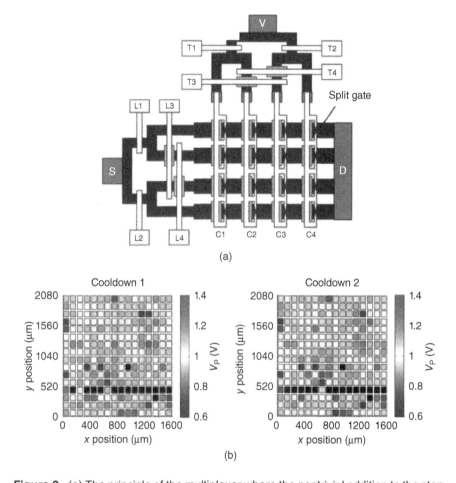

(a)

(b)

Figure 3. (a) The principle of the multiplexer where the nontrivial addition to the standard split-gate geometry is the layer of dielectric under some gates, shown as a gray area, that allows a bias to be applied to various lines without the underlying 2DEG being pinched off. (b) Normalized plots of the spatial distribution of the pinch-off voltage V_P of the split-gate transistors on two successive cooldowns showing the level of continuity of properties of each gate. Black areas denote failed devices. The vertical (horizontal) pitch of the actual devices is 130 μm (100 μm).

84% when illuminated) of actual ballistic quantization steps. This yield reflects something of the detailed distribution of donor atoms in the supply layer of the transistor, and various forms of disorder or interference that can mask the clear steps. We have been able to collect and evaluate the average and standard deviation of the pinch-off voltage (Fig. 3(b)) and the voltage at which the 1D/2D transition takes place. The average slope of the ladder of 1D steps is a measure of the local density of the 2D electron gas, and this has a standard deviation of 4% for devices that are typically a 100 μm apart. This places limitations on inferences from simple Hall bar measurements on the actual local electron density for 1D phenomena.

This initial multiplexer is used for gathering statistics of the device operation, which are now being used to try to distinguish between various explanations of the so-called 0.7 spin phenomenon.[9] A second generation of multiplexer allows each gate to be biased and charge to be held on each gate before an experiment is undertaken. The first application is to investigate phenomena in a 2×7 array of closely coupled quantum dots.[10]

For any real-world applications, such as in charge pumps for quantum metrology or specific quantum information processing systems, one needs a strategy to narrow the standard deviation of the device parameters. One intriguing possibility is to zero every gate to the pinch-off condition before starting any experiment, and an algorithm to achieve this has been developed. Of the attributes specific to manufacturability, reproducibility needs further improvement, simple simulations are available, and reliability and service life can be inferred from earlier cryogenic experiments.

5. Other nanoscale systems

Unlike the case of tunneling, consideration of quantum dot manufacturing has not advanced in the 3 years since the last FTM conference.[11] In the case of such applications as quantum key sources of single photons, one must search the surface for the particular dot at the required wavelength without guaranteeing its optimum positioning in any subsequent cavity.[12]

There are many examples of papers working on the new forms of carbon, fullerenes, nanotubes, and graphene, but these are still in the realms of one-offs in terms of individual or few devices, or very small and primitive integrated circuits. Issues such as yield and reproducibility to a prespecified performance remain as yet unstudied. The same applies to other quasi-monolayer electron systems.

6. Conclusion

In selected areas, there has been some progress toward the manufacturing of nanoscale artefacts. The remaining concern is living with the intrinsic fluctuations from one nanoscale artefact to the next, and whether there can be any design-around to cope these fluctuations.

Acknowledgments

This work is done with several teams of collaborators, listed as coauthors of the referenced papers, and funded by the Engineering and Physical Science Research Council of the United Kingdom over 20 years, and the Royal Society's Mercer Innovation Fund more recently.

References

1. M. J. Kelly, "Intrinsic top-down unmanufacturability," *Nanotechnology* **22** 234303 (2011).
2. M. J. Kelly and M. C. Dean, "A specific nanomanufacturing challenge," *Nanotechnology* **27**, 112501 (2016).
3. M. Missous, M. J. Kelly, and J. Sexton, "Extremely uniform tunnel barriers for low-cost device manufacture," *IEEE Electron Device Lett.* **36**, 543–545 (2015).
4. C. Shao, J. Sexton, M. Missous, and M. J. Kelly, "Achieving reproducibility needed for manufacturing semiconductor tunnel devices," *Electronics Lett.* **49**, 669–671 (2013).
5. V. A. Wilkinson, M. J. Kelly, and M. Carr, "Tunnel devices are not yet manufacturable," *Semicond. Sci. Technol.* **12**, 91–99 (1997).
6. R. K. Hayden, A. E. Gunnaes, M. Missous, R. Khan, M. J. Kelly, and M. J. Goringe, "Ex-situ re-calibration method for low-cost precision epitaxial growth of heterostructure devices," *Semicond. Sci. Technol.* **17**, 135–140 (2002).
7. Q. Z. Yang, M. J. Kelly, H. Beere, I. Farrer, and G. A. C. Jones, "The potential of split-gate transistors as one-dimensional electron waveguides revealed through the testing and analysis of yield and reproducibility," *Appl. Phys. Lett.* **94**, article no. 033502 (2009).
8. H. Al-Taie, L. W. Smith, B. Xu, *et al.*, "Cryogenic on-chip multiplexer for the study of quantum transport in 256 split-gate devices," *Appl. Phys. Lett.* **102**, 243102 (2013).
9. L. W. Smith, H. Al-Taie, F. Sfigakis, *et al.*, "Statistical study of conductance properties in one-dimensional quantum wires focusing on the 0.7 anomaly," *Phys. Rev. B* **90**, 045426 (2014).
10. R. K. Puddy, L. W. Smith, H. Al-Taie, *et al.*, "Multiplexed charge-locking device for large arrays of quantum devices," *Appl. Phys. Lett.* **107**, 143501 (2015).
11. M. J. Kelly, "Manufacturability and nanoelectronic performance," chapter in: S. Luryi, J. Xu, and A. Zaslavsky, eds., *Future Trends in Microelectronics: Frontiers and Innovations*, New York: Wiley, 2013, pp. 133–138.
12. M. C. Dean and M. J. Kelly, "The manufacturability of self-assembled quantum-dot-based devices," unpublished.

Part II

New Materials and New Physics

What could be more powerful in propelling the advances in our profession than the development of new materials and findings of new physics? While graphene still gets most of the spotlight on the stage, other atom-thin materials might bring even greater rewards, whereas the ever-present surface waves may have begun to gain more attention, thanks in part to the pioneering contributions of Dyakonov. Under all the waves lie interesting interplays of electrons and atoms that give rise to extraordinary transitions among metal, insulator, localization, distributed and correlated, and even computational states. No less intriguing and surprising are the reports of light emission from brand new material systems, as readers will find in this chapter.

Contributors

2.1

Surface Waves Everywhere

M. I. Dyakonov
*Laboratoire Charles Coulomb, Université Montpellier, CNRS,
Montpellier, France*

1. Introduction

By "surface waves" one means a special kind of waves that propagate at the inter-face between two different media. There exists a large variety of such waves, which are interesting on their own, and sometimes have also practical importance and technological applications. This article presents a brief and nonexhaustive review of this vast subject, designed as an introduction to the field for nonspecialists. Similarities between surface waves of completely different origin are outlined. I am concerned with the physical picture and avoid math as strongly as possible (no differential equations and no boundary conditions), as well as many inter-esting details. Sometimes I omit numerical coefficients. My own contributions to this field are also presented.

It should be understood that the material in each section of this article is a subject of many books and hundreds, if not thousands, of journal publications, which the interested reader should address for more information.

2. Water waves

These are certainly the first kind of surface waves that mankind has encountered. However, our prehistoric ancestors did not yet realize that waves are character-ized by the frequency ω and the wave vector $k = 2\pi/\lambda$, where λ is the wavelength. The relation between ω and k is called the *dispersion law*. We now establish the dispersion laws for water waves from considerations of dimensionality only.

However, first we must distinguish between several kinds of water waves, dif-fering in the nature of the return force that drives the oscillatory motion: this can be either *gravity* or *surface tension*.

Also we must distinguish between the cases of *deep* and *shallow* water. To do this, we must compare the water depth h with some other length. There are two characteristic lengths in our problem: the wavelength λ and the wave amplitude.

Future Trends in Microelectronics: Journey into the Unknown, First Edition.
Edited by Serge Luryi, Jimmy Xu, and Alexander Zaslavsky.
© 2016 John Wiley & Sons, Inc. Published 2016 by John Wiley & Sons, Inc.

If we are concerned with the linear regime, the amplitude is irrelevant (so long as it is small compared to the wavelength). Thus, in the linear regime we must compare h and λ, so that one has deep water when $h \gg \lambda$ and shallow water if $h \ll \lambda$.

- *Gravitational waves*

The return force is provided by gravity. In this case, we have the following material for constructing the link between frequency ω with dimension $[\text{s}^{-1}]$ and the wave vector k with dimension $[\text{cm}^{-1}]$: the freefall acceleration g $[\text{cm/s}^2]$ and water depth h [cm].

Deep-water waves $(h \gg \lambda)$. In this case, the depth h is irrelevant. Then the only possibility for satisfying the correct dimensionality is the following dispersion law: $\omega = (gk)^{1/2}$. The phase velocity $v = \omega/k$ is then given by $v \sim (g\lambda)^{1/2}$, so that longer wavelengths propagate faster than shorter ones. Beneath the surface, the water is perturbed on a depth $\sim\lambda$ (not the wave amplitude!); thus, a diver does not care about the storm at the surface, so long as he stays at depths greater than the typical wavelength.

Shallow-water waves $(h \ll \lambda)$. Now, independently of the wavelength, the wave velocity is $v \sim (gh)^{1/2}$ and the dispersion law is linear: $\omega = vk$, as for sound waves. This formula gives a simple estimate for the time needed for a perturbation of the ocean surface caused by an earthquake to cross the Pacific and cause a tsunami on a distant coast.

The so-called "shallow earthquakes" have their focus up to 70 km deep in the Earth crust (say 50 km), which is much greater than the ocean depth ($\sim$3 km). Thus, the ocean surface will be perturbed on a scale of 50 km and the resulting surface wave will be a *shallow water wave*, however strange it may sound when applied to the Pacific Ocean. This wave will propagate with a velocity $\sim(gh)^{1/2}$, where $h \sim 3$ km. This is about the velocity of an aircraft!

Note that gravity waves do not depend on the density of the liquid, because, as Galileo showed by dropping stones from the Pisa tower, the motion in the gravity field does not depend on the mass of the object.

- *Capillary waves*

Here, the return force is due to surface tension σ. As before, we construct the dispersion law from dimensionality considerations only. This time our ingredients are surface tension σ $[\text{g/s}^2]$ and water density ρ $[\text{g/cm}^3]$. Then the only possible form of the dispersion law is $\omega \sim (\sigma k^3/\rho)^{1/2}$, which leads to the wave velocity $v \sim (\sigma k/\rho)^{1/2} \sim (\sigma/\rho\lambda)^{1/2}$. Note that in contrast to gravitational waves, now the wave velocity *decreases* with increasing wavelength.

Thus, generally the dependence of velocity on wavelength is nonmonotonic: there exists a minimal velocity v_{MIN} at a wavelength λ_{MIN} at which the roles of

(a)   (b)

Figure 1. The V-shaped trail (the wake) behind (a) a duck. (Source: https://commons
.wikimedia.org/wiki/File:Bodensee_at_Lindau_-_DSC06962.JPG. Public domain).
(b) A small boat. (Source: https://commons.wikimedia.org/wiki/File:Wake.avon.gorge
.arp.750pix.jpg. Public domain).

gravity and surface tension are roughly equal. For water, λ_{MIN} is about 2 cm, and
the corresponding velocity v_{MIN} is about 30 cm/s.

- *Excitation of water waves by wind*

The wind can excite water waves with phase velocities that are inferior to the wind
velocity. The previous analysis shows that if the wind velocity is less than v_{MIN},
it cannot excite any waves at all. This is the rare case when the water surface is
absolutely flat and mirror-like.

  On the other hand, when the wind speed exceeds v_{MIN}, it will excite both the
long gravity waves and the short capillary waves; as a consequence, we see long
waves covered with capillary ripples.

- *Kelvin wake*

A duck swimming in a pond leaves a V-shaped trail behind. The same wake exists
for a small fishing boat and for an enormous aircraft carrier. The amazing fact
is that the wake angle is universal: it is the same for the duck and the aircraft
carrier, and it depends neither on the shape and speed of the vessel nor on the
density of the liquid (provided that the excited waves are deep water gravitational
waves).

  Kelvin[1] showed that the wake angle θ is determined by the simple equation
$\sin(\theta/2) = 1/3$, giving $\theta = 19.5°$. This celebrated result, shown in Fig. 1, is a direct
consequence of the dispersion law for deep-water gravitational waves: $\omega = (gk)^{1/2}$.

- *Rogue waves*

For many centuries, these waves, also known as freak waves, monster waves, killer
waves, and extreme waves, have been considered as mythical. However, they are

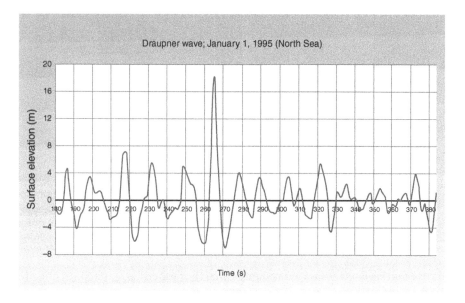

Figure 2. The "Draupner wave," a giant wave measured on New Year's Day 1995 in the North Sea, finally confirmed the existence of freak waves.[2] (Source: https://commons.wikimedia.org/wiki/File:Draupner_wave%282048x1270px%29.png. Public domain).

a real phenomenon, consisting in a sudden appearance, on the background of a mild ~5 m wave pattern, of a huge ~30 m single wave (the height of a 10-story building!), looking like a "wall of water" with a deep trough in front.

The first scientific registration of rogue wave is presented in Fig. 2.[2] With increasing maritime traffic such extreme waves are seen more and more often.

So far, there does not exist any theoretical understanding of rogue waves, and they still remain a mystery. The problem can be formulated as that of the statistics of extremely rare events, which apparently is not Gaussian: large fluctuations occur more frequently than the conventional theory predicts. There are indications[3, 4] that similar phenomena exist in optical and electrical noise.

3. Surface acoustic waves

In contrast to liquids and gases, in solids there are two types of sound: longitudinal and transverse, the speed of the latter being greater because of the contribution of shear stress. In 1885, Rayleigh predicted[5] that the longitudinal and transverse acoustic waves existing in the bulk of a solid can combine to produce a hybrid surface mode (the Rayleigh wave) with a velocity inferior to the velocities of both longitudinal and transverse sound (this is a necessary condition for their existence; otherwise, they would excite bulk acoustic waves and disappear). This hybridization makes it possible to satisfy the boundary conditions at the surface. The amplitude of the Rayleigh wave decreases exponentially away from the surface.

Such waves are used for nondestructive testing and in some electronic devices.

4. Surface plasma waves and polaritons

These are electromagnetic waves that exist when the dielectric constant is positive on one side of the interface and negative on the other side.[6] Negative dielectric constant $\varepsilon(\omega)$ is a well-known property of plasma below the so-called plasma frequency $\omega_P = 4\pi ne^2/m$, where n, e, and m are the electron density, charge, and mass, respectively. The dielectric constant of plasma is given by

$$\varepsilon(\omega) = 1 - \left(\frac{\omega_P}{\omega}\right)^2. \tag{1}$$

Electromagnetic waves cannot propagate in a plasma if $\omega < \omega_P$. This explains the possibility of radio communication with the opposite side of the Earth, a fact that was puzzling at the dawn of the radio era. Later, it was understood that the ionosphere acts as a mirror for frequencies below ω_P. In contrast, for television one needs much higher frequencies (because of the necessity of having a broad band), which are greater than the characteristic plasma frequency in the ionosphere. This is why, until the invention of cable television, one had to build high towers with emitting antennas and television could be received only in areas within the horizon viewed from the top of the tower.

Surface plasma waves can propagate along the interface between media with positive and negative dielectric constant (e.g., air and metal). Similarly, surface electromagnetic waves (polaritons) may exist at the air/dielectric crystal interface, where $\varepsilon < 0$ for frequencies between the limiting frequencies of the optical and acoustical phonons.

5. Plasma waves in two-dimensional structures

There exist two distinct situations: an ungated two-dimensional electron gas (2DEG), in a quantum well or heterostructure, and a 2DEG with a metallic gate, like in a field-effect transistors (FETs). The theoretical analysis of electromagnetic waves in such structures was done in Refs 6–8, whereas the first experimental results were obtained in Refs 9–11.

- *Ungated electron gas*

The relevant properties of the 2D plasma are the electron concentration, n, the electron charge, e, and the electron effective mass, m^*. From dimensional considerations one obtains the following dispersion relation:

$$\omega(k) = \left(\frac{2\pi n e^2 k}{m^* \varepsilon} \right)^{1/2}. \tag{2}$$

(Obviously, the 2π numerical factor cannot be guessed from dimensionality arguments. One needs some math to obtain this factor correctly.)

Comparing this result with that of Section 2.1, one can notice a striking similarity with the dispersion law for *deep-water waves*! One of the consequences is that a phenomenon similar to the Kelvin wake should also exist in a 2DEG. Certainly, we cannot sail a boat in our heterostructure; however, we could make some inclusion or defect in the 2D gas and pass a current. Then, a static Kelvin wake with an angle of 19.5° will appear in the downstream direction and produce some measurable electrical disturbance at the lateral boundaries of the sample. So far such an experiment was never done.

- *Gated electron gas or FET*

In addition to the electron concentration, in a gated 2DEG we now have another important parameter: the gate-to-channel capacitance per unit area C. The local electron charge in the channel ne is related to the local gate-to-channel voltage V_G by the plane capacitor formula: $ne = C(V_G - V_T)$, where V_T is the threshold voltage. For gate voltages below V_T, there are no electrons in the channel. The difference $U = V_G - V_T$ is called *the gate voltage swing*.

Plasma waves in a FET consist in joint oscillations of local gate voltage and electron concentration in the channel. Dimensionality arguments lead to the existence of a characteristic velocity:

$$v = \left(\frac{ne^2}{m^* C} \right)^{1/2} = \left(\frac{eU}{m^*} \right)^{1/2}, \tag{3}$$

and hence the dispersion law for plasma waves in a FET should be $\omega(k) = vk$.

Note again the interesting similarity with shallow water waves (or sound waves) described in Section 2.1. One has only to replace the *gravitational* energy per unit mass gh for shallow water waves by the *electrostatic energy* per unit mass eU/m^* in the case of plasma waves in a FET!

This analogy has led Shur and me to the prediction of plasma wave instability[12] in a FET for sufficiently high source–drain current, provided the boundary conditions at the source and drain are asymmetric. A similar instability is responsible for the performance of wind musical instruments (the clarinet is asymmetric and no sound is produced unless one blows strongly enough).

6.   Electronic surface states in solids

In a crystal, the periodic structure of the lattice results in the band structure of the electron energy spectrum and the states are described by Bloch wavefunctions. However, the termination of the periodic structure at the boundary of the crystal will mix the wavefunctions belonging to different bands and this leads to the formation of specific electron surface states.

•   *Tamm and Shockley states*

In the early days of the quantum theory of solids, Tamm[13] and later Shockley[14] demonstrated the existence of such states within a model illustrated in Fig. 3. Tamm used the tight-binding approximation, while Shockley used the nearly free electron approximation. The physical nature of Tamm states is identical to that of Shockley states.

While being quite important as a matter of principle, these results do not have predictive value. In particular, the energy of Tamm–Shockley surface states crucially depends on the exact point chosen for termination of the periodic potential, which clearly is unphysical.

•   *Surface states in HgTe quantum wells*

It is well known that the electron spectrum in a quantum well consists of two-dimensional minibands. In a rectangular, infinitely deep quantum well, the textbook energy spectrum is given by $E_n(k) = \hbar^2 \pi^2 n^2/2ma^2 + \hbar^2 k^2/2m$, where $n = 1, 2, 3, \ldots$, a is the well width, m is the electron mass, and $\hbar k$ is the in-plane electron momentum.

Some 33 years ago, Khaetskii and Dyakonov[15] became interested in understanding the energy spectrum in a quantum well formed by a gapless semiconductor, such as HgTe. The bulk energy spectrum of this peculiar material is presented in Fig. 4(a). The conduction and the valence bands have the same origin as bands of light and heavy holes in Ge, GaAs, and other III–V semiconductors, in which there exist light and heavy holes with a degeneracy at $k = 0$. The specific feature of HgTe is that the light-hole band is inverted, becoming the *conduction* band. The hole is ~20 times heavier than the electron.

One could expect that in a quantum well, the holes and the electrons would be quantized independently according to the above simple formula. This is indeed

Figure 3. The electron potential energy in the vicinity of the boundary with vacuum.

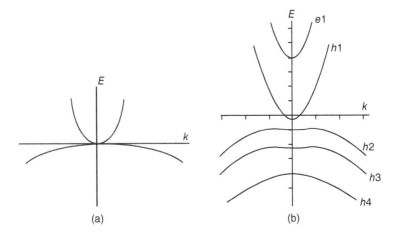

Figure 4. The bulk spectrum of the gapless semiconductor HgTe (a) and the two-dimensional spectrum in a HgTe quantum well[15] (b).

the case at $k = 0$ (see Fig. 4(b)). However, at finite k we have a surprise: the first hole subband ($h1$) is inverted (it becomes electronic-like). The electron subbands ($e1, e2, \dots$) as well as other hole subbands ($h2, h3, \dots$) are "normal."

Moreover, while the wavefunction in the $h1$ band at $k = 0$ has a maximum in the middle of the well (which is normal), for large k the maxima of the wavefunction appear near the well boundaries. This means that at large k the $h1$ band is formed by *surface states*! If the Fermi level is located in this band, but below the $e1$ band, only the surface states should contribute to conductivity.

We also predicted that similar surface states should exist at the boundary of vacuum with a bulk HgTe crystal.[16] However, in this case the band of surface states is superimposed on the bulk band spectrum.

The subject was further explored in Refs 17 and 18. In particular, Ref. 17 considered a quantum well of HgTe sandwiched between thick CdTe layers and made the first prediction of "time-reversal symmetry protected edge states," which were confirmed experimentally exactly 20 years later.[19]

• *Topological insulators*

The theoretical work of Pankratov *et al.*[17] was the first prediction belonging to the now very popular field of *topological insulators*. The exciting property is that in such materials the *surface* states (in 3D materials) or the *edge* states in 2D sandwich-like quantum wells are *time-reversal symmetry protected*, which means that they are immune to potential scattering and spin–orbit interaction (but can be destroyed by scattering by magnetic impurities, i.e., by spin scattering).

The 3D topological insulator was predicted by Fu and Kane[20] for binary compounds involving bismuth. The predicted symmetry-protected surface states were discovered in BiSb.[21] The surface states of a 3D topological insulator is a new type of a 2DEG, where the electron's spin is locked to its linear momentum.

There is a huge recent literature on this subject so we will not dwell on it further.

7. Dyakonov surface waves (DSWs)

Inspired by Rayleigh results for surface acoustic waves,[5] some time ago[22] I started a search for surface waves in another situation, where there exist two kinds of waves in the bulk: the ordinary and extraordinary electromagnetic waves in a bire-fringent crystal with eigenvalues of the dielectric tensor $\varepsilon_{\parallel}$ and $\varepsilon_{\perp}$. (It is interesting to note that all the important facts about optics of anisotropic crystals were estab-lished by Fresnel[23] long before Maxwell's equations were written.) The simplest configuration in which these new surface waves can exist is presented in Fig. 5.

In order to describe waves decaying away from the interface ($x = 0$), the x-components of their wave vectors should be imaginary: ik_1, $-ik_2$, and $-ik_3$ (for the wave in the isotropic media at $x > 0$, the extraordinary wave in the crystal at $x < 0$, and for the ordinary wave in the crystal, respectively). The dispersion laws for the three waves are

$$q^2 - k_1^{\,2} = \varepsilon,$$

$$\frac{(q^2\sin^2\phi - k_2^{\,2})}{\varepsilon_{\parallel}} + \frac{(q^2\cos^2\phi)}{\varepsilon_{\perp}} = 1, \qquad (4)$$

$$q^2 - k_1^{\,2} = \varepsilon_{\perp},$$

where all the wave vectors are dimensionless (written in units of ω/c).

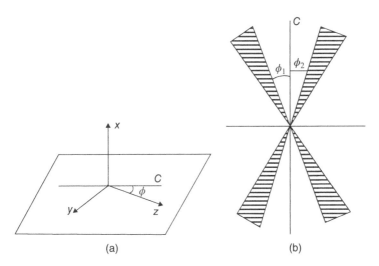

(a)   (b)

Figure 5. (a) The interface between an isotropic medium ($x > 0$) and a birefringent crystal ($x < 0$). The optical axis of the crystal is denoted by C. The direction of the surface wave phase velocity (z) makes an angle ϕ with the optical axis. (b) The angular intervals where the surface waves can propagate.

These equations should be complemented by the boundary conditions requiring the continuity of the tangential components of the electric and magnetic fields at the interface. After some tedious algebra, one can obtain the fourth equation[23]:

$$(k_1 + k_2)(k_1 + k_3)(\varepsilon k_3 + \varepsilon_\perp k_2) = (\varepsilon_\parallel - \varepsilon)(\varepsilon - \varepsilon_\perp)k_3. \tag{5}$$

Now, we can determine the four unknowns: k_1, k_2, k_3, and q, which can be done only numerically. However, from the above equation, one can see that a solution can exist only if $\varepsilon_\parallel > \varepsilon > \varepsilon_\perp$ or $\varepsilon_\perp > \varepsilon > \varepsilon_\parallel$. Further analysis shows that only the first case is acceptable; thus, the condition for existence of DSW is

$$\varepsilon_\parallel > \varepsilon > \varepsilon_\perp, \tag{6}$$

which means that the birefringent crystal must be *positive* ($\varepsilon_\parallel > \varepsilon_\perp$).

Another possibility was considered by Averkiev and Dyakonov.[24] Consider two identical positive birefringent crystals with optical axes parallel to their surfaces. Put one upon the other and make some angle α between their optical axes C_1 and C_2, as shown in Fig. 6(a). For such a configuration, we found that surface waves can exist within two sectors around the bisectrixes of the two angles formed by C_1 and C_2, as shown in Fig. 6(b).

The previously known electromagnetic surface waves (surface plasmons discussed in Section 4) exist under the condition that the permittivity of one of the materials forming the interface is negative, while the other one is positive. In contrast, the DSWs can propagate when both materials are transparent;

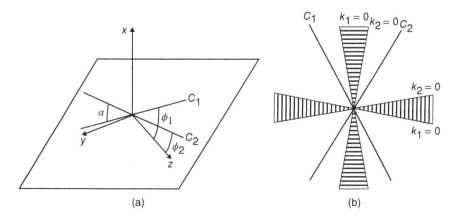

(a)   (b)

Figure 6. Surface electromagnetic waves at the interface between two identical birefringent crystals with optical axes C_1 and C_2 parallel to the interface and making an angle α between them: (a) geometry and (b) the angular intervals where surface waves can propagate.[21]

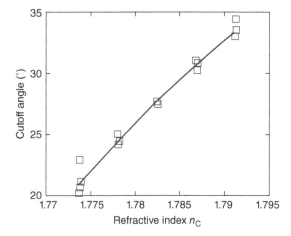

Figure 7. The measured upper cutoff angle as a function of the refractive index of the matching fluid. Line – theoretical, points – experimental. Note the extremely fine tuning. (Source: Takayama *et al.*[27] Reproduced with permission of the American Physical Society).

hence they are virtually lossless, which is their most attractive property. A large number of theoretical works appeared discussing these, or similar, waves under various conditions, in particular, in metamaterials[25] (see the review in Ref. 26).

The first experimental observation of DSW was reported only in 2009 by Takayama and coworkers,[27, 28] using the Otto–Kretschmann configuration and an index-matching fluid as an isotropic medium satisfying the condition $\varepsilon_{\parallel} > \varepsilon > \varepsilon_{\perp}$. The cutoff propagation angle was measured as a function of the permittivity of the fluid (see Fig. 7).

It is believed that the extreme sensitivity of DSWs to anisotropy, and thereby to stress, along with their low-loss (long-range) character render them particularly attractive for enabling high sensitivity tactile and ultrasonic sensing for next-generation, high-speed transduction and read-out technologies.

This subject continues to be of considerable interest, recent work can be found in Refs 29–34.

References

1. Lord Kelvin (W. Thomson), "Deep sea ship-waves," *Math. Phys. Papers* **4**, 303 (1910).
2. See en.wikipedia.org/wiki/Draupner_wave (accessed on Apr. 12, 2016).
3. D. R. Solli, C. Ropers, and P. Koonath, and B. Jalali, "Optical rogue waves," *Nature* **450**, 1054–1057 (2007).

4. X. Hu, P. Du, and C.-K. Cheng, "Exploring the rogue wave phenomenon in 3D power distribution networks," *Proc. IEEE 19th Conf. Electrical Perf. Electronic Packaging Syst.* (2010), pp. 57–60.

5. L. Rayleigh, "On waves propagated along the plane surface of an elastic solid," (1885), see http://plms.oxfordjournals.org/content/s1-17/1/4.full.pdf (accessed on Apr. 12, 2016).

6. R. H. Ritchie, "Plasma losses by fast electrons in thin films," *Phys. Rev.* **106**, 874–881 (1957).

7. F. Stern, "Polarizability of a two-dimensional electron gas," *Phys. Rev. Lett.* **18**, 546–548 (1967).

8. M. Nakayama, "Theory of surface waves coupled to surface carriers," *J. Phys. Soc. Japan* **36**, 393–398 (1974).

9. A. Eguiluz, T. K. Lee, J. J. Quinn, and K. W. Chiu. "Interface excitations in metal–insulator–semiconductor structures," *Phys. Rev. B* **11**, 4989–4993 (1975).

10. S. J. Allen, D. C. Tsui, and R. A. Logan, "Observation of the two-dimensional plasmon in silicon inversion layers," *Phys. Rev. Lett.* **38**, 980–983 (1977).

11. D. C. Tsui, E. Gornik, and R. A. Logan, "Far infrared emission from plasma oscillations of Si inversion layers," *Solid State Commun.* **35**, 875–877 (1980).

12. M. I. Dyakonov and M. S. Shur, "Shallow water analogy for a ballistic field effect transistor: New mechanism of plasma wave generation by dc current," *Phys. Rev. Lett.* **71**, 2465–2468 (1993).

13. I. E. Tamm, "On the possible bound states of electrons on a crystal surface," *Phys. Z. Soviet Union* **1**, 733–735 (1932).

14. W. Shockley, "On the surface states associated with a periodic potential," *Phys. Rev.* **56**, 317–323 (1939).

15. M. I. Dyakonov and A. V. Khaetskii, "Size quantization of the holes in a semiconductor with a complicated valence band and of the carriers in a gapless semiconductor," *Sov. Phys. JETP* **55**, 917–920 (1982).

16. M. I. Dyakonov and A. V. Khaetskii, "Surface states in a gapless semiconductor," *JETP Lett.* **33** 110–113 (1981).

17. O. A. Pankratov, S. V. Pakhomov, and B. A. Volkov, "Supersymmetry in heterojunctions: Band-inverting contact on the basis of $Pb_{1-x}Sn_xTe$ and $Hg_{1-x}Cd_xTe$," *Solid State Commun.* **61**, 93–96 (1987).

18. I. G. Gerchikov and A. V. Subashiev, "Interface states in subband structure of semiconductor quantum wells," *Phys. Stat. Sol. (b)* **160**, 443–457 (1990).

19. M. König, S. Wiedmann, C. Brüne, *et al.*, "Quantum spin Hall insulator state in HgTe quantum wells," *Science* **318**, 766–770 (2007).

20. L. Fu and C. L. Kane, "Topological insulators with inversion symmetry," *Phys. Rev B* **76**, 045302 (2007).

21. D. Hsieh, D. Qian, Y. Xia, L. Wray, R. J. Cava, Y. S. Hor, and M. Z. Hasan, "A topological Dirac insulator in a quantum spin Hall phase," *Nature* **452**, 970–974 (2008).

22. M. I. Dyakonov, "New type of electromagnetic waves propagating at an interface," *Sov. Phys. JETP* **67**, 714–716 (1988).
23. A.-J. Fresnel, *Œuvres Complètes*, vol. **1**, Paris: Imprimerie impériale, 1868.
24. N. S. Averkiev and M. I. Dyakonov, "Electromagnetic waves localized at the interface of transparent anisotropic media," *Opt. Spectrosc. (USSR)* **68**, 653–655 (1990).
25. D. Artigas and L. Torner, "Dyakonov surface waves in photonic metamaterials," *Phys. Rev. Lett.* **94**, 013901 (2005).
26. O. Takayama, L. C. Crassovan, D. Mihalache, and L. Torner, "Dyakonov surface waves: A review," *Electromagnetics* **28**, 126–145 (2008).
27. O. Takayama, L. Crassovan, D. Artigas, and L. Torner, "Observation of Dyakonov surface waves," *Phys. Rev. Lett.* **102**, 043903 (2009).
28. L. Torner, D. Artigas, and O. Takayama, "Dyakonov surface waves," *Opt. Photonics News* **20**, 25 (2009).
29. Z. Jacob and E. Narimanov, "Optical hyperspace for plasmons: Dyakonov states in metamaterials," *Appl. Phys. Lett.* **93**, 221109 (2008).
30. O. Takayama, D. Artigas, and L. Torner, "Practical dyakonons," *Opt. Lett.* **37**, 4311–4313 (2012).
31. O. Takayama, D. Artigas, and L. Torner, "Coupling plasmons and dyakonons," *Opt. Lett.* **37**, 1983–1985 (2012).
32. C. J. Zapata-Rodriguez, J. J. Miret, S. Vukovič, and Z. Jakšič, "Dyakonons in hyperbolic metamaterials," *Photonics Lett. Poland* **5**, 63–65 (2013).
33. M. A. Noginov, "Steering Dyakonov-like waves," *Nature Nanotechnol.* **9**, 414–415 (2014).
34. O. Takayama, D. Artigas, and L. Torner, "Lossless directional guiding of light in dielectric nanosheets using Dyakonov surface waves," *Nature Nanotechnol.* **9**, 419–424 (2014).

2.2

Graphene and Atom-Thick 2D Materials: Device Application Prospects

Sungwoo Hwang, Jinseong Heo, Min-Hyun Lee, Kyung-Eun Byun, Yeonchoo Cho, and Seongjun Park
Device Laboratory, Samsung Advanced Institute of Technology, Suwon 443-803, South Korea

1. Introduction

The electronic properties of low-dimensional systems have long been an interesting topic in both physics and engineering societies. Silicon MOSFET and III–V HEMT devices, for example, not only have been the most essential elements in micro- and nanoelectronics since their early days but have also provided wonderful playgrounds for correlated 2D electron systems, leading to such fundamental discoveries as the integer and fractional quantum Hall effects. These traditional 2D electron systems have now reappeared in the form of atomic sheets, shedding light on entirely new physics originating from the orbital confinement, as well as new materials science aspects. The new physics, in turn, enables innovation in the 2D device realm. In this chapter, we consider these new opportunities and accompanying challenges, examining various aspects of potential device applications of graphene and atom-thick 2D materials, including optoelectronic devices, new types of transistors, and possible CMOS integration. Direct growth is the key technology to make all these applications realistic, so we will also address the prospects of wafer-scale graphene and 2D materials growth.

2. Conventional low-dimensional systems

Back in 1967,[1] Stern and Howard considered the two-dimensional electron gas (2DEG) at the Si/SiO_2 interface, where the electron wavefunction is confined within a few nanometers in the direction perpendicular to the interface. High-mobility 2DEG was then created in III–V HEMTs by Stormer *et al.* in 1979.[2] Physicists utilized these 2DEGs as a wonderful playground for new types of electronic states. The integer quantum Hall effect,[3] a manifestation of

Future Trends in Microelectronics: Journey into the Unknown, First Edition.
Edited by Serge Luryi, Jimmy Xu, and Alexander Zaslavsky.
© 2016 John Wiley & Sons, Inc. Published 2016 by John Wiley & Sons, Inc.

magnetic field-induced localization–delocalization transition, was observed at the Si/SiO$_2$ interface. Subsequently, the fractional quantum Hall effect[4] arising in strongly correlated 2D electron liquids was found in high-mobility HEMTs. These observations were only made possible by superb quality of the devices. In 1981, when integer quantum Hall effect was first observed, Intel was already manufacturing processors with more than 100,000 transistors. Advances in large-scale CMOS circuits have been driving modern information technology to this date.

If the electronic wavefunction of a 2DEG is further confined, one can realize a one-dimensional electron gas (1DEG). In 1988, van Wees *et al.*[5] fabricated 1DEGs by placing split gates on a III–V HEMT and defining a narrow electron channel by negatively biasing these gates. The mobility of the HEMT wafer was 10^6 cm^2/V·s, resulting in the mean free path of several microns at 4 K making it possible to observe conductance quantization whenever the gate voltage opened a new conductance channel. Scott-Thomas *et al.*[6] fabricated silicon 1DEGs by the double-gate technique and observed Coulomb blockade oscillations. These early 1DEGs studied for fundamental physics have now reappeared in such devices as fin-FETs[7] and gate all-around (GAA) Si nanowire FETs,[8] illustrated in Figs. 1 and 2, respectively, as examples of ultimate highly scaled Si MOSFETs.

Figure 2 also shows the conductance measured as a function of gate bias V_G in GAA Si nanowire FETs with length ranging from 400 to 20 nm. It shows conductance overshoot when the channel length is shorter than 40 nm. This length scale is comparable to the mean free path of electrons in Si at room temperature, providing strong evidence of ballistic transport. One well-known but important message is that the device performance will not be improved much in these nodes even though we scale the channel length.

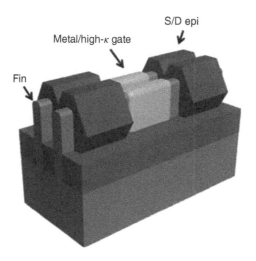

Figure 1. Schematic structure of a fin-FET.

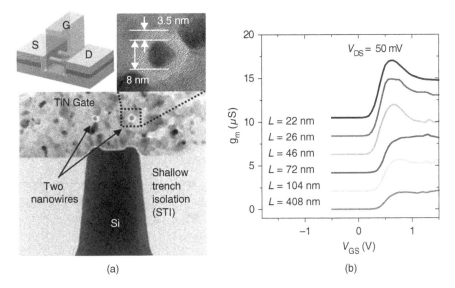

Figure 2. (a) Gate-all-around Si nanowire MOSFET; and (b) transconductance as a function of gate length. (Source: Cho et al.[8] Reproduced with permissions of the American Institute of Physics).

3.   New atomically thin material systems

Newly discovered atom-thick 2D systems, including graphene, can be thought of as a traditional 2DEG shrunk to atomic thickness. The extreme orbital confinement manifests itself in new physics originating from material aspects and dimensional confinement on the atomic scale. The new physics is giving us many new opportunities for creating new device functionalities and innovating conventional 2D devices.

Graphene, one isolated sheet of graphite or a honeycomb 2D lattice of carbon atoms, was discovered by Geim and Novoselov in 2004.[9] In layered graphite, each carbon atom has sp^2 hybridization, leading to strong intralayer covalent bonding and relatively weak van der Waals interlayer interaction, with an interlayer distance of 0.34 nm. Such weak bonding between layers leads to simple mechanical exfoliation using adhesive tape, resulting in atomically thin monolayers of carbon atoms (see Fig. 3(a) and (b)). Having linear energy–momentum dispersion relation, shown in Fig. 3(c), low energy excitations of carriers in graphene mimic relativistic Dirac particles with a reduced velocity of $c/300$, called massless Dirac fermions.[10] A simple consequence of this unique dispersion is the absence of backscattering, leading to mobility up to 200,000 $cm^2/V \cdot s$ at room temperature. More excitingly, many novel physical effects have been explored experimentally: half-integer quantum Hall effect,[9] Coulomb drag (frictional coupling between electric currents flowing in spatially separated

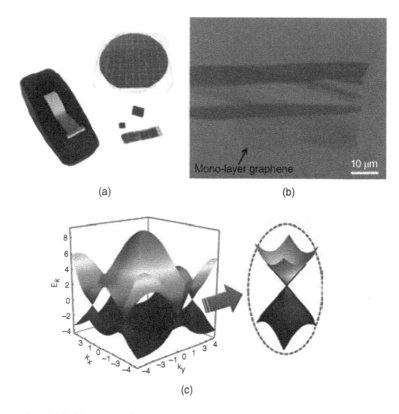

Mono-layer graphene

10 μm

(a)   (b)

(c)

Figure 3. (a) Highly oriented pyrolysis graphite (HOPG) with adhesive tape; (b) exfoliated graphene on SiO₂/Si substrate; and (c) electronic band structure of graphene and linear dispersion at K points in the Brillouin zone.

conducting layers due to interlayer electron–electron interactions),[11] and Hofstadter's butterfly effect,[12] when 2D electrons in both a magnetic field and a periodic electrostatic potential exhibit a self-similar recursive energy spectrum.

In addition to new fundamental physics, graphene has a number of superior technological properties in terms of transparency, bulk resistivity, current density, chemical inertness, surface area, and thermal conductivity. Graphene has high transparency of 97.7% over a wide spectral range, bulk resistivity of $10\ \mu\Omega$ cm, and can carry current densities of 10^6 A/cm², compared to 1.68 μΩ cm and 4020 A/cm² for Cu of monolayer thickness, respectively. It also has chemical inertness from aromatic bonds of 518 bond enthalpies kJ/mol compared to 348 for C—C bond; a high surface area of 2630 m²/g compared to 500 for activated carbon; and thermal conductivity of 5300 W/m·K, compared to 2320 for diamond. These properties can be combined and exploited in transparent electrodes, interconnects, barrier materials, supercapacitor, and heat sink applications – see Fig. 4. For example, transparent electrodes require both

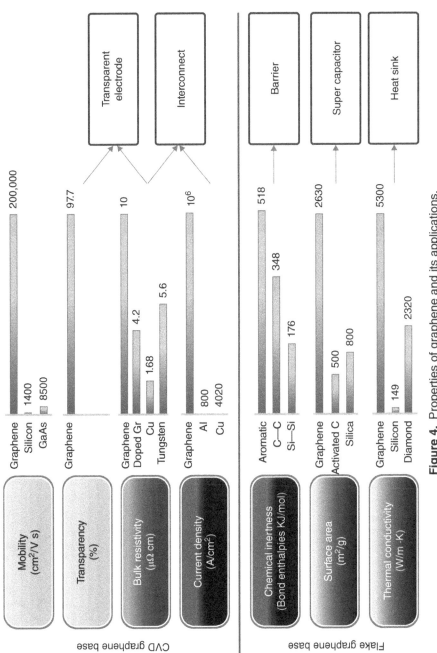

Figure 4. Properties of graphene and its applications.

high transparency and low sheet resistance. Doped four-layer graphene has $30\,\Omega/\square$ sheet resistance at 90% transparency, superior to commercial indium tin oxides.[13] Moreover, for flexible applications, graphene-based transparent electrode can withstand strains of more than 6%, whereas ITO cracks under 2–3% strain. As another example, heat management in modern integrated chips or CPUs is becoming increasingly important because of ever-larger heat dissipation as device density increases. Given a thermal conductivity of around 40 times larger than Si, graphene is expected to provide viable solutions for heat management in modern integrated chips when incorporated in the form of composites or inks.

Since its discovery in 2004, graphene has been at the center of new materials research both in academia and industry for disruptive innovation. However, for applications such as flexible digital electronics, new functional nanomaterials beyond graphene are required, with such properties as an electronic bandgap or atomically clean dielectric interfaces. The search for other layered materials that can be easily exfoliated has produced a large number of 2D materials besides graphene. These include monoatomic phosphorene and diatomic materials such as hexagonal boron nitride (h-BN) and transition metal dichalcogenides (TMDCs)[14] – see Table 1. While graphene is a semimetal, some TMDCs are semiconducting, whereas h-BN is a dielectric with an ultraclean surface. The most studied semiconducting TMDCs are MoS_2 (see Fig. 5 for bandstructure), $MoSe_2$, WS_2, and WSe_2. From the optoelectronic application viewpoint, these

Semimetal/superconductor	Graphene, WTe_2, $PtSe_2$, TiS_2	NbS_2, $NbSe_2$
Topological insulator/insulator	Bi_2S_3, Bi_2Se_3, Bi_2Te_3	h-BN
Semiconductor	Black phosphorus, MoS_2, $MoSe_2$, $MoTe_2$, WS_2, WSe_2, GaS, GaS_2, HfS_2, $HfSe_2$, In_2Se_3, ReS_2, $ReSe_2$, $ZrSe_2$, SnS_2, $SnSe_2$, $TiSe_2$, TaS_2, GeS	

Table 1. Some examples of layered 2D materials.

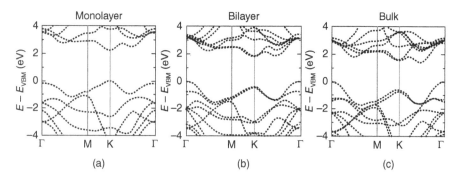

Figure 5. Band structure of (a) monolayer; (b) bilayer; and (c) bulk MoS_2.

2D materials span the range from zero bandgap graphene to a large bandgap semiconductor such as GaS ($\sim$3 eV), to the insulator h-BN. In addition to aforementioned basic components, there is also a new class of materials such as 2D superconductors, for example, NbS_2 and $NbSe_2$. With a whole library of 2D materials and their heterostructures, one can envisage a new atomic system with a broad range of functionalities.[15]

4.   Device application of new material systems

- *Photodetectors*

Two-dimensional materials provide strong light-matter interaction, such that even a single monolayer can have up to 10% absorption, depending on wavelength and material bandgap.[16] With large absorption coefficients of 10^5–10^6 cm^{-1} of 2D materials in the visible range, almost two orders of magnitude larger than Si, photodetectors based on 2D materials can be made as thin as a few tens of nanometers. One of the photodetector figures of merit is responsivity or external quantum efficiency, which measures the number of electrons generated per incident photon.[17] Focusing on vertically stacked photodetectors with a transparent electrode/photoactive region/electrode sequence, a responsivity of 20 A/W was measured in graphene/MoS_2/graphene heterostructures (see Table 2).

Material stack	Measurement conditions	R (A/W)	t_R (ms)	References
G/WSe_2/G	$V_D = 0.5$ V, $V_G = 0$, $\lambda = 514$ nm, $P = 10^{-5}$ μW/μm^2	0.33 (1.1)	<30 (28×10^{-3})	18
G/MoS_2/G	$V_D = 5$ V, $V_G = 30$ V, $\lambda = 633$ nm, $P = 0.24$ μW/μm^2	20	>10^4	18
G/WS_2/G	$V_D = 0$, $V_G = -20$ V, $\lambda = 488$ nm, $P = 10$ μW/μm^2	0.1	NA	19
G/MoS_2/G	$V_D = 0$ V, $V_G = -60$ V, $\lambda = 514$ nm, $P = 80$ μW/μm^2	0.06	NA	20
G/WSe_2/MoS_2/G	$V_D = 0$, $V_G = 0$, $\lambda = 532$ nm, $P = 3$–7 μW/μm^2	0.12	NA	21

Table 2. Performance of various 2D heterojunction photodetectors.

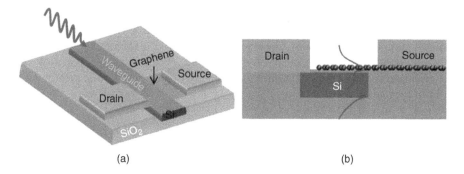

(a)   (b)

Figure 6. (a) Schematic of graphene–Si waveguide photodetector and its cross section with field intensity; (b) By design, the optical field is maximized at the interface between Si and graphene.

Another figure of merit of photodetectors is response time or bandwidth, a critical measure for optical communication. Graphene has the highest mobility and ultrabroadband photoresponse but only absorbs 2.3% of light.[22] The weak absorption of graphene can be overcome by integrating it with a Si waveguide in order to enhance the detection area along the propagation direction[23–25] (see Fig. 6).

Graphene/Si hybrid photodetectors showed a 3 dB bandwidth of 20 GHz at 1550 nm wavelength. Monolayer deposition of graphene on a Si waveguide offers simpler and cheaper integration than heteroepitaxial growth of III–V or Ge layers on Si for upcoming all-optical chips.

• *New types of transistors*

Graphene FETs showed unprecedented mobility of over 100,000 cm^2/V·s at room temperature,[26] two orders of magnitude larger than Si CMOS. As a result, graphene was considered a candidate for future electronics in post-Si era. But graphene lacks a bandgap of at least a few hundred millielectron volts, making it difficult to switch off the FET current. A number of graphene-based alternatives have been proposed – graphene nanoribbons,[27] bilayer graphene with tunable bandgap,[28] and graphene nanomeshes,[29] – but all have suffered from either lithographically patterned irregular edges, the absence of uniformity and controllability over large area, or small on/off current ratios.

The challenges of using standalone graphene for digital applications shifted the focus to graphene hybrid systems. In order to attain an appropriate energy barrier to turn off charge flow, semiconducting materials were brought in contact with graphene. For example, graphene–Si hybrid junctions make Schottky diodes where the barrier height of the junction can be tuned by gate voltage as the work-function of graphene is modulated (see Fig. 7).[30] This solid-state triode system,

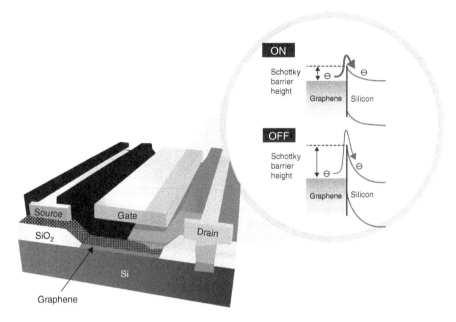

Figure 7. Graphene barristor and its operation principle.

called graphene barristor, enabled high on/off ratio of 10^5 and a half-adder circuit was demonstrated upon its integration.

The work on graphene barristors also demonstrated a new finding: the absence of pinning in graphene on Si. This unique property can be applied to reduce source–drain contact resistance, which is discussed in Section 5.2. Graphene/2D semiconductor heterotunnel junctions were also demonstrated to show large on/off ratio[31] and atomically thin semiconductors, such as WS_2, sandwiched between graphene layers, led to transparent and flexible tunneling transistors.[32] Recent results in this area are summarized in Table 3. Graphene–semiconductor–metal tunnel junctions on 8-in. wafer scale provide a versatile platform to be tailored by varying materials and deposition thicknesses for low-power electronics, as in Fig. 8.[35]

• *Challenges*

It should be noted that many conceptually new device concepts that appeared in the ITRS, including carbon nanotubes, single electron transistors, and spin MOSFETs, have not to date succeeded in displacing Si technology. In that regard, the new graphene- and/or other atom-thick 2D material-based optoelectronic or electronics devices discussed above could require many years before showing up in the market.[36] Even though they show higher photoresponsivity and

Materials	I_{ON} at 0.5 V ($\mu A/\mu m^2$)	I_{ON}/I_{OFF}	SS (mV/dec)	References
G/IGZO/Mo	10	10^6	30–1,000	18
G/MoS$_2$/Ti	10	10^3	20,000	33
G/WS$_2$/G	5	10^6	10,000	32
Pt/Al$_2$O$_3$/G/Si	10^{-6}	10^6	250	34

Table 3. Performance of vertical tunneling transistors based on 2D materials.

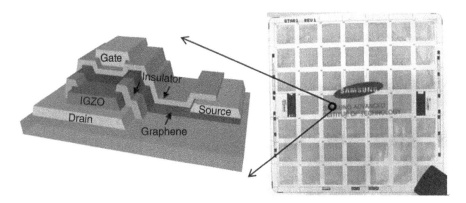

Figure 8. Graphene–semiconductor–metal tunneling junction transistors (SAIT).

optical communication speeds, as well as on/off ratios comparable to conventional devices, there are still many challenges to overcome.

As far as heterojunctions are concerned, there is the perennial question of interfaces. First of all, large-scale graphene grown on germanium[37] or catalytic metal surfaces[38] has to be transferred to the target substrate, including exposed Si or other 2D materials. The transfer process usually involves adhesives, polymers, or chemical etchants,[39] which in turn introduce organic, ionic, or metal residues. Spatial variation of these residues degrades device uniformity and controllability. Second, there is the 2D semiconductor and metal contact issue. With lack of precise doping control for both p- and n-type 2D semiconductors, true Ohmic contact is difficult to achieve. So far, only limited chemical doping has shown some reduction of contact resistance,[40] but the results are still inferior to state-of-the-art Si technology. Third, there is no thorough study on scaling of two-dimensional devices including both channel and contact area. Even though the best I_{ON} of 10 $\mu A/\mu m^2$ at 0.5 V for 2D heterojunction tunnel devices is comparable to state-of-the-art Si contact resistivity, all aforementioned issues result in nonuniform distribution and degradation of on-current for electronic devices.

5.   Components in Si technology

So far, we discussed devices based on active functions of graphene or related 2D materials for optoelectronic or electronic applications that might be adopted in 5–10 years. More realistic application of those materials is implementation as a component layered material in Si device process, enhancing device performance.

The first MOSFET, patented in 1960 by D. Kang[41] and shown in Fig. 9, was a flat device using only Si, SiO_2, and Al as constituent materials. Continued performance improvement and dimensional scaling of MOSFETs during last half century have been made possible only by continuously adopting new type of structures, materials, and processing technologies. Compared to Fig. 9, a modern CMOS device architecture, shown in several chapters of this book, is replete with tremendous complications and adaptations. The Al interconnect was replaced by Cu for the lower parasitic resistance in the late 1990s, in spite of the high diffusivity of Cu into active device regions and the deep defect level of Cu in Si. Also, the damascene process was adopted to solve the difficulty in etching Cu.[42] Then, SiO_2/poly-Si gate was replaced by high-κ/metal gate stack for scaling the equivalent oxide thickness (EOT) from the 45 nm technology node onward, regardless of the difficulties in threshold voltage (V_T) alignment and complex new process steps, such as atomic layer deposition (ALD).[43] Two-dimensional atom-thick materials hold potential promise as passive components in following two possible scenarios in sub-10 nm technology node.

•   *Interconnects*

The exponential increase in the number of interconnects and their increasing resistance with device downscaling have always comprised a major limiting factor for fabricating more complex electronic device. Until the recent sub-20 nm nodes, additional Cu metal layers and the damascene process provided reasonable resistance and manufacturability in the back-end-of-line (BEOL) process

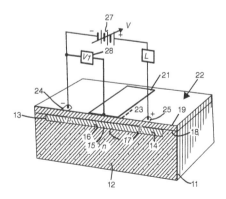

Figure 9. The original Si MOSFET[41] patent.

flow. However, shrunken Cu interconnects have to carry higher current densi-ties, resulting in migration, deformation, or diffusion to other regions because of increased heat dissipation. Moreover, electron scattering at the surfaces and grain boundaries, as well as line edge roughness (LER), start to play an increasing role in the conduction properties of metal, and the effective resistivity of Cu more than doubles as the line width decreases from 100 to 10 nm.[44] Furthermore, the thickness of the liner/barrier layer has remained almost constant and the portion of liner/barrier layer in interconnect has increased from 1% to 50%.[45]

There are few possible ways to overcome these issues: improvement of the effective resistivity of Cu by increasing grain size and decreasing LER, hybridization or replacement of interconnect material with lower resistivity materials, and introduction of an atomically thin liner/barrier layer. Chemical vapor deposition of graphene on catalytic metals[38] is a well-known method for large-scale and high-quality graphene film production. Among other metals, Cu offers the best quality of monolayer graphene because of near-zero solubility and self-terminated growth mechanism.[46] As a result, Cu–graphene hybrid interconnect appears a promising alternative to Cu alone, and graphene-covered Cu brings a lot of benefits. Since graphene is both a good electrical and thermal conductor, and a good barrier material, graphene–Cu interconnects can prevent electromigration and provide increased current-carrying capacity – see Fig. 10. These concepts were patented[47] by IBM, Toshiba, and others and published in a few of papers.[48, 49] For example, Yeh et al. reported the current-carrying capacity of 22 nm-node interconnect was increased from 1.1×10^8 to 1.1×10^9 A/cm^2 when Cu was replaced by graphite–Cu grown at 400 °C.[49]

Finally, full replacement of Cu by graphene can be expected at line widths of a few nanometers.[50] Experimentally, graphene shows a peak current capacity on the order of 10^8 A/cm^2 and the comparable resistivity to Cu lines scaled below

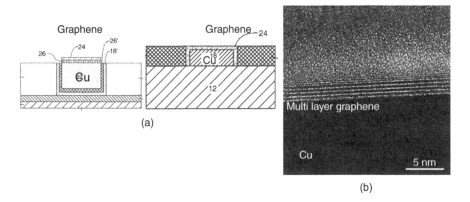

(a)

(b)

Figure 10. (a) Graphene–Cu interconnect patent;[47] and (b) TEM image for graphene–Cu interconnect structure (SAIT). (Source: http://www.google.co.uk/patents/US3102230. Public domain).

40 nm.[51, 52] Also, multilayer graphene intercalated with iron chloride may out-perform copper in a current capacity and resistivity ($>10^7$ A/cm^2, 3.2 $\mu\Omega$ at 8 nm width).[48, 53]

- *Source–drain contacts*

Scaling issues also challenge the front-end-of-line (FEOL) process. The Si/metal contact resistance increases as the inverse square of the device size, unlike other resistances that increase linearly as the device size decreases. This rapid increase in the contact resistance has led to it to exceed the channel resistance below 32 nm node[54] (see Fig. 11). Larger contact resistance in scaled devices prevents optimal scaling of the supply voltage, resulting in greater power consumption.

Theoretically, the contact resistance is strongly dependent on the Schottky barrier height, formed by the misalignment between the metal workfunction and the semiconductor affinity of electrons or holes. Thus, low workfunction (~4.0 eV) metals for n-type Si and high workfunction (~5.2 eV) metals for p-type Si are supposed to guarantee Ohmic contact behavior and low contact resistance. However, most metals show high measured Schottky barrier heights of about 0.5 eV due to pinning effects at the semiconductor interface. There have been a few candidates for resolving this contact issue by introducing a highly doped Si S/D layers or thin insertion layers between highly doped Si and contact metal. The highly doped Si results in a sufficiently narrow depletion width for effective charge tunneling, and this technique was widely used in current Si devices. However, this method is approaching the dopant solubility

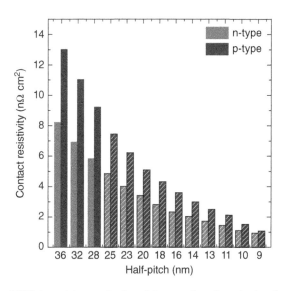

Figure 11. ITRS target for contact resistance at various technology nodes.

limit in semiconductor ($<10^{21}$ cm^{-3}), and faces challenges in shallow junction formation, high source–drain leakage currents, and low thermal stability. The insertion layer prevents Fermi level pinning between metal and semiconductor, and at the same time, the layer is thin enough that tunneling probability of charges approaches unity. For example, Agrawal *et al.* reported that 1 nm-thick TiO$_{2-x}$ layer inserted between Ti and *n*-Si reduces contact resistance by an order of magnitude to 9.1×10^{-9} Ω cm^{2}.[55] However, the insulating layer is hard to control when its thickness falls below 1 nm and the thermal stability on the Si layer may be insufficient.

On the other hand, 2D materials with monolayer thickness less than 1 nm can be an optimal candidate for depinning the interface and maximizing tunneling probability. As mentioned in Section 4.2, the Schottky barrier height between graphene and Si was changed by the same amount as the workfunction of graphene by a gate voltage, proving depinning of the graphene–Si interface.[30] This is an attractive property for the Si/metal contact formation. Also, the graphene can reduce the metal workfunction by charge transfer, similar to surface doping.[56] This could prove helpful for n-type semiconductor contacts that need a low workfunction below 4 eV (see Fig. 12).

In addition, 2D materials have other attractive features for insertion into contacts: nonpermeability, high thermal stability, and digital thickness control with low predicted Schottky barrier heights, as shown in Fig. 13.[57]

To test the effect of a graphene insertion layer at metal/Si junction, a Si AFM tip was contacted to graphene/Ni and Ni surfaces for comparison in a vacuum chamber.[58] As shown in Fig. 14, the Si/Ni junction exhibited rectifying

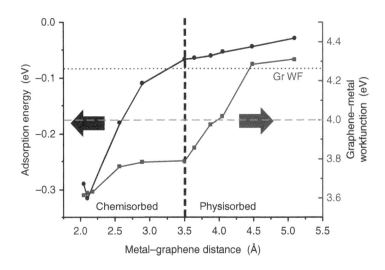

Figure 12. Workfunction of metal-doped graphene and its formation energy as function of distance calculated using density functional theory.

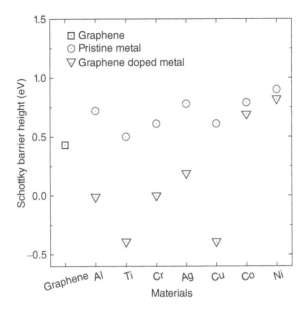

Figure 13. Predicted Schottky barrier heights for n-type Si of graphene, metals, and graphene-doped metals.

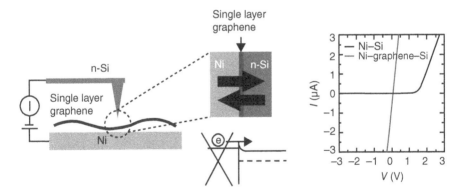

Figure 14. Schematic of conducting AFM experiment for measuring contact resistance of a metal/Si junction with and without graphene (left); band diagram with inserted graphene layer (middle); corresponding *I–V* curves (right).

diode-like $I-V$ characteristics, whereas the graphene-inserted junction was Ohmic with low resistance. The measured minimum contact resistance ranged from 10^{-8} to $10^{-9}\,\Omega\cdot cm^2$, comparable to state-of-the-art Si technology.

6. Graphene on Ge

For most of device applications, direct growth of high-quality graphene and 2D materials on various substrates is the most important issue. Typically, graphene has been grown on metallic catalyst substrates such as Ni[38] or Cu.[46] For interconnect applications, growth of graphene on insulators such as SiO_2 or SiN will be essential. Contact applications require the growth of high-quality graphene or 2D materials directly on the semiconductor surface. Recent reports on the growth of graphene on Si[59] and Ge[37, 60] and growth of MoS_2 on fused silica[61] are therefore of great interest. Another important aspect is the crystallinity of graphene. Polycrystalline graphene was usually obtained in earlier works on metallic catalysts. Great progress on single-crystal graphene growth has been made using a single seed,[62] and recent work using multiple aligned seeds[37] represents an interesting breakthrough.

7. Conclusion

Newly emergent 2D materials including graphene and TMDCs began as playgrounds for observing exotic physics, such as unconventional quantum Hall effect and quantum transport phenomena. Just as the fundamental understanding and exploration of conventional 2DEG in an earlier era became the cornerstone of modern IT industry, these atom-thick 2D materials may lead to an industrial revolution in the post-Si era. In particular, for electronic devices, atomically thin tunnel transistors can offer additional transparency and flexibility with comparable performance to Si, whereas for photodetectors 2D materials can provide wider bandwidth and cheaper integration with Si for data communication. They may also provide atomically thin, flexible photodetectors compatible with any surface. However, these device applications requiring active functions of 2D materials may take 5–10 years to reach the marketplace due to challenges and issues of material quality and integration. For more realistic near-term applications, we have discussed Cu–graphene hybrid interconnects and graphene-inserted source–drain contacts that may help with the critical challenges in further scaling of Si technology.

References

1. F. Stern and W. E. Howard, "Properties of semiconductor surface inversion layers in the electric quantum limit," *Phys. Rev.* **163**, 816–835 (1967).

2. H. L. Stormer, R. Dingle, A. C. Gossard, W. Wiegmann, and M. D. Sturge, "Two-dimensional electron gas at a semiconductor–semiconductor interface," *Solid State Commun* **29**, 705–709 (1979).

3. K. von Klitzing, G. Dorda, and M. Pepper, "New method for high-accuracy determination of the fine-structure constant based on quantized Hall resistance," *Phys. Rev. Lett.* **45**, 494–497 (1980).

4. D. C. Tsui, H. L. Stormer, and A. C. Gossard, "Two-dimensional magnetotransport in the extreme quantum limit," *Phys. Rev. Lett.* **48**, 1559–1562 (1982).

5. B. J. van Wees, H. van Houten, C. W. J. Beenakker, *et al.*, "Quantized conductance of point contacts in a two-dimensional electron gas," *Phys. Rev. Lett.* **60**, 848–850 (1988).

6. J. H. F. Scott-Thomas, S. B. Field, M. A. Kastner, H. I. Smith, and D. A. Antoniadis, "Conductance oscillations periodic in the density of a one-dimensional electron gas," *Phys. Rev. Lett.* **62**, 583–586 (1989).

7. V. Moroz, "TCAD eases FinFET design and variability analysis," *EE Times Asia* (2012).

8. K. H. Cho, K. H. Yeo, Y. Y. Yeoh, *et al.*, "Experimental evidence of ballistic transport in cylindrical gate-all-around twin silicon nanowire metal-oxide-semiconductor field-effect transistors," *Appl. Phys. Lett.* **92**, 052102 (2008).

9. K. S. Novoselov, A. K. Geim, S. V. Morozov, *et al.*, "Two-dimensional gas of massless Dirac fermions in graphene," *Nature* **438**, 197–200 (2005).

10. A. H. Castro Neto, F. Guinea, N. M. R. Peres, K. S. Novoselov, and A. K. Geim, "The electronic properties of graphene," *Rev. Mod. Phys.* **81**, 109–162 (2009).

11. R. V. Gorbachev, A. K. Geim, M. I. Katsnelson, *et al.*, "Strong Coulomb drag and broken symmetry in double-layer graphene," *Nature Phys.* **8**, 896–901 (2012).

12. C. R. Dean, L. Wang, P. Maher, *et al.*, "Hofstadter's butterfly and the fractal quantum Hall effect in Moire superlattices," *Nature* **497**, 598–602 (2013).

13. S. Bae, H. Kim, Y. Lee, *et al.*, "Roll-to-roll production of 30-inch graphene films for transparent electrodes," *Nature Nanotechnol.* **5**, 574–578 (2010).

14. K. S. Novoselov, D. Jiang, F. Schedin, *et al.*, "Two-dimensional atomic crystals," *Proc. Natl. Acad. Sci.* **102**, 10451–10453 (2005).

15. A. K. Geim and I. V. Grigorieva, "Van der Waals heterostructures," *Nature* **499**, 419–425 (2013).

16. M. Bernardi, M. Palummo, and J. C. Grossman, "Extraordinary sunlight absorption and one nanometer thick photovoltaics using two-dimensional monolayer materials," *Nano Lett.* **13**, 3664–3670 (2013).

17. F. H. L. Koppens, T. Mueller, P. Avouris, A. C. Ferrari, M. S. Vitiello, and M. Polini, "Photodetectors based on graphene, other two-dimensional materials and hybrid systems," *Nature Nanotechnol.* **9**, 780–793 (2014).

18. J. Heo, *et al.*, unpublished (2015).

19. L. Britnell, R. M. Ribeiro, A. Eckmann, *et al.*, "Strong light-matter interactions in heterostructures of atomically thin films," *Science* **340**, 1311–1314 (2013).

20. W. J. Yu, Y. Liu, H. Zhou, *et al.*, "Highly efficient gate-tunable photocurrent generation in vertical heterostructures of layered materials," *Nature Nanotechnol.* **8**, 952–958 (2013).

21. C.-H. Lee, G.-H. Lee, A. M. van der Zande, *et al.*, "Atomically thin p–n junctions with van der Waals heterointerfaces," *Nature Nanotechnol.* **9**, 676–681 (2014).

22. R. R. Nair, P. Blake, A. N. Grigorenko, *et al.*, "Fine structure constant defines visual transparency of graphene," *Science* **320**, 1308 (2008).

23. K. Kim, J.-Y. Choi, T. Kim, S.-H. Cho, and H.-J. Chung, "A role for graphene in silicon-based semiconductor devices," *Nature* **479**, 338–344 (2011).

24. X. Gan, R.-J. Shiue, Y. Gao, *et al.*, "Chip-integrated ultrafast graphene photodetector with high responsivity," *Nature Photonics* **7**, 883–887 (2013).

25. A. Pospischil, M. Humer, M. M. Furchi, *et al.*, "CMOS-compatible graphene photodetector covering all optical communication bands," *Nature Photonics* **7**, 892–896 (2013).

26. K. I. Bolotin, K. J. Sikes, J. Hone, H. L. Stormer, and P. Kim, "Temperature-dependent transport in suspended graphene," *Phys. Rev. Lett.* **101**, 096802 (2008).

27. X. Li, X. Wang, L. Zhang, S. Lee, and H. Dai, "Chemically derived, ultrasmooth graphene nanoribbon semiconductors," *Science* **319**, 1229–1232 (2008).

28. J. B. Oostinga, H. B. Heersche, X. Liu, A. F. Morpurgo, and L. M. K. Vandersypen, "Gate-induced insulating state in bilayer graphene devices," *Nature Mater.* **7**, 151–157 (2008).

29. J. Bai, X. Zhong, S. Jiang, Y. Huang, and X. Duan, "Graphene nanomesh," *Nature Nanotechnol.* **5**, 190–194 (2010).

30. H. Yang, J. Heo, S. Park, *et al.*, "Graphene barristor, a triode device with a gate-controlled Schottky barrier," *Science* **336**, 1140–1143 (2012).

31. L. Britnell, R. V. Gorbachev, R. Jalil, *et al.*, "Field-effect tunneling transistor based on vertical graphene heterostructures," *Science* **335**, 947–950 (2012).

32. T. Georgiou, R. Jalil, B. D. Belle, *et al.*, "Vertical field-effect transistor based on graphene-WS_2 heterostructures for flexible and transparent electronics," *Nature Nanotechnol.* **8**, 100–103 (2013).

33. W. J. Yu, Z. Li, H. Zhou, *et al.*, "Vertically stacked multi-heterostructures of layered materials for logic transistors and complementary inverters," *Nature Mater.* **12**, 246–252 (2013).

34. C. Zeng, E. B. Song, M. Wang, *et al.*, "Vertical graphene-base hot-electron transistor," *Nano Lett.* **13**, 2370–2375 (2013).

35. J. Heo, K.-E. Byun, J. Lee, *et al.*, "Graphene and thin-film semiconductor heterojunction transistors integrated on wafer scale for low-power electronics," *Nano Lett.* **13**, 5967–5971 (2013).

36. K. S. Novoselov, V. I. Fal'ko, L. Colombo, P. R. Gellert, M. G. Schwab, and K. Kim, "A roadmap for graphene," *Nature* **490**, 192–200 (2012).
37. J.-H. Lee, E. K. Lee, W.-J. Joo, *et al.*, "Wafer-scale growth of single-crystal monolayer graphene on reusable hydrogen-terminated germanium," *Science* **344**, 286–289 (2014).
38. K. S. Kim, Y. Zhao, H. Jang, *et al.*, "Large-scale pattern growth of graphene films for stretchable transparent electrodes," *Nature* **457**, 706–710 (2009).
39. L. Gao, G.-X. Ni, Y. Liu, B. Liu, A. H. Castro Neto, and K. P. Loh, "Face-to-face transfer of wafer-scale graphene films," *Nature* **505**, 190–194 (2014)
40. L. Yang, K. Majumdar, H. Liu, *et al.*, "Chloride molecular doping technique on 2D materials: WS_2 and MoS_2," *Nano Lett.* **14**, 6275–6280 (2014).
41. K. Dawon, "Electric field controlled semiconductor device," patent US3102230 A (1963).
42. L. Zuckerman, "I.B.M. to make smaller and faster chips," Sept. 22, 1997, available at www.nytimes.com/1997/09/22/business/ibm-to-make-smaller-and-faster-chips.html.
43. K. Mistry, C. Allen, C. Auth, *et al.*, "A 45 nm logic technology with high-κ + metal gate transistors, strained silicon, 9 Cu interconnect layers, 193 nm dry patterning, and 100% Pb-free packaging," *Tech. Digest IEDM* (2007), pp. 247–250.
44. S. Im, N. Srivastava, K. Banerjee, and K. E. Goodson, "Scaling analysis of multilevel interconnect temperatures for high-performance ICs," *IEEE Trans. Electron Devices* **52**, 2710–2719 (2005).
45. G. C. Schwartz and K. V. Srikrishnan, eds., *Handbook of Semiconductor Interconnection Technology*, 2nd ed., Boca Raton, FL: CRC Press, 2006.
46. X. Li, W. Cai, J. An, *et al.*, "Large-area synthesis of high-quality and uniform graphene films on copper foils," *Science* **324**, 1312–1314 (2009).
47. J. A. Ott and A. A. Bol, "Use of graphene to limit copper surface oxidation, diffusion and electromigration in interconnect structures," patent US8610278 B1 (2013).
48. D. Kondo, H. Nakano, B. Zhou, *et al.*, "Intercalated multi-layer graphene grown by CVD for LSI interconnects," *IEEE Intern. Interconnect Technol. Conf. (IITC)* (2013), pp. 1–3.
49. C.-H. Yeh, H. Medina, C.-C. Lu, *et al.*, "Scalable graphite/copper bishell composite for high-performance interconnects," *ACS Nano* **8**, 275–282 (2014).
50. A. Naeemi and J. D. Meindl, "Conductance modeling for graphene nano-ribbon (GNR) interconnects," *IEEE Electron Device Lett.* **28**, 428–431 (2007).
51. R. Murali, Y. Yang, K. Brenner, T. Beck, and J. D. Meindl, "Breakdown current density of graphene nanoribbons," *Appl. Phys. Lett.* **94**, 243114 (2009).
52. R. Murali, K. Brenner, Y. Yang, T. Beck, and J. D. Meindl, "Resistivity of graphene nanoribbon interconnects," *IEEE Electron Device Lett.* **30**, 611–613 (2009).

53. D. Kondo, H. Nakano, B. Zhou, *et al.*, "Sub-10-nm-wide intercalated multi-layer graphene interconnects with low resistivity," *IEEE Intern. Interconnect Technol. Adv. Metallization Conf. (IITC/AMC)* (2014), pp. 189–192.

54. A. M. Noori, M. Balseanu, P. Boelen, *et al.*, "Manufacturable processes for 32-nm-node CMOS enhancement by synchronous optimization of strain-engineered channel and external parasitic resistances," *IEEE Trans. Electron Devices* **55**, 1259–1264 (2008).

55. A. Agrawal, J. Lin, M. Barth, *et al.*, "Fermi level depinning and contact resistivity reduction using a reduced titania interlayer in n-silicon metal–insulator–semiconductor Ohmic contacts," *Appl. Phys. Lett.* **104**, 112101 (2014).

56. G. Giovannetti, P. A. Khomyakov, G. Brocks, V. M. Karpan, J. van den Brink, and P. J. Kelly, "Doping graphene with metal contacts," *Phys. Rev. Lett.* **101**, 026803 (2008).

57. M.-H. Lee, C.-Y. Cho, H.-R. Kim, *et al.*, unpublished (2015).

58. K.-E. Byun, H.-J. Chung, J. Lee, *et al.*, "Graphene for true Ohmic contact at metal–semiconductor junctions," *Nano Lett.* **13**, 4001–4005 (2013).

59. P. Thanh Trung, F. Joucken, J. Campos-Delgado, J.-P. Raskin, B. Hackens, and R. Sporken, "Direct growth of graphitic carbon on Si(111)," *Appl. Phys. Lett.* **102**, 013118 (2013).

60. G. Wang, M. Zhang, Y. Zhu, *et al.*, "Direct growth of graphene film on germanium substrate," *Nature Sci. Rep.* **3**, article no. 2465 (2013).

61. K. Kang, S. Xie, L. Huang, *et al.*, "High-mobility three-atom-thick semiconducting films with wafer-scale homogeneity," *Nature* **520**, 656–660 (2015).

62. Y. Hao, M. S. Bharathi, L. Wang, *et al.*, "The role of surface oxygen in the growth of large single-crystal graphene on copper," *Science* **342**, 720–723 (2013).

2.3

Computing with Coupled Relaxation Oscillators

N. Shukla and S. Datta
Department of Electrical Engineering, The Pennsylvania State University, University Park, PA 16802, USA ; Department of Electrical Engineering, University of Notre Dame, Notre Dame, IN 46556, USA

A. Parihar and A. Raychowdhury
School of Electrical and Computer Engineering, Georgia Institute of Technology, Atlanta, GA 30332, USA

1. Introduction

Complementary metal-oxide-semiconductor (CMOS) transistors supporting the Boolean computational framework have been the backbone of modern computation, and have fueled the information technology revolution for the past four decades. However, there is a class of computing tasks, such as associative processing, which involves the determination of a degree of match (or association) between two quantities in multidimensional space, for which the Boolean approach consumes a significant amount of computing resources. Mathematically, the quantification of the degree of match between two quantities requires the calculation of a distance (using a suitable distance norm) between two points that represents the two quantities in a high-dimensional space. It turns out that the so-called "Boolean bottleneck" arises from the requirement of a significant number of power-intensive multiply–accumulate (MAC) operations involved in the calculation of the distance norm.

Associative computing-based applications such as pattern recognition, visual saliency, and image segmentation are becoming increasingly pervasive as we aim to design intelligent video analytic systems[1] with the eventual goal of realizing a "visual cortex on a silicon platform," to mimic human vision on a hardware platform. Consequently, the computational inefficiency of CMOS in associative computing, along with the increasing relevance of associative processing-based applications, motivates the development of an alternate non-Boolean computational approach.

In this chapter, we focus on our recent work in experimentally demonstrating tunable coupled relaxation oscillators, based on the phenomenon of insulator–metal transition in vanadium dioxide (VO_2), as an efficient non-Boolean computational primitive capable of performing the visual saliency task. Our approach is an embodiment of "let the physics do the computing" paradigm and judiciously exploits the phase synchronization dynamics of pairs of coupled oscillators to compute a fractional distance norm that is suitable for associative computing.

2. Vanadium dioxide-based relaxation oscillators

VO_2 is a transition metal oxide that exhibits an abrupt insulator-to-metal transition (IMT) at $\sim$340 K in its bulk single-crystal form. The insulating state (under no strain) has a monoclinic M1 crystal structure with an optically measured bandgap of 0.6 eV, which abruptly collapses into a metallic phase, accompanied by a structural transformation to the rutile crystal structure (with no bandgap).[2, 3] The abrupt collapse of the bandgap amplifies the free-carrier concentration and manifests itself as a sharp change in resistivity (up to five orders in magnitude in bulk single crystal). Further, the IMT in VO_2 can be induced using temperature, as in Fig. 1(a), strain,[4] doping,[5] and electrical stimulus, as in Fig. 1(b). The phase transition can be triggered on ultrafast timescales ($\sim$75 fs) as experimentally demonstrated by time-resolved optical pumping.[6] However, the exact physics and the origin of the phase transition have been topics of debate for the past several decades; various models have been proposed to explain the IMT in VO_2, attributing it to varying degrees of

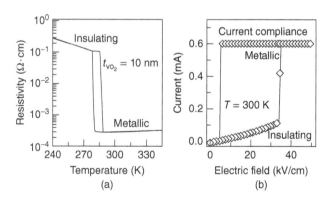

Figure 1. (a) Typical resistivity versus temperature characteristics of 10 nm thick VO_2 films epitaxially grown on (001) TiO_2; (b) Typical current–voltage characteristics of two-terminal VO_2 devices (VO_2 channel of 2 μm length, 10 μm width).

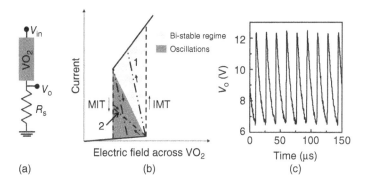

Figure 2. (a) Schematic of the two-terminal VO_2 oscillator circuit; (b) Schematic of the I–V characteristics of VO_2 device without R_S (short dashed line). A series resistor modifies the I–V dynamics across the phase transition in VO_2 (NDR load lines 1 and 2), leaving the circuit in either bistable or unstable oscillatory mode, depending on R_S; (c) Typical time-domain waveform of the VO_2 relaxation oscillator.

contribution from a Mott–Hubbard type correlated phase transition as well as from a Peierls-like structural instability.[7, 8]

We can harness the electrically induced large, abrupt, and hysteretic change in resistivity of the VO_2 to implement VO_2-based relaxation oscillators by connecting a resistive load R_S in series, as shown in Fig. 2(a). When a voltage is applied across the circuit so that the electric field across the VO_2 device exceeds the critical field, $E > E_2$, the IMT is triggered resulting in a negative differential resistance (NDR) characteristic.[9] The NDR is characterized by a decrease in resistivity and a simultaneous reduction in the electric field across the VO_2 device. If, as a result, the electric field across the VO_2 drops below a critical value, $E < E_1$, the metallic phase becomes unstable and the VO_2 channel undergoes the metal–insulator transition (MIT) and returns to the insulating state with a corresponding increase in the electric field across the VO_2. Subsequently, the process repeats itself in a self-sustaining manner, resulting in sustained electrical oscillations.

Representing this scenario through an electrical load line, as in Fig. 2(b), a VO_2 device operating on load line 2 will oscillate, since the electric field across the VO_2 channel drops below the critical field for MIT, whereas a device operating on load line 1 will not exhibit oscillatory behavior.[9] The resulting representative time-domain waveform of the relaxation oscillator is shown in Fig. 2(c).

The properties of the oscillator such as frequency cannot be tuned with a constant R_S, thereby constraining its utility in computational circuits. The tuning capability can be added by replacing the resistor R_S with a metal-oxide semiconductor field-effect transistor (MOSFET) (drain of the MOSFET is connected to the VO_2) and utilizing it as a voltage-controlled variable

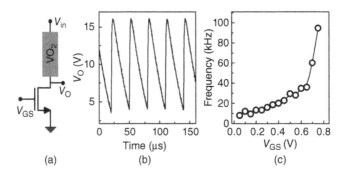

Figure 3. (a) Schematic of a VO$_2$ relaxation oscillator consisting of a two-terminal VO$_2$ device (channel length/width of 4/6 µm) in series with the source–drain of a MOSFET; (b) Time-domain waveform of the HVFET oscillator. (c) Oscillation frequency versus V_{GS}, enabling VCO operation.

resistor, thereby creating a hybrid-vanadium dioxide metal-oxide semiconductor field-effect transistor (HVFET) oscillator,[10] illustrated in Fig. 3. Modulating the gate-to-source voltage V_{GS} of the MOSFET changes the resistance in series with the VO$_2$ and changes the operating frequency of the oscillator. The voltage-controlled oscillator (VCO) function is shown in Fig. 3(c).

3.   Experimental demonstration of pairwise coupled HVFET oscillators

To make such oscillators relevant to associative computing applications, we explore in detail their synchronization dynamics. Figure 4(a) illustrates the schematic of the capacitively coupled HVFET oscillators. The use of capacitive coupling is motivated by its high-pass filtering characteristics that block any DC interaction while simultaneously allowing for the exchange of reactive power among the oscillator pair, allowing the oscillators to synchronize.[9] The resulting time-domain waveforms of the synchronized oscillators are shown in Fig. 4(b), while the corresponding power spectrum is shown in Fig. 4(c). Further, as shown in Fig. 4(d), the resonant frequency of the coupled oscillators can be tuned by modulating the input gate voltage difference $\Delta V_{GS} = V_{GS,2} - V_{GS,1}$.[10]

4.   Computing with pairwise coupled HVFET oscillators

For performing computation, we investigate using simulations, the phase dynamics of the synchronized HVFET oscillators using the equivalent circuit illustrated in Fig. 5(a). Figure 5(b) shows the simulated voltage flow diagrams for the coupled oscillators (at different input gate voltages), as the individual oscillators in the coupled configuration traverse between their respective insulating and the metallic states.[11, 12] The simulations reveal that the shape of the periodic orbit

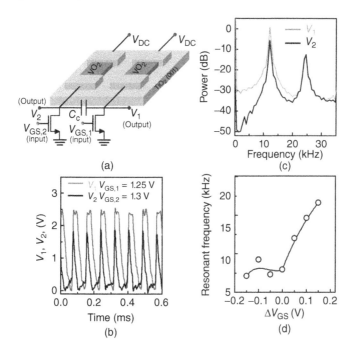

Figure 4. (a) Schematic of the capacitively coupled HVFET oscillators (coupling $C_C = 2.2$ nF); (b) Time-domain waveform of the synchronized coupled HVFET oscillators after eliminating DC offsets ($V_{GS,1} = 1.25$ V, $V_{GS,2} = 1.3$ V); (c) Power spectrum of the coupled waveform in (b); (d) Variation of coupled frequency as a function of $\Delta V_{GS} = V_{GS,2} - V_{GS,1}$.

in the phase space (i.e., the relative phase difference between the oscillators) is a strong function of the difference between the input voltages $V_{GS,1}$ and $V_{GS,2}$. For instance, when the gate voltage inputs to the two oscillators are equal, say $V_{GS1} = V_{GS2} = 0.3$ V, the oscillators have a high out-of-phase locking character, with only a small fraction of the periodic orbit lying in the phase space that represents in-phase locking (shown in gray in Fig. 5(b)). On the other hand, as the input voltage difference between the oscillators increases, the periodic orbit evolves to having a higher percentage of in-phase locking character.[13]

This sensitivity of the periodic orbit of the HVFET oscillators can be harnessed for associative computing. By encoding the two quantities that need to be compared in the gate input voltages, the evolving periodic orbit of the coupled oscillators in phase space can be exploited to calculate the degree of match between them. To quantify the periodic orbit, we define a time-averaged *xor* measure by (i) thresholding the analog output to a binary stream, (ii) applying the *xor* operation on these binary values at every time instance, and, finally, (iii) averaging the *xor* output for a finite time duration (over at least three to

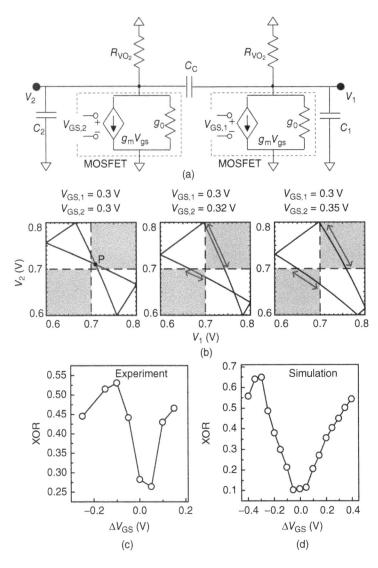

Figure 5. (a) Equivalent circuit of the capacitively coupled oscillators, where C_1 and C_2 are the equivalent capacitances of the VO_2 devices combined with the respective MOSFET output circuits; (b) Effect of input gate voltages on the shape of the steady-state periodic orbit of the coupled oscillators; (c) Experimental; and (d) simulated time-averaged *xor* output of the oscillations as a function of ΔV_{GS}.

four complete periodic orbits). This metric quantifies the degree of in-phase locking between the oscillators, and a higher value indicates higher in-phase locking in the periodic orbit. It is discernible from Fig. 5(b) that the time-averaged *xor* value is minimized when the two inputs are similar and the value subsequently

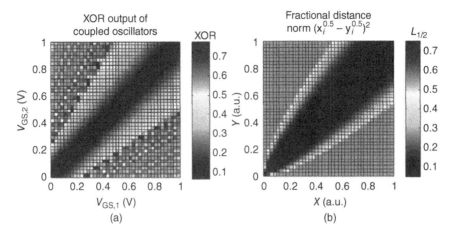

Figure 6. (a) Two-dimensional contour map of the time-averaged *xor* output of the coupled oscillators as function of $V_{GS,1}$ and $V_{GS,2}$; (b) The fractional distance norm $L_{0.5} = (x_i^{1/2} - y_i^{1/2})^2$ suggesting that *xor* output of the oscillators computes a distance norm close to $L_{0.5}$.

increases with increased separation ΔV_{GS} between the gate voltage inputs. This is confirmed through both experiments and simulations, compared in Fig. 5(c) and (d), indicating that a quantitative determination of the degree of match can indeed be performed using the pairwise coupled HVFET oscillators.

The pronounced similarity between the 2D contour map of the *xor* output of the oscillators with the function $(x_i^{1/2} - y_i^{1/2})^2$, illustrated in Fig. 6, indicates that the specific distance norm calculated by the oscillators is the $L_{0.5}$ "fractional distance norm."

5. Associative computing using pairwise coupled oscillators

We investigate the inherent ability of the synchronization dynamics of the coupled oscillators to compute a distance norm for visual saliency determination. Visual attention/saliency is the property of the human brain by which certain aspects of an image are more important aka *salient* than other aspects of the same image.

Usually, objects in the image that have a high degree of contrast with their neighborhood are salient to the human brain. Figure 7(a) shows a sample image wherein the squirrel and the bark of the tree are more prominent in comparison to the details in the white background. To replicate this functionality with the coupled oscillators, edge detection is performed using an array of pairwise oscillators to approximate the degree of dissimilarity between a given image pixel and its immediate neighbors (distance computation) – see Fig. 7(b). Different edges such as vertical, horizontal, and diagonal are detected based on the selection

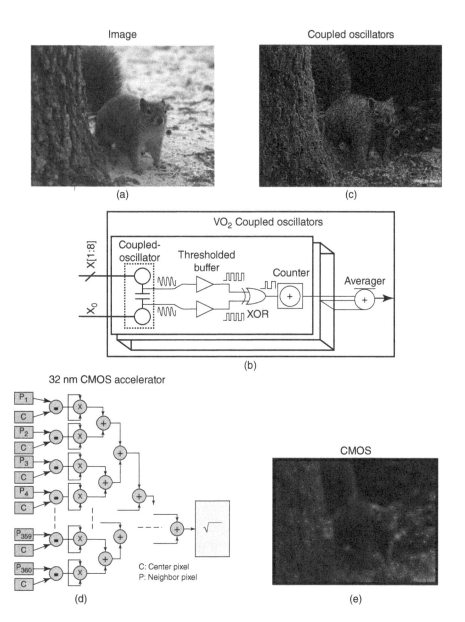

Figure 7. (a) Input image processed by the oscillators; (b) Schematic of the processing scheme for visual saliency using coupled HVFET oscillators; (c) Output of the coupled oscillators; (d) Functional block diagram for the 32 nm CMOS accelerator-based processing scheme for finding the visual saliency; (e) Visual saliency output of the 32 nm CMOS accelerator.

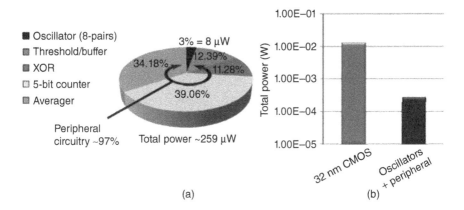

Figure 8. (a) Power distribution among various functional blocks in the HVFET coupled-oscillator-based associative computing scheme for visual saliency; (b) Comparison of the power requirements with a CMOS implementation. The total power consumption is calculated to be 13.1 mW and 259 μW, respectively.

of neighboring pixels for comparison. As this concept is expanded to include the comparison of pixels within a larger neighborhood (pixels surrounding reference pixel; a 3 × 3 neighborhood is used here), the output approximates the visual saliency (see Fig. 7(c)). For comparison, the same image is also evaluated using a 32 nm CMOS accelerator, as shown in Fig. 7(d) and (e).

The power-performance projection of the HVFET synchronized oscillator-based processing scheme was calculated to be ∼259 μW, as shown in Fig. 8(a). It can be observed that the oscillators performing as the fundamental distance computing functional block consume only 8 μW of power; the remaining power is consumed by the external peripheral circuitry required to read out the *xor* output of the oscillators. Thus, designing new power-efficient read-out schemes will enable further reduction in power. To evaluate if the oscillator-based computing approach provides a power-performance advantage, the total power requirements of the 32 nm CMOS-based accelerator in performing the same task were also evaluated and the total power consumption is measured to be ∼13.1 mW. Comparing the power requirements for the two computing approaches in Fig. 8(b), we find a substantial advantage (∼50× power reduction) in performance can be attained in associative computing applications using coupled HVFET oscillators.

6. Conclusion

We have experimentally demonstrated coupled relaxation oscillators with input programmable synchronization that harness the insulator–metal transition in VO_2. The phase dynamics of the synchronized oscillators can be used to calculate a fractional distance norm, thereby providing an alternative non-Boolean route to associative computing applications such as visual saliency and pattern matching. Coupled oscillator systems have the potential to achieve a significant advantage in power consumption over their CMOS counterparts in future video analytic applications.

References

1. V. Narayanan, S. Datta, G. Cauwenberghs, D. Chiarulli, S. Levitan, and P. Wong, "Video analytics using beyond-CMOS devices," *Proc. Design Automation Test Europe (DATE) Conf.* (2014), pp. 1–5.
2. L. A. Ladd and W. Paul, "Optical and transport properties of high quality crystals of V_2O_4 near the metallic transition temperature," *Solid State Commun.* **7**, 425–428 (1969).
3. J. H. Park, J. M. Coy, T. S. Kasirga, *et al.*, "Measurement of a solid-state triple point at the metal–insulator transition in VO_2," *Nature* **500**, 431–434 (2013).
4. J. Cao, E. Ertekin, V. Srinivasan, *et al.*, "Strain engineering and one-dimensional organization of metal–insulator domains in single-crystal vanadium dioxide beams," *Nature Nanotechnol.* **4**, 732–737 (2009).
5. X. Tan, T. Yao, R. Long, *et al.*, "Unraveling metal–insulator transition mechanism of VO_2 triggered by tungsten doping," *Nature Sci. Rep.* **2**, article no. 466 (2012).
6. A. Cavalleri, T. Dekorsy, H. Chong, J. Kieffer, and R. Schoenlein, "Evidence for a structurally-driven insulator-to-metal transition in VO_2: A view from the ultrafast timescale," *Phys. Rev. B* **70**, 161102 (2004).
7. R. M. Wentzcovitch, "VO_2: Peierls or Mott–Hubbard? A view from band theory," *Phys. Rev. Lett.* **72**, 3389–3392 (1994).
8. T. M. Rice and J. P. Pouget, "Comment on 'VO_2: Peierls or Mott–Hubbard? A view from band theory'," *Phys. Rev. Lett.* **73**, 3042 (1994).
9. N. Shukla, A. Parihar, E. Freeman, *et al.*, "Synchronized charge oscillations in correlated electron systems," *Nature Sci. Rep.* **4**, article no. 4964 (2014).
10. N. Shukla, A. Parihar, M. Cotter, *et al.*, "Pairwise coupled hybrid vanadium dioxide-MOSFET (HVFET) oscillators for non-Boolean associative computing," *Tech. Dig. IEDM* (2014), pp. 28.7.1–28.7.4.
11. A. Parihar, N. Shukla, S. Datta, and A. Raychowdhury, "Synchronization of pairwise-coupled, identical, relaxation oscillators based on metal–insulator phase transition devices: A model study," *J. Appl. Phys.* **117**, 054902 (2015).
12. S. Datta, N. Shukla, M. Cotter, A. Parihar, and A. Raychowdhury, "Neuro-inspired computing with coupled relaxation oscillators," *Proc. Design Automation Conf. (DAC)* (2014), pp. 1–6.
13. A. Parihar, N. Shukla, S. Datta, and A. Raychowdhury, "Exploiting synchronization properties of correlated electron devices in a non-Boolean computing fabric for template matching," *IEEE J. Emerg. Sel. Top. Circ. Syst.* **4**, 450–459 (2014).

2.4

On the Field-Induced Insulator–Metal Transition in VO$_2$ Films

Serge Luryi
Department of Electrical and Computer Engineering, Stony Brook University, Stony Brook, NY 11794, USA

Boris Spivak
Department of Physics, University of Washington, Seattle, WA 98195, USA

1. Introduction

First-order metal–insulator transitions (MITs) in crystalline materials have been known for many years[1] and correspond to a transformation between states with a dielectric (semiconductor) and a metallic types of conductivity. These transitions occur under the influence of certain external parameters, such as temperature and pressure, as well as with varying material composition. Materials exhibiting these phenomena include many transition metal oxides, of which over 40 are known to possess MITs.[2, 3]

Among the best-known MIT materials are vanadium oxides. Being able to combine with oxygen in 2-, 3-, 4-, and 5-valent states, vanadium forms a series of oxides of which at least eight exhibit MITs.[4] Phase transition in V$_2$O$_3$ occurs at a critical temperature $T_C = 150$ K and in VO$_2$ at $T_C = 340$ K, with the electrical conductivity changing by up to 10 and 5 orders of magnitude, respectively. The MIT in vanadium oxides is also accompanied by a discontinuous variation of other than electrical properties, such as optical and magnetic. Vanadium dioxide, VO$_2$, is of particular interest for technology because its transition occurs near room temperature and furthermore its T_C is tunable over a wide range by doping with impurities such as tungsten. A recent review of vanadium oxide electronics[5] exhaustively cites the relevant literature. Another recent review[6] has a somewhat different focus and perspective.

Despite numerous attempts, no commercially viable applications of VO$_2$ have been found thus far. One known application, in infrared-vision systems, employs a closely related but nontransitioning, nonstoichiometric VO$_x$ rather than VO$_2$. The reason for choosing VO$_x$ over VO$_2$ had to do with a difficulty of

Future Trends in Microelectronics: Journey into the Unknown, First Edition.
Edited by Serge Luryi, Jimmy Xu, and Alexander Zaslavsky.
© 2016 John Wiley & Sons, Inc. Published 2016 by John Wiley & Sons, Inc.

dealing with the hysteretic nature of a transition in VO_2. A recently proposed VO_2 sensor for infrared-vision systems utilizes a spectacular bolometric effect that is of nonhysteretic nature,[7] even though it is within the hysteretic loop of the transition. This proposal has not found commercial use either.

The dramatic difference in the electronic spectrum on the metallic and the dielectric side of the transition opens the possibility for *switching* applications triggered by an external electric field applied to a thin transitioning film in a gated structure. Since the field penetrates very differently into the dielectric and metallic phases, the field energy in the two phases is different (a possibility to the best of our knowledge first suggested by Valiev *et al.*[8]). As a result, application of an external electric field shifts the critical temperature of the transition.[8-21]

We had previously estimated[14] such a shift δT_C for a transistor-like structure with a thin VO_2 film sandwiched between two metallic plates insulated from the film by dielectric layers (cf. Fig. 1). One of these plates is the ground plane and to the other a gate voltage V_G is applied, resulting in

$$\delta T_C = -\frac{(C_M - C_D)V_G^{\,2}}{2(S_M - S_D)}, \tag{1}$$

where S_M and S_D denote the entropy densities and C_M and C_D the electrostatic capacitances per unit area of the film in the metallic and the dielectric phase, respectively. In this estimate, it is assumed that the entire film undergoes the transition. As suggested in Ref. 14, this may be "true for a thin enough film, where the formation of in-plane transition boundary is energetically unfavorable." The negative sign in Eq. (1) implies that application of a gate voltage drives the system into its metallic state.* For an applied voltage $V_G = 1$ V in an exemplary gated

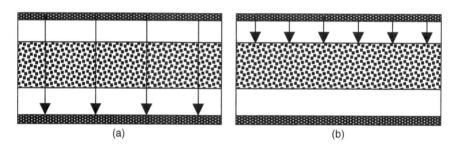

(a)   (b)

Figure 1. Schematic diagram of an experimental arrangement. Grounded VO_2 film is sandwiched between two insulated metallic gates. When the film is in the dielectric state (a), the electric field lines penetrate through the film and terminate on the bottom metallic plate. When the film is metallic (b), the electric field lines terminate on the film surface.

---

* Because of a wrong sign in an equation leading to Eq. (1), the opposite conclusion was erroneously stated in Ref. 14.

structure, we found $\delta T_C \approx -1$ K. This is a substantial shift, but it is hardly adequate for switching applications.

The electrostatic effect of Eq. (1) is based on thermodynamics and is independent of a physical origin of the MIT. However, this mechanism is rather weak, as it allows shifting T_C by at most several degrees. A much stronger effect, albeit model-dependent, is introduced in the next section and will form the basis for our subsequent discussion. Here we wish to stress that the idea of inducing insulator-to-metal transition in a thin vanadium oxide film by applying an external electric field is a sound idea. The effect surely exists!

Nevertheless, despite multiple reports that observe such an effect and the sound theoretical reasons for its existence (see Refs 8–21 and above-cited reviews[5, 6]), the status of this problem is unsatisfactory and we offer our reflections on the reason why. We believe that for a sufficiently large applied field the transition indeed occurs, but it is confined to an infinitesimal sliver at the surface of the sample. In its "insulator" phase, the VO$_2$ sample is, in fact, a semiconductor with a relatively narrow bandgap and a rather high conductivity at typical experimental temperatures. As a result, the field-induced transition is never observed in a gated bulk sample. The transition has been reported only in double-gated films, but even there the effect is rather weak because most of the film remains semiconducting and shunts the emergent metallic sliver.

We discuss the conditions under which the transition can be induced in the entire film, rather than its top sliver. We show that this favorable situation can be realized when the film is sufficiently thin, so that the energy cost of converting the entire film into the metallic phase (which is "thermodynamically wrong" in the part of the film away from the surface) is smaller than the would-be cost of creating a domain boundary between the two phases. At this time, we cannot predict how thin the film should be for the proposed stabilization of the entire-film transition by the domain boundary energy, but we discuss the physical quantities that should be determined to make this prediction.

Finally, we discuss the need for the ground plane, which does not sit well with the technological reality that the best and thinnest VO$_2$ films are grown on insulating monocrystalline sapphire substrates. In large transistor-like structures with gate lengths of a micron and longer, the ground plane appears indispensible, for otherwise the field configuration would be accommodated with tiny domain boundaries across the film. The phase boundary must be comparable to the film area for its energy to count. Whether or not this can be accomplished with deep submicron gate lengths is an open question.

2. Electron concentration-induced transition

There is no definitive theory of the metal–insulator transition in vanadium oxides. Possible mechanisms under discussion are the electron–phonon or Peierls mechanism and electron correlation mechanism of Mott–Hubbard type.[1, 22, 23] Ultrafast experiments,[24-27] where the transition is excited by femtosecond optical pulses, support the electron correlation model, at least in VO_2. Nowadays, this seems to be the prevalent view of most practitioners.[5, 6, 28]

The correlation model naturally leads to the assumption that the transition is controlled by the concentration of mobile carriers. This implies a model of the transition triggered by the total electron concentration in the dielectric phase, irrespective of whether it results from doping, thermal excitation, photo excitation – or is induced by an applied electrostatic field. We are interested, of course, in the latter case. In an intrinsic semiconductor in the absence of an applied field, the chemical potential μ is approximately in the middle of the bandgap E_G. As the field F is applied, μ moves toward one of the allowed bands and the concentration of electrons (or holes) in the vicinity of the surface increases. At some critical concentration we can expect a transition to the metallic state.

Although there is yet no quantitative theory to describe such a transition, the very existence of induced concentration effect should be regarded as likely. The critical carrier concentration may be estimated to be of the order of the thermal carrier concentration that arises in the dielectric phase at the transition temperature T_C due to the thermal excitation across the gap. That concentration is of the order 10^{18} cm^{-3} and should be easily achievable by field effect. The induced-carrier effect is much "stronger" than the electrostatic field energy effect and should work in a wide range of temperatures. The expected "strong" effect is very attractive for switching applications.[14]

Qualitatively, we assume that the transition driven by electron concentration is controlled by the position of the chemical potential μ relative to the conduction band E_C (in a homogeneous system), see Fig. 2. We shall denote by $\varphi(T)$ the critical position of μ at a given temperature T that corresponds to the transition. We regard the temperature as fixed and vary the concentration. Dielectric phase corresponds to $\mu < \varphi(T)$.

We shall continue describing the system by its dielectric diagram even when the semiconductor phase is unstable. This provides the meaning to $\mu > \varphi$ as the level of the chemical potential that would be present if the dielectric phase were stabilized externally.

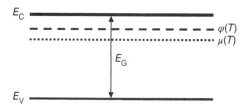

Figure 2. Schematic band diagram of VO_2 in its semiconductor phase. At a fixed temperature T, the position of chemical potential μ varies with the induced electron concentration. The metal–insulator transition occurs at a critical position $\mu = \varphi$.

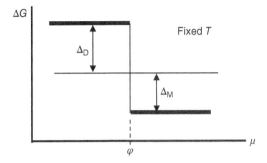

Figure 3. Difference $\Delta G = G_M - G_D$ in the free energy densities between the metallic and dielectric phases. For $\mu > \varphi$, the dielectric phase is unstable and the meaning of μ is the level the chemical potential would have if the phase were stabilized externally.

For the difference ΔG in the free energy density per unit volume between the metallic and dielectric phases we assume the model illustrated in Fig. 3. Here in this diagram, the quantity Δ_D is the penalty the system pays for keeping a unit volume metallic in the region where $\mu < \varphi$, that is, where the volume should be dielectric; similarly, Δ_M is the penalty for maintaining a dielectric unit volume in the region where it should be metallic. The sum of these quantities, $\Delta_D + \Delta_M = \Delta$, is the total discontinuity in the free energy density upon transition, but their ratio is completely unknown at this time.

3. Field-induced transition in a film

Consider a dielectric VO$_2$ film of thickness d in an external electric field F. To determine what happens with the film, when part of it, a layer of thickness x, is subject to $\mu > \varphi$ (while stabilized in the dielectric phase), we must tally the penalties. The reason the layer may not transition to metal is associated with the boundary energy b between the two phases. The quantity b is per unit area and its value should be also considered unknown at this time. The value of x depends only on the electric field and increases with the field. There is a well-defined equation $x = x(F)$ that is determined by electrostatics of the accumulation layer.

Three possible configurations are illustrated in Fig. 4. In the configuration (a), the film remains dielectric, paying the penalty $x\Delta_M$.

In the configuration (b), a thin layer x_M goes metallic and screens the applied field. The remaining portion $(x - x_M)$ is no longer under metallic condition. Since the field does not penetrate beyond x_M, the film underneath has $\mu < \varphi$ and no penalty is extracted except, of course, for the boundary energy b, which is the only penalty in this configuration.

We can safely assume that x_M is very small, on the order of $\hbar v_F / E_G$ where $v_F \approx 10^8$ cm/s is the Fermi velocity of electrons in the metallic phase, and

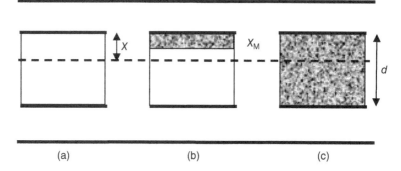

Figure 4. Possible configurations of the transitioning thin film in an external electric field: (a) the entire film remains dielectric; (b) a sublayer of thickness x_M goes metallic and screens the applied field; (c) the entire film goes metallic. The strength of the field is characterized by the thickness x of a sublayer where $\mu > \varphi$ (while the film is stabilized in the dielectric phase). The relation $x = x(F)$ is well defined by the electrostatics and need not be specified analytically.

$E_G \approx 0.5$ eV is the bandgap energy in the dielectric phase. The actual extent of x_M should be determined from the Ginzburg–Landau equation that describes the spatially inhomogeneous transition. The combination $x_M = \hbar v_F / E_G$ appears reasonable on dimensional considerations and gives $x_M \approx 10$ Å.

In the configuration (c), the entire film goes metallic and there is no boundary penalty. However, most of the film, $(d - x)$, is in the regime $\mu < \varphi$ and hence pays the penalty for being metallic, $\Delta_D(d - x)$.

For *thick enough films*, only configurations (a) and (b) are competitive. The transition occurs in thick films at large enough fields, when $x > b/\Delta_M$. Due to the smallness of x_M, the effect of the transition is not very strong and may be hard to see in resistance measurements. The reason is that the conductivity of a thick VO_2 film in its "dielectric" phase (where the intrinsic free-carrier concentration is considerable at room temperature, $n \approx 10^{18}$ cm^{-3}) is rather high and it would not be much changed by the conversion to metal of a surface sublayer of thickness x_M.

The critical field F_C for the transition in thick films is defined by the electrostatic relation $x_C = x(F_C)$ with a finite thickness of penetration x_C. For thick films, the critical field penetration thickness is given by $x_C = b/\Delta_M$.

In the opposite limit of *thin films*, where $d\Delta_D < b$, configuration (c) in Fig. 4 becomes competitive, $\Delta_D(d - x) < x\Delta_M$, and the entire film undergoes transition. This happens at large enough fields, $x > d\Delta_D/(\Delta_D + \Delta_M)$, and the critical field now corresponds to

$$x_C = \frac{d\Delta_D}{\Delta_M + \Delta_D}. \tag{2}$$

Equation (2) together with the electrostatic relation $x_C = x(F_C)$ determines the critical field for driving the transition in a thin film.

Note that in light of the $d\Delta_D < b$ condition for thin films, the thin-film critical field of Eq. (2) is *smaller* than the corresponding field, $x_C = b/\Delta_M$, for thick films.

4.   Need for a ground plane

The above estimates necessarily assume a double-gated structure (cf. Figs 1 and 4). In this case, the domain boundary is *planar* and the boundary energy *per unit area* may compete on equal footing with the phase energy densities corresponding to different configurations. However, the best thin films of VO$_2$ are produced on insulating crystalline substrates, such as sapphire. In these structures, a meaningful ground plane is hard to fabricate. The ground plane is "meaningful" if it attracts most of the field lines emanating from the gate electrode; therefore, it must be positioned close to the film, much closer than the source–drain spacing. Otherwise, instead of the configurations of Fig. 4, the electric field lines would terminate on the source and drain contacts and the domain walls would be "vertical" (across the film). Vertical walls may also tangibly contribute to the energy balance; thus, they possess enough energy to stabilize nonhysteretic branches in temperature excursions within the hysteresis loop of the metal–insulator transition in thin VO$_2$ films.[7] However, for long gates (e.g., on the order of a micron or longer), the vertical-wall area would be negligible compared to the area of the film and the boundary energy would hence not be significant.

The situation may be more favorable for deep submicron gates. It seems reasonable that for short enough gates the vertical wall energy may be sufficient to stabilize the transition in the entire film. Philosophically, however, the short-gate quest seems to undermine perhaps the main advantage of the three-terminal switch based on a metal–insulator transition: namely, that its speed is independent of the gate length. As discussed earlier,[14] unlike field-effect transistors, the speed of such a switch is not limited by carrier drift time under the gate.

Therefore, we believe the main research effort would be best deployed in the development of a viable technology for the deposition of a high-quality thin VO$_2$ film on a substrate that could incorporate a *meaningful insulated ground plane*.

5.   Conclusion

First-order metal insulator phase transitions in thin VO$_2$ films can be controlled by an applied electric field. The effect can be used for the implementation of useful devices such as a three terminal ultrafast switch. For the successful development of such a switch, it is imperative that the thin film be double-gated and moreover be thin enough to transition as a whole, rather than separating into two planar domains, the metallic and the semiconducting. We have argued that in thin enough films, the entire-film transition can be stabilized by the domain boundary energy.

It would be very worthwhile to develop a quantitative theory that microscopically describes the spatially inhomogeneous first-order transition. Such a theory should be able to provide an estimate for the key quantities required to assess the needed film thickness. First and foremost, this is the domain boundary energy b, about which we presently know next to nothing, quantitatively. Another parameter one would like to know is the breakdown of the free-energy discontinuity upon transition, $\Delta = \Delta_D + \Delta_M$, into its components describing the "penalties" on the metallic and the dielectric sides of the transition. While the Δ itself can be estimated from the transition latent heat data (about 1 kcal/mol, according to Ref. 1), nothing is known about its breakdown into Δ_D and Δ_M. Still another interesting (and unknown at this time) quantity is the thickness x_M of a metallic sliver that forms on the surface bulk (or thick-film) VO_2 in response to an applied electric field. The smallness of this quantity according to our estimate, $x_M = \hbar v_F / E_G \approx 10\,\text{Å}$, should account for the unsatisfactory status of the problem today.

References

1. N. Mott, *Metal–Insulator Transitions*, London: Taylor & Francis, 1997.
2. P. A. Cox, *Transition Metal Oxides*, Oxford: Clarendon Press, 1995.
3. J. M. Honig, "Transitions in selected transition metal oxides," in: P. P. Edwards and C. R. N. Rao, eds., *The Metallic and the Non-Metallic States of Matter*, London: Taylor & Francis, 1985, p. 261.
4. F. A. Chudnovskiy, "Metal–semiconductor phase transition in vanadium oxides and its technical applications," *Sov. Phys. Tech. Phys.* **20**, 999–1012 (1976).
5. A. L. Pergament, G. B. Stefanovich, and A. A. Velichko, "Oxide electronics and vanadium dioxide perspective: A review," *J. Sel. Top. Nanoelectron. Comput.* **1**, 24–43 (2013).
6. Y. Zhou and S. Ramanathan, "Correlated electron materials and field effect transistors for logic: A review," *Crit. Rev. Solid State Mater. Sci.* **38**, 286–317 (2013).
7. M. Gurvitch, S. Luryi, A. Polyakov, and A. Shabalov, "Nonhysteretic phenomena in the metal–semiconductor phase-transition loop of VO_2 films for bolometric sensor applications," *IEEE Trans. Nanotechnol.* **9**, 647 (2010). See also, "Nonhysteretic behavior inside the hysteresis loop of VO_2 and its possible application in infrared imaging," *J. Appl. Phys.* **106**, 104105 (2009).
8. K. A. Valiev, Yu. V. Kopaev, V. G. Mokerov, and A. V. Rakov, "Electron structure and phase transitions in lower vanadium oxides in an electric field," *Zh. Eksp. Teor. Fiz.* **60**, 2175–2187 (1971) [*Soviet Physics – JETP* **33**, 1168–1174 (1971)].
9. V. V. Mokrousov and V. N. Kornetov, "Field effects in vanadium dioxide films," *Fiz. Tverd. Tela* **16**, 3106–3107 (1974).

10. G. P. Vasil'ev, I. A. Serbinov, and L. A. Ryabova, "Switching in the VO$_2$-dielectric-semiconductor system," *Pis'ma Zh. Tekhn. Fiz.* **3**, 342–344 (1977).

11. E. V. Babkin, G. A. Petrakovskii, and A. A. Charyev, "Anomalous features of the conductivity and of the galvanomagnetic properties of vanadium dioxide in strong electric fields," *Pis'ma Zh. Eksp. Teor. Fiz.* **43**, 538–540 (1986) [*JETP Letters* **43**, 697–700 (1986)].

12. C. Zhou, D. M. Newns, J. A. Misewich, and P. C. Pattnaik, "A field-effect transistor based on the Mott transition in a molecular layer," *Appl. Phys. Lett.* **70**, 598–600 (1997).

13. D. M. Newns, J. A. Misewich, C. C. Tsuei, A. Gupta, B. A. Scott, and A. Schrott, "Mott transition field effect transistor," *Appl. Phys. Lett.* **73**, 780–782 (1998).

14. F. Chudnovskiy, S. Luryi, and B. Spivak, "Switching device based on first-order metal–insulator transition induced by external electric field," in S. Luryi, J. M. Xu, and A. Zaslavsky, eds., *Future Trends in Microelectronics: The Nano Millennium*, New York: Wiley, 2002, pp. 148–155.

15. H. T. Kim, B. G. Chae, D. H. Youn, S. L. Maeng, G. Kim, K.-Y. Kang, and Y.-S. Lim, "Mechanism and observation of Mott transition in VO$_2$-based two- and three-terminal devices," *New J. Phys.* **6**, article no. 52 (2004).

16. H. T. Kim, B. G. Chae, D. H. Youn, *et al.*, "Raman study of electric-field-induced first-order metal–insulator transition in VO$_2$-based devices," *Appl. Phys. Lett.* **86**, 242101 (2005).

17. C. Ko and S. Ramanathan, "Observation of electric field-assisted phase transition in thin film vanadium oxide in a metal-oxide-semiconductor device geometry," *Appl. Phys. Lett.* **93**, 252101 (2008).

18. Z. Yang, C. Ko, V. Balakrishnan, G. Gopalakrishnan, and S. Ramanathan, "Dielectric and carrier transport properties of vanadium dioxide thin films across the phase transition utilizing gated capacitor devices," *Phys. Rev. B* **82**, 205101 (2010).

19. D. Ruzmetov, G. Gopalakrishnan, C. Ko, V. Narayanamurti, and S. Ramanathan, "Three-terminal field effect devices utilizing thin film vanadium oxide as the channel layer," *J. Appl. Phys.* **107**, 114516 (2010).

20. M. A. Belyaev, A. A. Velichko, P. P. Boriskov, N. A. Kuldin, V. V. Putrolaynen, and G. B. Stefanovitch, "The field effect and Mott transistor based on vanadium dioxide," *J. Sel. Top. Nanoelectron. Comput.* **2**, 26–30 (2014).

21. D.-H. Qiu, Q.-Y. Wen, Q.-H. Yang, Z. Chen, Y.-L. Jing, and H.-W. Zhang, "Electrically-driven metal–insulator transition of vanadium dioxide thin films in a metal-oxide–insulator–metal device structure," *Mater. Sci. Semicond. Process.* **27**, 140–144 (2014).

22. T. M. Rice, H. Launois and J. P. Pouget, "Comment on 'VO$_2$: Peierls or Mott–Hubbard? A view from band theory'," *Phys. Rev. Lett.* **73**, 3042 (1994).

23. H. Nakatsugawa and E. Iguchi, "Electronic structures in VO$_2$ using the periodic polarizable point-ion shell model and DV-Xα method," *Phys Rev. B* **55**, 2157–2163 (1997).

24. M. F. Becker, A. B. Buckman, R. M. Walser, T. Lepine, P. Georges, and A. Brun, "Femtosecond laser excitation of the semiconductor–metal phase transition in VO$_2$," *Appl. Phys. Lett.* **65**, 1507–1509 (1994).

25. A. Cavalleri, Cs. Tóth, C. W. Siders, J. A. Squier, F. Ráksi, P. Forget, and J. C. Kieffer, "Femtosecond structural dynamics in VO$_2$ during an ultrafast solid–solid phase transition," *Phys. Rev. Lett.* **87**, 237401 (2001).

26. G. I. Petrov, V. V. Yakovlev, and J. Squier, "Raman microscopy analysis of phase transformation mechanisms in vanadium dioxide," *Appl. Phys. Lett.* **81**, 1023–1025 (2002).

27. R. Yoshida, T. Yamamoto, Y. Ishida, *et al.*, "Ultrafast photoinduced transition of an insulating VO$_2$ thin film into a nonrutile metallic state," *Phys. Rev. B* **89**, article no. 205114 (2014).

28. S. Hormoz and S. Ramanathan, "Limits on vanadium oxide Mott metal–insulator transition field-effect transistors," *Solid State Electron.* **54**, 654–659 (2010).

2.5

Group IV Alloys for Advanced Nano- and Optoelectronic Applications

Detlev Grützmacher
Peter Grünberg Institut-9 and JARA-FIT, Forschungszentrum Jülich GmbH, 52425 Jülich, Germany

1. Introduction

Mainstream semiconductor technology builds on elements of group IV within the periodic table. Crystalline silicon remains the principal base material, whereas germanium and carbon have entered the mainstream in the embedded source/drain technology, as well as in heterojunction bipolar transistors (HBTs) used in BiCMOS technology. Tin, the next group IV element in the periodic table, is a semimetal in its α-Sn phase with a negative bandgap of about 0.4 eV. Most interestingly, group IV alloys containing Sn, in particular $Ge_{1-x}Sn_x$ alloys, have been predicted to be direct bandgap semiconductors.[1]

Recently, it has been shown that alloying Ge with Sn enables the fabrication of fundamental direct bandgap group IV semiconductors, as well as optically pumped GeSn lasers grown on Si(001).[2] This achievement might pave the route toward efficient and monolithically integrated group IV light emitters, that is, lasers, for electronic–photonic integrated circuits (EPICs) that could solve the emerging power consumption crisis in complementary metal-oxide semiconductor (CMOS) technology by enabling optical on-chip and chip-to-chip data transfer. The clock distribution via copper lines takes about 30% of the energy consumption of modern CPU's and the limited bandwidths and delay times are problematic for further scaling. The large parasitic capacitances introduced by various layers of Cu interconnects demand high transistor I_{ON} currents, which could be reduced if some of the Cu lines are replaced by optical interconnects. Changing from electrons to photons for the data transfer would lead to a tremendous reduction in energy consumption of ICs.[3]

In addition, reducing the parasitic capacitances and thus relaxing the demands for high I_{ON} currents might be beneficial for the implementation of steep slope devices operating at very low supply voltages $V_{DD} < 0.4$ V. To this end, GeSn being a direct bandgap semiconductor with a small ~ 0.5 eV

Future Trends in Microelectronics: Journey into the Unknown, First Edition.
Edited by Serge Luryi, Jimmy Xu, and Alexander Zaslavsky.

bandgap might be attractive as well, due to the small effective electron masses and predicted electron mobility and injection velocity exceeding that of InAs.

2. Epitaxial growth of GeSn layers by reactive gas source epitaxy

Chemical vapor deposition (CVD) is the method of choice to grow alloys in the group IV Si–Ge–C material system. Alloys in the Si–C–Ge–Sn material system are of substantial technological relevance since they allow for strain and bandgap engineering based on the mature Si technology. Alloys such as GeSn[1], SiGeSn[4], and CSiGeSn[5] have been predicted to have a direct bandgap for Sn and C concentrations exceeding the solid solubility limit. In particular, Sn concentrations above 8–10% are required for relaxed $Ge_{1-x}Sn_x$ alloys in order to achieve a direct bandgap. At these high Sn concentrations, it becomes increasingly difficult to avoid the formation of Sn precipitates. The large >4% lattice mismatch between GeSn and SiGeSn alloys and the technologically relevant Si substrates increases the challenge to design suitable growth processes.

To avoid phase separation via the formation of precipitates, low substrate temperatures are required.[6] These precipitates can form either via Sn diffusion in the bulk of the crystal or via surface diffusion during the epitaxial growth. To avoid the latter, high growth rates proved to be successful, since it shortens the time available for surface diffusion of adatoms that are rapidly incorporated into the crystal. It has been shown that Sn precipitates occur in molecular beam epitaxy (MBE) grown films if the growth temperatures exceed 150 °C.[7] The combination of very low growth temperatures and high growth rates in conventional CVD systems typically yields compromised crystal quality and even though GeSn alloys with high Sn concentration have been fabricated, no efficient photoluminescence was reported until recently. To overcome this problem, reactive gas source epitaxy (RGSE) has been developed. The basic idea is to use highly reactive molecules and radicals, thereby leading to strongly exothermic reactions at the crystal surface, resulting in a "hot" substrate surface without affecting the low substrate temperatures in the bulk. The combination of halogenides with hydride radicals has proved promising for the epitaxial growth of a wide range of materials combining low temperature processing, high growth rates, and excellent material properties. Here we focus on the growth of Sn-containing alloyed materials with compositions far beyond the solid solubility limit. The RGSE technique combines three key ingredients: highly reactive components leading to exothermic surface reactions, controlled suppression of gas phase reactions, and low substrate temperatures.

We use Ge_2H_6 and $SnCl_4$ as precursors to deposit GeSn alloys. The reactor base pressure was kept at 200 mbar. The growth temperature was adjusted between 425 and 325 °C, depending on the desired Sn concentration; higher Sn concentrations require lower deposition temperatures. Figure 1 shows the growth rate of Ge on Si(100) substrates as a function of deposition temperature in an Arrhenius plot to elucidate the basic principles of the growth process. The standard Ge growth rate using GeH_4 as a precursor is also shown for comparison. The GeH_4 precursor rate shows a strong dependence on the temperature, typical for a CVD process controlled by surface kinetics, with an activation energy of 1.3 ± 0.1 eV. Replacing the GeH_4 with a Ge_2H_6 precursor leads to a reduction of the activation energy to 0.7 ± 0.1 eV and consequently to substantially increased growth rates at low temperatures – see Fig. 1. Remarkably, if the Ge_2H_6 partial pressure is reduced from 15 to 5 Pa, the growth rate drops by a factor of 3 (triangles), despite the fact that in the kinetically controlled CVD regime the growth rate dependence usually does not depend on the precursor partial pressure. This leads to the assumption that the growth rate in the kinetically controlled regime for this growth process depends on the amount of GeH_x ($x = 1, 2, 3$) radicals on the surface. Since Ge_2H_6 is a rather unstable molecule, it is rather likely that Ge_2H_6 dissociates in the gas phase, similar to earlier experiments performed using Si_2H_6.[8] However, it was also shown[8] that in the presence of an H_2 ambient, these radicals will react to form SiH_4 in the gas phase, which is a rather stable molecule and not suitable for low temperature growth. Accordingly, we can assume that also in the case discussed here the following reactions may occur:

$$Ge_2H_6 + H_2 => 2GeH_3^{\cdot} + H_2 => 2GeH_4$$

$$Ge_2H_6 + H_2 => GeH_4 + GeH_2^{\cdot} + H_2 => 2GeH_4. \qquad (1)$$

To suppress the reactions forming GeH_4 and to prove that GeH_x radicals play an important role in the growth process the H_2 ambient was replaced by N_2. The result is presented in Fig. 1. The activation energy is further reduced to ~0.5 eV and the growth rate increases substantially below 400 °C. These findings indicate that GeH_x radicals play a dominant role in the growth process. The radicals are supplied to the surface and deliver excess energy to overcome energy barriers in the surface kinetic reactions. It is important to realize that the reactor geometry with a gas supply via a showerhead very close to the substrate surface is crucial to the efficient supply of radicals to the surface due to the short retention time of the molecules in the gas phase. Also, in order to achieve high growth rates, a large density of radicals is necessary, that is, a high Ge_2H_6 partial pressure. At the same time, a high dilution is beneficial to avoid unwanted gas reactions. These conditions require minimizing the gas phase making the process substantially different from conventional CVD. Similar experiments using conventional CVD showed much less impact on the growth rate, that is, an activation energy of 1.3 eV for a Ge_2H_6 process and little dependence on the carrier gas (H_2 or N_2).[9]

In the next step, $SnCl_4$ is added to the gas phase as a precursor for Sn allowing the formation of GeSn alloys. Previously, $SnCl_4$ was shown to be a suitable precursor for the growth of GeSn alloys by conventional CVD with Sn concentrations up to 8%.[9] In combination with GeH_x radicals on the surface, $SnCl_4$ might be of particular interest, because exothermic reaction releasing HCl and chlorine appear very attractive. Excess energy of the exothermic reactions at

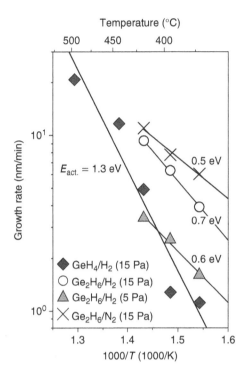

Figure 1. Arrhenius plots for the growth rate of Ge on Si using GeH$_4$ and Ge$_2$H$_6$ precursors at different partial pressures, and using H$_2$ as well N$_2$ as carrier gas.

the surface may allow for high growth rate at low deposition temperatures while maintaining good crystallinity. Moreover, chlorine-mediated chain reactions with hydride radicals might be beneficial to achieve high growth rates at low temperatures. Again it will be crucial to limit the reactions to the surface and to avoid reactions in the gas phase in order to make use of the excess energy at the surface.

Figure 2 shows the dependence of the GeSn growth rate and of the Sn concentration in the GeSn alloy on the deposition temperature for two different SnCl$_4$ partial pressures, $p_{SnCl4} = 0.6$ and 1.3 Pa. At both partial pressures, a monotonic decrease in growth rate and an increase in Sn concentration are observed with decreasing deposition temperatures. The growth rate at $p_{SnCl4} = 1.3$ Pa is generally lower and the temperature dependence weaker than at $p_{SnCl4} = 0.6$ Pa. Apparently, a higher SnCl$_4$ partial pressure leads to an increased HCl concentration at the surface, which in turn etches the GeSn film at a rate that increases with temperature. At 375 °C, the etch rate is negligible and a remarkable growth rate of about 40 nm/min in observed, almost independent of the SnCl$_4$ partial pressure. However, the Sn concentration in the films grown at 375 °C with $p_{SnCl4} = 1.3$ Pa is 10% compared to 4% at $p_{SnCl4} = 0.6$ Pa. The Sn concentration of the films grown at $p_{SnCl4} = 1.3$ Pa drops much faster with increasing temperatures than those grown at 0.6 Pa. At 450 °C, the Sn

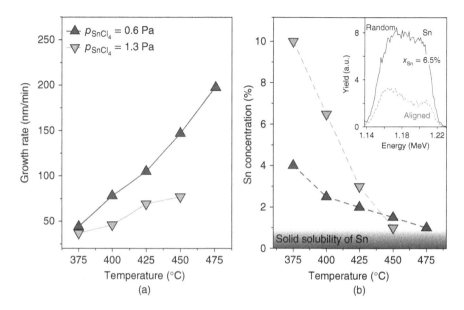

Figure 2. Dependence of growth rate (a) and Sn concentration (b) on the deposition temperature for GeSn films grown on Si with a $SnCl_4$ partial pressure of 0.6 Pa (-▲-) and 1.3 Pa (-▼-). Inset: Rutherford backscattering (RBS) indicating a high density of lattice defects in GeSn grown directly on Si.

concentration is even smaller for the film grown at the higher p_{SnCl4} of 1.3 Pa, indicating that Sn is etched faster by HCl than Ge.

Since the films were grown directly on Si(100) substrates, they contain a large number of misfit dislocations degrading the crystalline quality. Accordingly, the Rutherford backscattering (RBS) data shown in the inset of Fig. 2 shows a poor minimum scattering yield as a result of a high defect concentration. Thus, in the next optimization step, the GeSn films were deposited on strain-relaxed Ge buffer layers deposited on Si(100) substrates. The growth of GeSn on these virtual substrates followed the same growth kinetics as directly on Si films. However, the RBS data exhibits a drastically improved minimum yield, manifesting pristine crystalline quality. The TEM image in Fig. 3 shows no extended defects in the bulk of a GeSn film with 12.6% Sn content, whereas we do observe a dense array of misfit dislocations at the Ge/GeSn interface, as well as dislocations penetrating into the Ge buffer layer. A close inspection of the misfit dislocations at the interface reveals that these are 90° Lomer dislocations, indicated by arrows in Fig. 3, that are expected to efficiently relax the strain.

A further reduction of the growth temperature to 325 °C enabled the deposition of films with Sn concentrations up to 14.5%. By adjusting the total and partial pressures carefully, a growth rate of around 30 nm/min could be maintained even at these low deposition temperatures. As a result, the chosen growth methodology, using highly reactive gas sources that form radicals at the surface

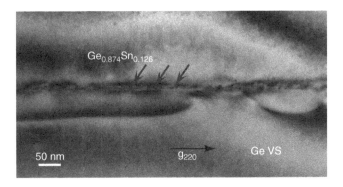

Figure 3. Cross-sectional transmission electron microscope (TEM) image of a GeSn film deposited on a relaxed Ge buffer, with 90° Lomer partial dislocations (angled arrows) relaxing the strain at the interface.

leading to subsequent chain reactions, can be fully employed to grow high-quality crystalline GeSn films with Sn concentration far beyond the solid solubility limit.

3.   Optically pumped GeSn laser

The prediction of a direct bandgap in GeSn dates back to 1982.[1] However, there was considerable uncertainty about the amount of Sn required in a GeSn alloy to obtain a direct bandgap. Recently, it was demonstrated that about 8.5% Sn is required to obtain a direct bandgap for relaxed GeSn alloys,[2, 10] whereas about 20% would be needed for a completely strained film pseudomorphically grown on Ge. In our approach, the GeSn films of several hundred nanometer thickness were grown on Ge virtual substrates. Our GeSn films are predominantly relaxed with some residual strain on the order of 10–20%, depending on the film thickness and Sn concentration. For these conditions, the indirect to direct bandgap transition was found to occur around ~10% Sn content.[2, 10] Optically pumped lasers were fabricated from GeSn films with 12.6% of Sn, using the simplest geometry of etching mesa structures into the GeSn. The mirror and waveguide losses, as well as the limited mode overlap, resulted in rather high pumping power threshold values and lasing was only observed at low temperatures up to 90 K was feasible to obtain lasing.

In order to improve the device, the Ge buffer layer was partly removed underneath the GeSn mesa structure as shown in Fig. 4. This conferred several substantial advantages. First, the GeSn film will further relax in the underetched area, leading to an increase of the energy separation between the direct and indirect bandgap for a given Sn concentration (>8.5%) since the direct bandgap shrinks. Second, due to the reduced bandgap, the optical mode is confined in the underetched area, leading to an improved mode overlap of $\Gamma = 95\%$. Finally, the dislocated GeSn/Ge interface is removed and waveguide losses are reduced.

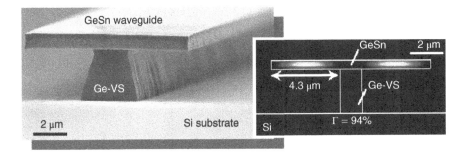

Figure 4. SEM micrograph of an underetched Fabry–Perot waveguide laser (left) and the calculated intensity of the fundamental TE mode (right).

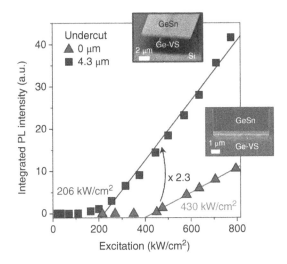

Figure 5. Integrated PL intensity versus optical excitation power (L–L curve) for etched mesa Fabry–Perot laser before (triangles) and after (squares) partial underetching of the mesa. The insets show SEM micrographs illustrating the two configurations.

As a result, the performance of the optically pumped laser is substantially improved, as shown in Fig. 5 that compares the characteristics of an optically pumped laser prior and subsequent to the underetching of the mesa. Clearly, both devices show a distinct threshold behavior with an increase by more than a factor of 8000 in intensity as soon as the threshold pumping power is surpassed. At the same time, the linewidth of the emission spectra is reduced by a factor of 10. Detailed analysis of the spectra in Fig. 5 unambiguously proves lasing. After underetching, the pump power required to reach the lasing threshold was reduced by more than a factor of 2 from 430 to 206 kW/cm^2 and the efficiency is

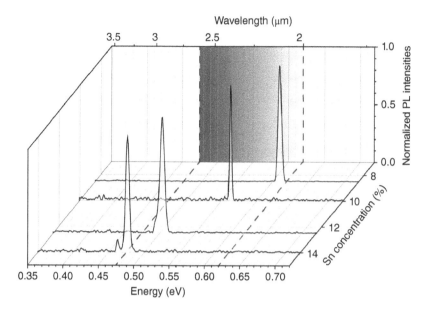

Figure 6. Emission spectra of optically pumped GeSn laser. Fabry Perot cavities were fabricated by defining mesa structures in the GeSn film and subsequent under-etching. The emission wavelength shifts from 2 to 2.6 μm for increasing the Sn concentration from 8.5% to 14%.

improved by a factor of 2.3. Lasing was obtained up to 135 K, compared to 90 K for the simple mesa structure.

The latter is remarkable since the underetched mesa structure has an inferior thermal contact to the substrate due to the narrow GeSn film and the remaining part of the Ge virtual substrate (marked as Ge-VS in Fig. 4). Rough estimates of the actual temperature of the GeSn laser due to the high pumping power suggest a realistic temperature of more than 100 K above the measured temperature, indicating operation close to room temperature.

Due to the lattice relaxation of the underetched mesa structures, lasing was obtained for GeSn films having a Sn concentration of 8.5%. Changing the Sn concentration from 8.5% to 14% gives a tuning range of the emission wavelength from 2 to 2.6 μm, as illustrated in Fig. 6. Generally, the emission spectra of the underetched devices are slightly shifted toward longer wavelengths compared to those of the simple mesa type Fabry–Perot waveguide devices. This shift is attributed to relaxation of the residual strain in the freestanding parts of the underetched structures. The lattice relaxation was also detected in micro-Raman measurements.

The temperature dependence of the emission intensity and of emission wavelength has been measured.[2, 11] The band structure around the Γ point was calculated using the 8-band $k \cdot p$ method including strain effects. Details can be

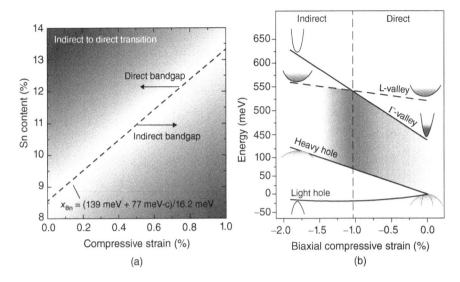

Figure 7. Image (a) shows the dependence of the Sn concentration required to obtain a direct bandgap on the residual compressive strain in the GeSn film. Image (b) shows the calculated electronic band structure as function of compressive strain for a $Ge_{0.875}Sn_{0.125}$ alloy.

found in the supplementary information of Ref. 2. The temperature dependence of the PL intensity was simulated, which includes the calculation of the joint density of states, with the band offset ΔE between the minima of Γ and L valleys being the key fitting parameter. By fitting a large amount of experimental PL data from GeSn samples with different Sn concentrations and residual strain, measured by RBS and X-ray diffraction (XRD), a set of parameters used in the 8-band $k \cdot p$ method was established. Figure 7 shows the key results of this self-consistent band structure calculation. Figure 7(a) shows the dependence of the Sn concentration required to obtain a direct bandgap on the residual compressive strain in the GeSn film. Accordingly, a Sn concentration of 8.5% is required in order to obtain a fundamental direct bandgap in fully relaxed GeSn films. Figure 7(b) shows the calculated electronic band structure as function of compressive strain for a specific $Ge_{0.875}Sn_{0.125}$ alloy. Here the maximum ΔE of about 80 meV is obtained for zero strain and the material becomes indirect once the compressive strain exceeds -1%.

4. Potential of GeSn alloys for electronic devices

The above-described fitting methodology to obtain band structure parameter for the 8-band $k \cdot p$ method was in turn used to calculate the effective mass of electrons in GeSn. Figure 7 shows that for relaxed GeSn with a direct bandgap

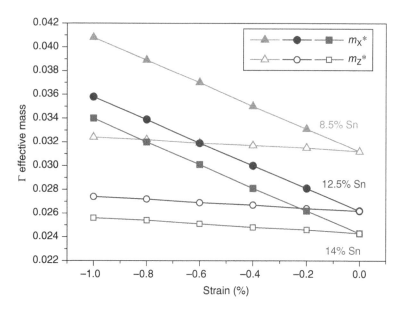

Figure 8. Effective mass of electrons in the Γ valley versus Sn concentration and compressive strain.

(Sn content >8.5%), the Γ valley forms the conduction band edge, which is expected to harbor low effective mass electrons and a low density of states. For relaxed GeSn the valence band is degenerate. Adding compressive strain shifts the heavy holes up in energy, whereas tensile strain does the same to the light holes. Tensile strain is accessible by applying a SiN_x stressor material to the GeSn film. Figure 8 depicts the calculated effective mass for electrons in the GeSn alloys as a function of residual compressive strain for three Sn concentrations: 8.5%, 12.5%, and 14%. In calculating the in-plane mass m_X^* (open symbols) and out-of-plane mass m_Z^*, we find the lowest masses for relaxed GeSn with higher Sn content. Interestingly, the m^* for relaxed $Ge_{0.86}Sn_{0.14}$ is $0.024m_0$, close to the effective electron mass of InAs.

Figure 9(a) shows the calculated room-temperature mobility of electrons in the Γ valley of relaxed GeSn with 12.5% Sn content, including band nonparabolicity as well as phonon scattering. The calculation reveals extremely high mobilities, up to 250,000 cm²/V·s for relaxed GeSn. However, due to the small separation of Γ and L valleys, electrons in both valleys have to be considered. The higher mass and lower mobility of electrons in the L valley, shown in Fig. 9(b), reduce the overall mobility to about 80,000 cm²/V·s – see Fig. 9(c). These calculations do not include alloy scattering, which may have an impact even at room temperature for very high-mobility electrons. Still, it appears likely that GeSn may have electron mobilities competitive with InAs and hole mobilities comparable to Ge. Hence GeSn may offer high-mobility p- and n-MOS circuitry

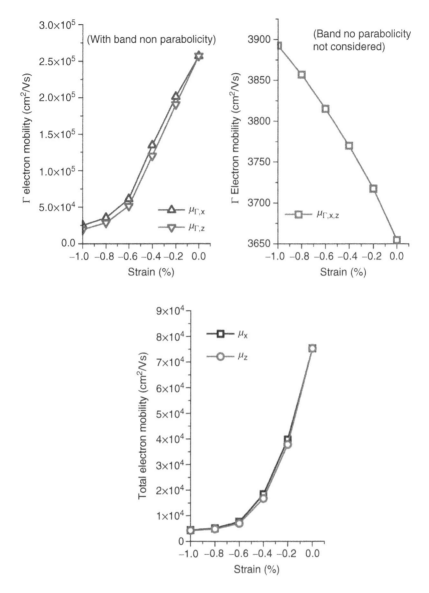

Figure 9. Electron mobility of electrons in the Γ and L valleys versus residual compressive strain for a GeSn alloy containing 12.5% Sn.

within the same layer and may thus be a strong alternative for the combination of III/V n-MOS with Ge p-MOS, especially since electronic and optoelectronic devices can be integrated as well. Finally, the combination of a group IV alloy having a direct bandgap and very low effective charge carrier masses might be a viable route for high-performance, ultralow-power TFET devices.[12] In addition to improved tunneling currents compared to Si TFETs,[13] the problem of deep traps due to anti-side defects in III/V compounds, which leads to trap-assisted tunneling, is not present in group IV alloys.

5. Conclusion

RGSE was employed to grow GeSn alloys up to 14%, thus far beyond the solid solubility limit of 1% Sn in Ge. The RGSE technique enables growth at high rates and low deposition temperatures while maintaining pristine crystallographic quality. Optically pumped lasers were fabricated from GeSn films deposited on Ge virtual substrates. Underetching of the waveguide structures led to drastic improvement of the laser performance, cutting the threshold pump power by half and increasing the efficiency by a factor of 2.3. Lasing was observed up to 135 K. The experimental data on the optical properties were used to fine-tune the parameter set for $k \cdot p$ bandstructure calculations. Relaxed GeSn alloys with Sn concentrations exceeding 8.5% exhibit a direct fundamental bandgap. The electrons in the Γ valley are predicted to have very low effective masses and high mobilities, making GeSn a potential alternative material for high-speed, low-power electronics. Hence, GeSn may offer a versatile platform for EPIC.

Acknowledgments

The above-presented work summarizes the in-depth collaboration with a large number of scientists. S. Wirths and N. von den Driesch (PGI-9, FZJ) grew the samples by RGSE. R. Geiger and H. Sigg (Paul Scherrer Institute, CH) and S. Wirths performed PL and optical pumping measurements. D. Stange and T. Stoica performed PL investigations at FZJ. J.-M. Hartmann (CEA-LETI, France) supplied the Ge virtual substrates. M. Luysberg (ER-C and PGI-5, FZJ) performed TEM measurements. J. Faist (ETH-Zürich) calculated the optical waveguide properties. Z. Ikonic (Leeds University) led the effort in bandstructure calculations of GeSn alloys (Figs. 7–9). G. Mussler and B. Holländer (PGI-9, FZJ) did extensive XRD and RBS measurements. D. Buca and S. Mantl scientifically supervised and coordinated the GeSn team at PGI-9, FZJ.

References

1. C. H. L. Goodman, "Direct-gap group IV semiconductors based on tin," *IEE Proc. I: Solid State Electron Dev.* **129**, 189–192 (1982).
2. S. Wirths, R. Geiger, N. von den Driesch, *et al.*, "Lasing in direct-bandgap GeSn alloy grown on Si," *Nature Photonics* **9**, 88–92 (2015).
3. K.-H. Koo and K. C. Saraswat, "Study of performances of low-κ Cu, CNTs, and optical interconnects," chapter in: N. K. Jha and D. Chen, eds., *Nanoelectronic Circuit Design*, New York: Springer, 2011, pp. 377–407.

4. P. Moontragoon, R. A. Soref, and Z. Ikonic, "The direct and indirect bandgaps of unstrained $Si_xGe_{1-x-y}Sn_y$ and their photonic device applications," *J. Appl. Phys.* **112**, 073106 (2012)

5. R. Soref, "Direct-bandgap compositions of the CSiGeSn group-IV alloy," *Opt. Mater. Express* **4**, 836–842 (2014).

6. S. Wirths, D. Buca, G. Mussler, *et al.*, "Reduced pressure CVD growth of Ge and $Ge_{1-x}Sn_x$ alloys," *ECS J. Solid State Sci. Technol.* **2**, N99–N102 (2013).

7. Y. Shimura, N. Tsutsui, O. Nakatsuka, A. Sakai, and S. Zaima, "Low temperature growth of $Ge_{1-x}Sn_x$ buffer layers for tensile-strained Ge layers," *Thin Solid Films* **518**, S2–S5 (2010).

8. S. M. Gates and C. M. Chiang, "Dissociative chemisorption mechanisms of disilane on Si(100)-(2×1) and H-terminated Si(100) surfaces," *Chem. Phys. Lett.* **184**, 448–454 (1991).

9. F. Gencarelli, B. Vincent, L. Souriau, *et al.*, "Low-temperature Ge and GeSn chemical vapor deposition using Ge_2H_6," *Thin Solid Films* **520**, 3211–3215 (2012).

10. N. von den Driesch, D. Stange, S. Wirths, *et al.*, "Direct bandgap group IV epitaxy on Si for laser applications," *Chem. Mater.* **27**, 4693–4702 (2015).

11. D. Stange, S. Wirths, N. von den Driesch, *et al.*, "Optical transitions in direct bandgap $Ge_{1-x}Sn_x$ alloys," *ACS Photonics* **2**, 1539–1545 (2015).

12. S. Wirths, A. T. Tiedemann, Z. Ikonic, *et al.*, "Band engineering and growth of tensile strained Ge/(Si)GeSn heterostructures for tunnel field effect transistors," *Appl. Phys. Lett.* **102**, 192103 (2013).

13. L. Knoll, Q.-T. Zhao, A. Nichau, *et al.*, "Inverters with strained Si nanowire complementary tunnel field-effect transistors," *IEEE Electron Device Lett.* **34**, 813–815 (2013).

2.6

High Sn-Content GeSn Light Emitters for Silicon Photonics

D. Stange, C. Schulte-Braucks, N. von den Driesch, S. Wirths, G. Mussler, S. Lenk, T. Stoica, S. Mantl, D. Grützmacher, and D. Buca
Peter Grünberg Institut-9 and JARA-FIT, Forschungszentrum Jülich GmbH, 52425 Jülich, Germany

R. Geiger, T. Zabel, and H. Sigg
Laboratory for Micro- and Nanotechnology, Paul Scherrer Institut, 5232 Villigen PSI, Switzerland

J. M. Hartmann
Université Grenoble Alpes and CEA-LETI/MINATEC, 38054 Grenoble, France

Z. Ikonic
Institute of Microwaves and Photonics, University of Leeds, LS2 9JT Leeds, United Kingdom

1. Introduction

The continuous progress of computer technology, with a larger amount of data transfer and higher data processing speed, has strongly increased the energy needed in order to run large data centers. With every new processor generation, the power consumption increases with the square of the clock frequency.[1] Given the development of IT networks, energy consumption becomes one of the main bottlenecks. A large fraction of the energy consumption in data centers (38% in 2009) is coming from the cooling systems used to dissipate the heat mainly produced by copper interconnects linking devices between and on chips.[2]

The partial substitution of copper by optical interconnects would reduce the heat generation enormously. Indeed, attenuation of data transmission at the 10 Gb/s rate can be 1000 times lower optically, requiring much less power.[1]

Future Trends in Microelectronics: Journey into the Unknown, First Edition.
Edited by Serge Luryi, Jimmy Xu, and Alexander Zaslavsky.
© 2016 John Wiley & Sons, Inc. Published 2016 by John Wiley & Sons, Inc.

The present chip technology is based on silicon with increasing number of other materials integrated into electrical circuits. Due to the constant research in silicon photonics, nearly all the components that are necessary for the photonic part of a silicon-based electronic–photonic integrated circuit (EPIC) have already been developed. The only exception is the light source, which today is based on III–V materials, adding complexity to the fabrication process due to the need for bonding techniques on the Si.[3, 4] Besides being costly, the bonding process of toxic III–V materials onto Si chips faces various technical challenges. For example, layers can be damaged during annealing steps due to thermal mismatch between III–V and Si materials.[5] For industrial manufacturing of photonic systems, a straightforward integration of a light source would be preferable. Devices based only on group IV materials directly grown on Si would be ideal from that point of view.

Unfortunately, group IV semiconductors like Si and Ge exhibit an indirect bandgap. Radiative carrier recombination is then not efficient, making them unsuitable for laser applications. Different concepts like Si nanocrystals[6] or Ge under high tensile strain[7] are presently under investigation. Another approach consists in growing GeSn layers with high amounts of tin. This reduces the energy separation between conduction band valleys and the valence band as Sn content increases, with a stronger effect for the Γ-valley than for the L-valley,[8–10] leading to a transition from an indirect to a direct bandgap for sufficiently high Sn content. The transition point for unstrained GeSn compounds was calculated to occur at a Sn concentration between 5% and 12%.[11, 12]

Recently, the fundamental direct bandgap in partially relaxed GeSn alloys was demonstrated experimentally.[13] The transition point for fully relaxed layers occurred for a Sn concentration of ~9%. Lasing by optical pumping of waveguide structures was demonstrated and a differential gain of 0.4 cm/kW was determined.

The growth of strain-relaxed GeSn is not trivial. Indeed, the lattice mismatch between the Ge-on-Si substrate and the GeSn alloy leads to the presence of a compressive strain inside the GeSn layer, which shifts the transition point to really high Sn concentrations. This chapter will present a systematic photoluminescence (PL) study of compressively strained, direct-bandgap GeSn alloys, followed by the analysis of two different optical source designs. First, a direct bandgap GeSn light emitting diode (LED) will be characterized via power- and temperature-dependent electroluminescence (EL) measurements. Then, lasing will be demonstrated in a microdisk (MD) resonator under optical pumping.

The integration of direct-bandgap GeSn-based devices as a light source for on-chip communications offers the possibility to monolithically integrate the complete photonic circuit within mainstream silicon technology.[14] The fabrication of a compact electrically pumped laser source could thus become a reality in the coming years.[15]

2. Experimental details of the GeSn material system

The $Ge_{1-x}Sn_x$ alloys were grown on 2.5 μm thick Ge virtual substrates (Ge-VS)[16] on top of 200 mm Si(100) wafers in an industrial AIXTRON TRICENT reduced-pressure chemical vapor deposition (CVD) reactor[17, 18] from commercial Ge_2H_6 and $SnCl_4$ precursors,[19] with PH_3 and B_2H_6 used for n-type and p-type *in situ* doping, respectively.

The growth temperature was kept in the 340–350 °C range, with high growth rates in order to avoid Sn precipitation. A TEM micrograph of a typical 400 nm thick epitaxially grown $Ge_{0.875}Sn_{0.125}$ layer is shown in Fig. 1. The crystalline quality of that layer, which is partly relaxed, is very high. Misfit dislocations are confined at the interface between GeSn and the Ge-VS underneath.[19] The GeSn layer is thus free from threading dislocations (at least on the TEM scale).

Material analysis regarding crystallinity, Sn concentration, and strain was performed via Rutherford-backscattering-spectroscopy/channeling (RBS/C) and X-ray diffraction using reciprocal space mapping (XRD-RSM).

Here, we will succinctly describe material properties using $Ge_{0.875}Sn_{0.125}$ epilayers of various thicknesses. Up to a certain thickness, $Ge_{1-x}Sn_x$ grows pseudomorphically on Ge-VS (e.g., its in-plane lattice parameter is equal to that of the Ge-VS, while its out-of-plane lattice parameter is higher than its bulk value). For coherently grown $Ge_{0.875}Sn_{0.125}$, the compressive strain reaches −1.7%. Above a certain critical layer thickness, the alloy starts to plastically relax through the formation of misfit dislocations at the interface.

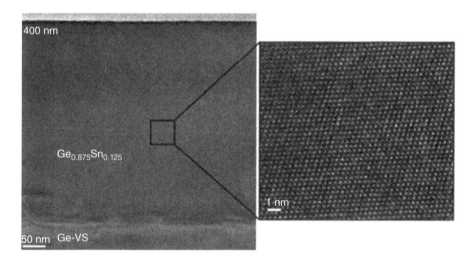

Figure 1. Cross-sectional TEM image of a 400 nm thick $Ge_{0.875}Sn_{0.125}$ layer on a Ge-VS. The inset shows the high crystallinity of the material.

The amount of strain inside the GeSn lattice strongly affects its band structure. The strain-dependent band structure for $Ge_{0.875}Sn_{0.125}$ in Fig. 2(a) is calculated via the 8-band $k \cdot p$ method,[21, 22] using conventional deformation potentials for indirect valleys. With increasing relaxation of the highly compressively strained layers, the conduction band is shifted toward lower

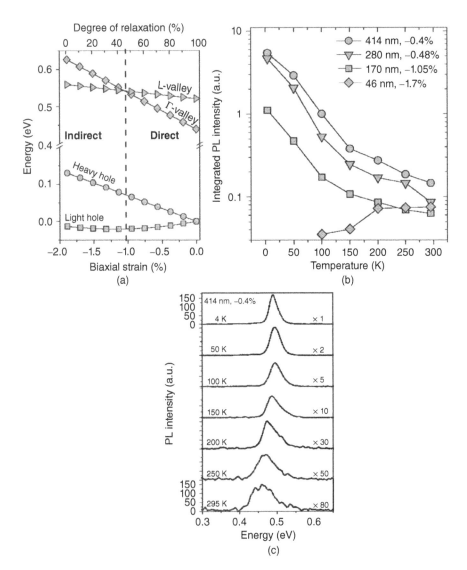

Figure 2. (a) Calculated $Ge_{0.875}Sn_{0.125}$ band structure versus biaxial compressive strain. Optical PL characterization of $Ge_{0.875}Sn_{0.125}$ layers; (b) temperature-dependent integrated PL intensity of $Ge_{0.875}Sn_{0.125}$ layers with a constant composition versus strain; (c) PL spectra of the 400 nm thick $Ge_{0.875}Sn_{0.125}$ layer.[20]

energies. However, the Γ-valley decreases faster in energy than the L-valley. The transition from an indirect to a direct bandgap semiconductor is predicted to occur at a compressive strain of about 1%. In addition, the valence band splitting is reduced with decreasing compressive strain, with the valence bands becoming degenerate at zero strain.

Temperature-dependent integrated PL intensity is a suitable method to determine whether a semiconductor has a direct or indirect fundamental bandgap.[13] The integrated PL intensity of four layers with thickness increasing from 46 to 400 nm (and hence decreasing compressive strain) is shown in Fig. 2(b). The corresponding strain is given in the figure inset.

The PL intensity of the pseudomorphically grown $Ge_{0.875}Sn_{0.125}$ alloy (e.g., under a compressive strain of -1.7%) decreases with temperature T. This is due to the decreasing Γ valley electron population as T decreases (reduction of thermal transfer of carriers). This indicates an indirect band structure of this alloy, confirming the calculations of Fig. 2(a).

For a thicker 170 nm layer, we find a residual strain of -1.05% and the integrated PL intensity increases continuously. According to calculations of Fig. 2(a), the material might be just at the transition from an indirect to a direct bandgap semiconductor as indicated by the constant value of integrated intensity for 300 K and 250 K. The trend of continuously increasing signal with decreasing T is observed for all GeSn layers with less than -1.05% compressive strain (see Fig. 2(b)). Given this trend, these materials can be classified as being direct bandgap semiconductors. With decreasing compressive strain, the energy difference between the Γ- and L-valleys increases, which results in increased intensities at a constant temperature. Full temperature-dependent PL spectra of a 400 nm thick GeSn layer are shown in Fig. 2(c). The strong luminescence reveals the high optical quality of the material, which is mandatory for the fabrication of light sources.

3. Direct bandgap GeSn light emitting diodes

LEDs made out of GeSn alloys have already been reported by various groups.[23–25] Room-temperature EL was associated with Γ–valence band recombination; however, the directness of these materials remained questionable. Tin concentrations of up to 10% were reached,[26, 27] but temperature dependence was not reported. The Sn content and the strain in the previously fabricated diodes suggested that the GeSn alloys had an indirect bandgap. As a consequence, the current density needed to inject a sufficient density of carriers remained very high, for example, in the kiloampere per square centimeter range.[24, 28, 29] Here we will give a short overview of light emission in a direct-bandgap $Ge_{0.89}Sn_{0.11}$ LED that could fulfill a major requirement for future optoelectronic devices with reduced power consumption.

The LEDs presented here are p–i–n structures based on GeSn. They contain a Sn concentration of about 11.5% with a residual strain of -0.8%.

In accordance with calculations regarding this configuration, the material should have a direct bandgap with a small energy offset between Γ- and L valley $\Delta E_{\Gamma L} = 2.4\,\text{meV}$. The carrier concentration in the p^+ boron-doped region reaches values of $2 \times 10^{18}\,\text{cm}^{-3}$, while the n+ phosphorus doping was $3 \times 10^{19}\,\text{cm}^{-3}$. The "intrinsic" region is unintentionally p-doped at $9 \times 10^{17}\,\text{cm}^{-3}$, possibly due to crystal defects.[30]

The mesa of the diode was fabricated by reactive ion etching using Cl_2 and Ar plasma and passivated by Al_2O_3/SiO_2 10:150 nm stacks deposited via atomic layer deposition (ALD)/plasma-enhanced chemical vapor deposition (PECVD). Contact windows were defined by optical lithography and opened by CHF_3 dry etch (in order to contact the p and n regions afterward). A schematic of the LED structure and a 3D SEM can be seen in Fig. 3(a) and (b).[31] The contacts were made by sputtering 10 nm of Ni and annealing the structure at 325 °C in a forming gas atmosphere to form NiGeSn.[32] As a last step, Al was deposited to facilitate wire bonding.

Electroluminescence was measured under a pulsed injection at a frequency of 1988 Hz. The temperature-dependent EL of a $Ge_{0.89}Sn_{0.11}$ p–i–n diode is shown in Fig. 4(a). At current densities of about 120 A/cm^2, a clear EL signal is obtained. The emission energy is blueshifted when the temperature

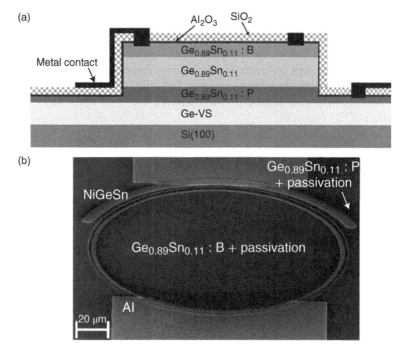

Figure 3. (a) Schematic of the p–i–n GeSn LED; (b) SEM of such a diode structure with 150 μm diameter.

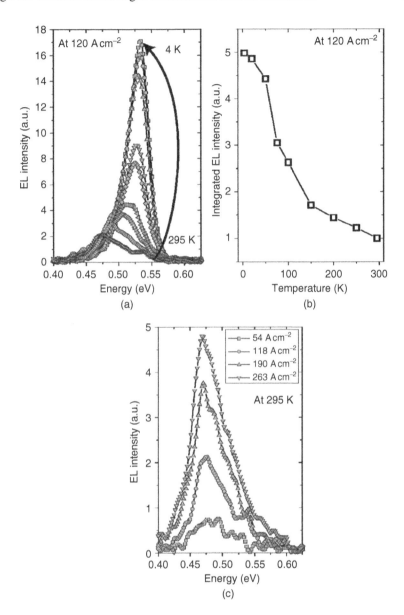

Figure 4. Electroluminescence of $Ge_{0.89}Sn_{0.11}$ LEDs with 150 μm diameter: (a) temperature dependence at a current density of 120 A/cm^2; (b) integrated intensity; and (c) power dependence at room temperature.

decreases, in agreement with the expected temperature dependence of the bandgap. At low temperatures, both thermal activation of charge carriers from Γ- into the L-valley and defect-related nonradiative recombination processes are suppressed, resulting in an EL increase. The continuously increasing integrated signal in Fig. 4(b) can then attribute to the direct bandgap of the alloy.

Power-dependent measurements at room temperature show the suitability of these group IV alloys as light emitters at low current densities of 55 A/cm^2 (see Fig. 4(c)). Reducing the temperature allows observation of a signal at even lower current densities.

It is, however, obvious that GeSn p–i–n homojunction LEDs do not provide a good confinement of carriers inside the active region (when targeting optimized LED structures and designing electrically pumped lasers). Also, heterostructures like Ge/GeSn/Ge do not offer high band offsets. The Ge layer sitting on top of the thick, nearly fully relaxed GeSn layer, is necessarily under tensile strain, making type I band alignment difficult to reach. Therefore, it seems mandatory to use SiGeSn as a cladding material. In addition, multiquantum well (MQW) structures may offer low threshold powers given their 2D density of states.[33]

4. Group IV GeSn microdisk laser on Si(100)

Both the integration of an on-chip light source and dense packaging require a compact resonator geometry. One possible design is the MD geometry, which is already used in III–V technology as a suitable approach for integrated light sources on a silicon chip.[34, 35]

Group IV microdisk experiments were also reported including SiC and pseudomorphic Ge/GeSn/Ge heterostructures. Whispering gallery modes (WGM) were observed in both cases; however, no evidence of lasing was reported.[36, 37] The same was true for Ge membranes under high tensile strain[38] or MDs that used SiN or other stressing methods.[39, 40]

The microdisks presented here are underetched, with an undercut of 3.6 μm using selective etching between Ge and GeSn. This leads to an increase of the effective refractive index and, therefore, improves the optical confinement of WGM modes inside the disk resonator. Another beneficial impact of this undercut is the strain relaxation that occurs at the edges of the disk. By partially etching the Ge buffer beneath the Ge$_{1-x}$Sn$_x$ layer, the latter is indeed able to fully relax. With a higher degree of relaxation of the GeSn lattice, the energy separation between Γ- and L-valleys becomes larger, yielding a direct bandgap alloy. The same effect was discussed in the previous section regarding relaxation in Ge$_{0.875}$Sn$_{0.125}$ layers. The alteration of the band structure due to the disk geometry (i.e., due to strain gradients) leads to a diffusion of the charge carriers toward the edges of the disk. This results in increased Γ-population at the disk edges, where the WGM are formed, which may help in reducing the lasing threshold. For a strain change from −0.4% to 0% (e.g., fully relaxed

$Ge_{0.875}Sn_{0.125}$ layers), the Γ-population changes from 10% to 30% of the total electron density in the conduction band, as calculated by the 8-band $k \cdot p$ and deformation potential method.

The MD laser mesa structure was defined by e-beam lithography followed by ~1 µm dry etching with Cl_2/Ar. Then, Ge was selectively dry-etched isotropically by a CF_4 plasma.[41] For the reduction of nonradiative surface recombination, a 10 nm thick Al_2O_3 layer is deposited at 300 °C by ALD.

The previously described strain relaxation process was investigated via Raman spectroscopy. A WiTec setup with a 532 nm laser diode was used to measure the signal of Ge–Ge vibrations in a GeSn lattice. The Raman modes in Ge bulk are seen at ~300 cm^{-1}. With an increasing amount of Sn in the alloy, the signal is shifted to lower wavenumbers, whereas strain shifts the Raman peaks toward higher wavenumbers.[41, 42] As a result, for a given Sn content, plastic relaxation in GeSn should lead to a wavenumber decrease. This is exactly what is seen in Fig. 5(a): a shift toward lower wavenumbers when moving from the center toward the edges of the disk. The complete distribution of the Raman peak's wavenumber inside the disk is presented in Fig. 5(b). A further proof of strain relaxation is given by locally resolved PL measurements. The PL peak from the edge of the disk is redshifted compared to the signal from the center, due to the lower bandgap initiated by strain relaxation at the edges of the disk (see Fig. 5(c)).

Strain variation along the radius of disk induces local band structure changes. The lower energy of the bands at the edges compared to the center leads to a built-in electric fields that results in a diffusion of electrons toward the edges of the disk. Because of the location of WGM at the disk edges, carrier diffusion is a boon to lasing action and reduces the lasing threshold.

Incident light coming from the top from a pulsed 1064 nm Nd:YAG laser diode with a pulse duration of 6 ns was used to optically pump the microdisk laser. The generated laser light is scattered out by imperfections at the microdisk sidewalls and guided to an InSb detector.

An 8 µm diameter disk with a Sn concentration of 12.5% was investigated. The compressive strain in the initial $Ge_{0.875}Sn_{0.125}$ layer was −0.4% according to X-ray diffraction (XRD). Band structure calculations give the Γ–L energy separation $\Delta E_{\Gamma L} = 73$ meV for $Ge_{0.875}Sn_{0.125}$ at 100 K. Spontaneous emission becomes visible for optical pumping of the microdisk at low power. With increasing pumping power, an exponential enhancement of emission is observed (see Fig. 6(a)). The lasing threshold at 20 K is 128 kW/cm^2. For pumping powers above 500 kW/cm^2, light emission slowly starts to saturate.

Figure 6(b) shows PL microdisk spectra below and above threshold. The broad spectrum coming from spontaneous emission reveals a more than 10 times larger full width half maximum (FWHM) compared to the narrow laser linewidth. These measurements demonstrate the typical laser characteristics of this novel group IV semiconductor laser based on GeSn. Details of the microdisk laser can be found in Ref. 43.

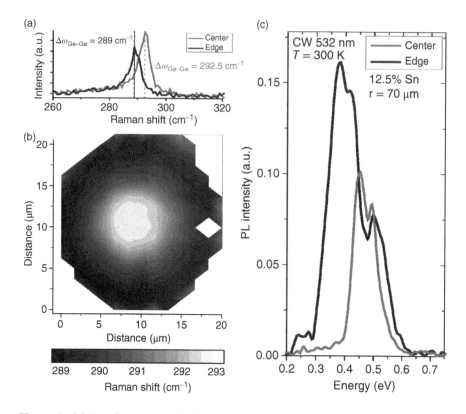

Figure 5. (a) Two Raman spectra belonging from the full map demonstrate the relaxation from the center to the edge; (b) full map of the Raman shift of $Ge_{0.875}Sn_{0.125}$ related to lattice relaxation in the underetched part of the MD; (c) locally separated PL measurements at the center and edge of the disk confirm the strain relaxation at the edges indicated by the Raman signal's redshift.

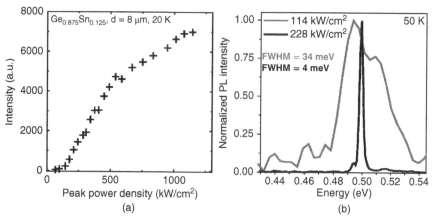

Figure 6. (a) Light-in light-out curve of a $Ge_{0.875}Sn_{0.125}$ microdisk laser at 20 K; (b) Spectra below and above threshold show the expected collapse of the linewidth above the onset of lasing.

5. Conclusion and outlook

In conclusion, we presented growth and optical characterization of high-quality GeSn alloys with very high Sn content. The change in band structure coming from plastic strain relaxation was discussed via PL analysis of different layers with various degrees of relaxation.

We have fabricated direct bandgap LEDs from our GeSn alloys, measured their luminescence, and discussed avenues to improve carrier confinement and obtain even more efficient LEDs.

Special attention was allocated to fabrication and characterization of underetched microdisk GeSn/Ge structures. They were investigated by Raman mapping and band structure calculations. Because of the undercut, nearly full relaxation was obtained at the edges. For a Sn concentration of 12.5%, lasing in those whispering gallery mode devices was achieved at a record-breaking $128\,kW/cm^2$ threshold.

Combining these two results, electrical injection into a resonator geometry coupled with the suggested improvements for LED structures should pave the way toward an electrically pumped GeSn laser grown directly on Si(100). This unique feature of GeSn alloys can be advantageous over bonded III–V light sources and could lead in the future to an affordable use of such heterostructures in complex EPICs.

References

1. UC-Santa Barbara Institute for Energy Efficiency (2015), see //iee.ucsb.edu/content/optical-interconnects-and-switching-reduce-power-consumption.
2. Emerson Network Power, "Energy logic: Reducing data center energy consumption by creating savings that cascade across systems," white paper (2009).
3. D. Liang, M. Fiorentino, T. Okumura, *et al.*, "Electrically-pumped compact hybrid silicon microring lasers for optical interconnects," *Opt. Express* **17**, 20355 (2009).
4. H. Park, A. W. Fang, S. Kodama, and J. E. Bowers, "Hybrid silicon evanescent laser fabricated with a silicon waveguide and III–V offset quantum wells," *Opt. Express* **13**, 9460–9464 (2005).
5. D. Pasquariello and K. Hjort, "Plasma-assisted InP-to-Si low temperature wafer bonding," *IEEE J. Sel. Top. Quantum Electron.* **8**, 118–131 (2002).
6. L. Pavesi, L. Dal Negro, C. Mazzoleni, G. Franzò, and F. Priolo, "Optical gain in silicon nanocrystals," *Nature* **408**, 440–444 (2000).
7. J. Liu, X. Sun, D. Pan, *et al.*, "Tensile-strained, n-type Ge as a gain medium for monolithic laser integration on Si," *Opt. Express* **15**, 11272–11277 (2007).
8. C. Eckhardt, K. Hummer, and G. Kresse, "Indirect-to-direct gap transition in strained and unstrained Sn_xGe_{1-x} alloys," *Phys. Rev. B* **89**, 165201 (2014).

9. P. Moontragoon, Z. Ikonić, and P. Harrison, "Band structure calculations of Si–Ge–Sn alloys: Achieving direct bandgap materials," *Semicond. Sci. Technol.* **22**, 742–748 (2007).

10. G. He and H. Atwater, "Interband transitions in Sn_xGe_{1-x} alloys," *Phys. Rev. Lett.* **79**, 1937–1940 (1997).

11. S. Gupta, B. Magyari-Köpe, Y. Nishi, and K. C. Saraswat, "Achieving direct band gap in germanium through integration of Sn alloying and external strain," *J. Appl. Phys.* **113**, 073707 (2013).

12. K. L. Low, Y. Yang, G. Han, W. Fan, and Y.-C. Yeo, "Electronic band structure and effective mass parameters of Sn_xGe_{1-x} alloys," *J. Appl. Phys.* **112**, 103715 (2012).

13. S. Wirths, R. Geiger, N. von den Driesch, *et al.*, "Lasing in direct-bandgap GeSn alloy grown on Si," *Nature Photonics* **9**, 88–92 (2015).

14. K. P. Homewood and M. A. Lourenço, "Optoelectronics: The rise of the GeSn laser," *Nature Photonics* **9**, 78–79 (2015).

15. R. Soref, "Group IV photonics: Enabling 2 µm communications," *Nature Photonics* **9**, 358–359 (2015).

16. J. M. Hartmann, A. Abbadie, N. Cherkashin, H. Grampeix, and L. Clavelier, "Epitaxial growth of Ge thick layers on nominal and 6° off Si(001); Ge surface passivation by Si," *Semicond. Sci. Technol.* **24**, 055002 (2009).

17. S. Wirths, Z. Ikonic, A. T. Tiedemann, *et al.*, "Tensely strained GeSn alloys as optical gain media," *Appl. Phys. Lett.* **103**, 192110 (2013).

18. S. Wirths, D. Buca, G. Mussler, A. T. Tiedemann, *et al.*, "Reduced pressure CVD growth of Ge and Sn_xGe_{1-x} alloys," *ECS J. Solid State Sci. Technol.* **2**, N99 (2013).

19. N. von den Driesch, D. Stange, S. Wirths, *et al.*, "Direct bandgap group IV epitaxy on Si for laser applications," *Chem. Mater.* **27**, 4693–4702 (2015).

20. D. Stange, S. Wirths, N. von den Driesch, *et al.*, "Optical transitions in direct bandgap $Ge_{1-x}Sn_x$ alloys," *ACS Photonics* **2**, 1539–1545 (2015).

21. T. Bahder, "Eight-band k·p model of strained zinc-blende crystals," *Phys. Rev. B* **41**, 11992–12001 (1990).

22. T. B. Bahder, "Erratum: Eight-band k·p model of strained zinc-blende crystals [*Phys. Rev. B* 41, 11992 (1990)]," *Phys. Rev. B* **46**, 9913–9913 (1992).

23. S. Q. Yu, S. A. Ghetmiri, W. Du, *et al.*, "Si based GeSn light emitter: Mid-infrared devices in Si photonics," *Proc. SPIE* **9367**, 93670R (2015).

24. M. Oehme, K. Kostecki, T. Arguirov, *et al.*, "GeSn heterojunction LEDs on Si substrates," *IEEE Photonics Technol. Lett.* **26**, 187–189 (2014).

25. J. P. Gupta, N. Bhargava, S. Kim, T. Adam, and J. Kolodzey, "Infrared electroluminescence from GeSn heterojunction diodes grown by molecular beam epitaxy," *Appl. Phys. Lett.* **102**, 251117 (2013).

26. J. D. Gallagher, C. L. Senaratne, P. Sims, T. Aoki, J. Menéndez, and J. Kouvetakis, "Electroluminescence from GeSn heterostructure pin diodes at the indirect to direct transition," *Appl. Phys. Lett.* **106**, 091103 (2015).

27. J. D. Gallagher, C. L. Senaratne, C. Xu, *et al.*, "Non-radiative recombination in $Ge_{1-y}Sn_y$ light emitting diodes: The role of strain relaxation in tuned heterostructure designs," *J. Appl. Phys.* **117**, 245704 (2015).

28. M. Oehme, J. Werner, M. Gollhofer, M. Schmid, M. Kaschel, E. Kasper, and J. Schulze, "Room-temperature electroluminescence from GeSn light-emitting pin diodes on Si," *IEEE Photonics Technol. Lett.* **23**, 1751–1753 (2011).

29. B. Schwartz, M. Oehme, K. Kostecki, *et al.*, "Electroluminescence of GeSn/Ge MQW LEDs on Si substrate," *Opt. Lett.* **40**, 3209–3212 (2015).

30. O. Nakatsuka, N. Tsutsui, Y. Shimura, S. Takeuchi, A. Sakai, and S. Zaima, "Mobility behavior of $Ge_{1-x}Sn_x$ layers grown on silicon-on-insulator substrates," *Japan. J. Appl. Phys.* **49**, 04DA10 (2010).

31. C. Schulte-Braucks, D. Stange, N. von den Driesch, *et al.*, "Negative differential resistance in direct bandgap GeSn p–i–n structures," *Appl. Phys. Lett.* **107**, 042101 (2015).

32. S. Wirths, R. Troitsch, G. Mussler, *et al.*, "Ternary and quaternary Ni(Si)Ge(Sn) contact formation for highly strained Ge p- and n-MOSFETs," *Semicond. Sci. Technol.* **30**, 055003 (2015).

33. D. Stange, N. von den Driesch, D. Rainko, *et al.*, "Study of GeSn-based heterostructures: Towards optimized group IV MQW LEDs," *Opt. Express* **24**, 1358–1367 (2016).

34. J. Van Campenhout, P. Rojo Romeo, P. Regreny, *et al.*, "Electrically pumped InP-based microdisk lasers integrated with a nanophotonic silicon-on-insulator waveguide circuit," *Opt. Express* **15**, 6744–6749 (2007).

35. G. Morthier, T. Spuesens, and P. Mechet, G. Roelkens, and D. Van Thourhout, "InP microdisk lasers integrated on Si for optical interconnects," *IEEE J. Sel. Top. Quantum Electron.* **21**, article no. 6975106 (2015).

36. M. Radulaski, T. M. Babinec, K. Mu, *et al.*, "Visible photoluminescence from cubic (3C) silicon carbide microdisks coupled to high quality whispering gallery modes," *ACS Photonics* **2**, 14–19 (2015).

37. R. Chen, S. Gupta, Y. Huang, *et al.*, "Demonstration of a Ge/GeSn/Ge quantum-well microdisk resonator on silicon: Enabling high-quality Ge(Sn) materials for micro- and nanophotonics," *Nano Lett.* **14**, 37–43 (2014).

38. A. Z. Al-Attili, S. Kako, M. K. Husain, *et al.*, "Whispering gallery mode resonances from Ge micro-disks on suspended beams," *Front. Mater.* **2**, 1–9 (2015).

39. A. Ghrib, M. El Kurdi, M. de Kersauson, *et al.*, "Tensile-strained germanium microdisks," *Appl. Phys. Lett.* **102**, 221112 (2013).

40. A. Ghrib, M. El Kurdi, M. Prost, *et al.*, "All-around SiN stressor for high and homogeneous tensile strain in germanium microdisk cavities," *Adv. Opt. Mater.* **3**, 353–358 (2015).

41. S. Gupta, R. Chen, Y. Huang, *et al.*, "Highly selective dry etching of germanium over germanium–tin ($Ge_{1-x}Sn_x$): A novel route for $Ge_{1-x}Sn_x$ nanostructure fabrication," *Nano Lett.* **13**, 3783–3790 (2013).

42. H. Lin, R. Chen, Y. Huo, T. I. Kamins, and J. S. Harris, "Raman study of strained $Ge_{1-x}Sn_x$ alloys," *Appl. Phys. Lett.* **98**, 261917 (2011).

43. D. Stange, S. Wirths, R. Geiger, *et al.*, "Optically pumped GeSn microdisk laser on Si," *ACS Photonics* **3**, 1279–1285 (2016).

2.7

Gallium Nitride-Based Lateral and Vertical Nanowire Devices

Y.-W. Jo, D.-H. Son, K.-S. Im, and J.-H. Lee
School of Electronics Engineering, Kyungpook National University, 80, Daehak-ro, Buk-gu, Daegu, South Korea

1. Introduction

GaN-based field-effect transistors (FETs) are widely used in power switching and amplifier applications due to the superior material properties of GaN, such as wide bandgap, high breakdown electric field, and high electron saturation velocity, which result in high current density and high breakdown voltage.[1, 2] However, GaN-based devices still suffer from challenges, such as surface and buffer-related trapping effects, that occasionally result in severe current collapse decrease and degradation of the off-state performances. While surface-related problems can be effectively addressed by appropriate surface treatment and passivation techniques,[3–6] trapping in the buffer layer cannot be easily solved because the buffer layer usually contains many defects originating from the heteroepitaxial growth of the GaN layer. Recently, AlGaN-/GaN-based triple-gate fin-shaped field-effect transistors (fin-FETs) have demonstrated superior off-state performance and also suggested a possible avenue for reducing the buffer-related trapping effects.[7–10] Even though the GaN-based gate-all-around (GAA) structure have not been realized with top-down approach thus far due to their complicated fabrication process, the GAA structure seems to exhibit greatly improved on- and off-state performances due to the enhanced gate controllability.[11–14] Furthermore, the GAA structure should also be more effective in eliminating buffer layer trapping because the entire active region of the device is surrounded by the gate metal and completely separated from the underlying thick GaN buffer layer.

In this chapter, we report on the first fabrication and characterization of GaN-based lateral and vertical nanowire (NW) FETs by using top-down approach, where we combined conventional e-beam lithography and dry etching techniques with strong anisotropic tetramethyl ammonium hydroxide (TMAH) wet etching.

Future Trends in Microelectronics: Journey into the Unknown, First Edition.
Edited by Serge Luryi, Jimmy Xu, and Alexander Zaslavsky.
© 2016 John Wiley & Sons, Inc. Published 2016 by John Wiley & Sons, Inc.

2. Crystallographic study of GaN nanowires using TMAH wet etching

Wet etching techniques play an important role in the semiconductor device fabrication process. Wet etching usually provides high etching selectivity that often offers an advantage in simplifying the fabrication process compared to the dry plasma etching. However, it is very difficult to etch the III-nitride semiconductors with the conventional wet solutions because these materials have strong chemical stability.[15] A few chemical solutions, such as hydroxide-based solutions (e.g., NaOH or KOH) were investigated to the etching of the III-nitride semiconductors due to their ability to react with Ga atoms to form Ga_2O_3 layer. Vertical and narrow nanorod arrays were formed using a plasma dry etch followed by an anisotropic wet etch in hydroxide solution for the purpose of high efficiency light emitting diodes (LEDs).[16] However, it is very rare to use NaOH or KOH solution as the GaN etchant in electronic device fabrication because it contains alkali metal ions that cause severe threshold voltage instability. On the other hand, TMAH does not suffer from the ion contamination and has been successfully utilized in GaN-based device fabrication.[3, 8–11, 17] When the GaN layer is exposed to the TMAH solution, Ga-polar (0001) plane is hardly etched (near-zero etch rate), whereas N-face plane is easily etched with the highest rate. This is because the dangling bond density (DBD) of N-face facet or other planes is larger than that of the Ga-polar (0001) facet.[16, 18, 19]

The GaN NW patterns along the $\langle 1\bar{1}00 \rangle$ and $\langle 11\bar{2}0 \rangle$ were then defined by e-beam lithography using a polymethyl methacrylate (PMMA) resist and followed by transformer-coupled plasma-reactive ion etching (TCP-RIE) using a BCl_3/Cl_2 gas mixture as shown in Fig. 1(a) and (b), respectively. The cross section of the initial etched GaN layer had a trapezoidal shape with sloped sidewall surfaces as shown in the left SEM image of Fig. 1(a) and the top SEM image of Fig. 1(b). After the dry etch, the NWs were etched in TMAH (5% solution at 90 °C) for 30 min. Due to its strong anisotropic etch characteristics, the etching proceeds only in the lateral directions but not in vertical direction along <0001>. Etching along $\langle 1\bar{1}00 \rangle$ direction makes the pattern narrower but still maintains the trapezoidal shape as shown in the right SEM image of Fig. 1(a). Etching along $\langle 11\bar{2}0 \rangle$ direction, on the other hand, changes the shape from a wide trapezoid with sloped angles to a narrow rectangle with very steep side-wall surfaces as shown in the bottom SEM image Fig. 1(b). Figure 2(b) and (c) shows the GaN crystal with wurtzite structure with $(1\bar{1}01)$ and $(11\bar{2}0)$ crystal planes, respectively. The $(1\bar{1}01)$ crystal plane has 58.4° slope (Figs 1(a) and 2(b)). On the other hand, the $(11\bar{2}0)$ crystal plane consists of both Ga and N atoms with 90° angle from the (0001) plane (Fig. 1(b) and (c)). The reason for the difference of the exposed plane along the wafer direction is due to the different etching rate between along $\langle 11\bar{2}0 \rangle$ and $\langle 1\bar{1}00 \rangle$ directions. When the rectangular GaN pattern is etched by dry etching, the (0001) planes at the inside and the outside of the pattern and sidewalls of $\{1\bar{1}01\}$ facets are exposed. The $\{1\bar{1}01\}$ facets have higher DBD (16.0 nm^{-2}) than the (0001) plane (11.4 nm^{-2}), which results in a higher TMAH etch rate,[18, 19] so

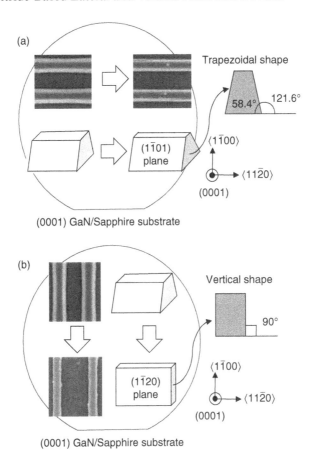

Figure 1. Schematic illustrations of the fin pattern directions and etching profile. SEM images of the fin test structure after TMAH etching along the (a) $\langle 11\bar{2}0 \rangle$ and (b) $\langle 1\bar{1}00 \rangle$.

the TMAH-etched GaN layer has a hexagonal shape due to its wurtzite crystal structure. As shown in the schematic of Fig. 1(a), the pattern along $\langle 1\bar{1}00 \rangle$ direction consists of the bottom hexagonal (0001) plane and the sidewall $\{1\bar{1}01\}$ facets with the trapezoidal shape. On the other hand, along the $\langle 11\bar{2}0 \rangle$ direction as shown in the schematic of Fig. 1(b), the sloped sidewall is first etched with relatively fast rate and when the vertical $(11\bar{2}0)$ crystal plane is exposed, the etching slows because the DBD of $(11\bar{2}0)$ plane (14.0 nm^{-2}) is smaller than that of $(1\bar{1}01)$ plane.[18, 19] Further etching widens the $(11\bar{2}0)$ plane along $\langle 1\bar{1}00 \rangle$ direction.

To form the vertical GaN NW structure, the plasma-enhanced chemical vapor deposition (PECVD) SiO$_2$ mask was first patterned by e-beam lithography with width and pitch of 300 nm. Dry etching with TCP-RIE was performed to obtain the initial trapezoidal NWs with height of 300 nm as shown in Fig. 3(a).

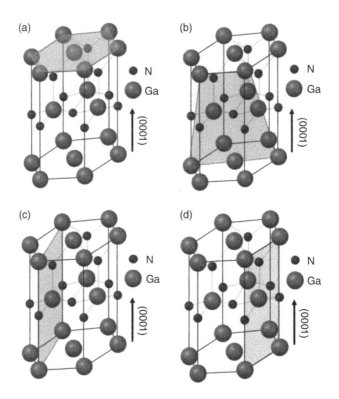

Figure 2. GaN crystal structure with Ga-face polarity showing various crystal planes: (a) (0000) *c*-plane; (b) (1$\bar{1}$01) *r*-plane; (c) (11$\bar{2}$0) *a*-plane; and (d) (1$\bar{1}$00) *m*-plane.

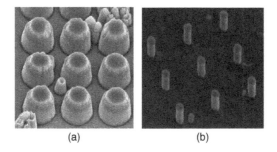

Figure 3. SEM images of GaN vertical NW structure after dry etching (a) and after subsequent TMAH wet etching for 11 min (b). The top SiO_2 mask layer was deposited by PECVD.

Anisotropic wet etching in TMAH solution (5% concentration at $80\,°C$) was then applied for 11 min to reduce the diameter of the nanowire. The TMAH solution etches the nanowire under the SiO_2 mask layer only in the lateral direction but not in the vertical direction, which results in NWs with average diameter of 100 nm and very steep sidewalls as shown in Fig. 3(b). The initial vertical GaN nanowire structure has two different planes: top (0001) c-plane and semipolar $\{1\bar{1}01\}$ planes formed at the sidewalls. The TMAH solution can selectively etch the $\{1\bar{1}01\}$ N-polar sidewalls while leaving the (0001) Ga-polar top surface unaffected. The $\{1\bar{1}01\}$ N-polar facets of the GaN vertical nanowires were removed gradually accompanied with the exposure of the $\{1\bar{1}01\}$ facets. This is because the DBD of $\{1\bar{1}01\}$ facets ($12.1\,nm^{-2}$) is smaller than that of $\{1\bar{1}01\}$ facets ($16.0\,nm^{-2}$).[18] The GaN nanowire with $\{1\bar{1}01\}$ facets has the nonpolar characteristics.[20] The corresponding atomic models matched to these crystal planes are shown in Fig. 2(b) and (d), respectively.

3. Ω-shaped-gate lateral AlGaN/GaN FETs

To fabricate the AlGaN/GaN Ω-shaped-gate nanowire FET, the GaN epitaxial layers were first grown on c-plane sapphire substrate by MOCVD. The structure consisted of 30 nm-thick low temperature-grown GaN as a nucleation layer, 2 µm-thick undoped GaN semi-insulating layer in growth sequence, followed by an 80 nm-thick undoped GaN layer and 30 nm-thick $Al_{0.3}Ga_{0.7}N$ layer. Hall effect measurements showed that the two-dimensional electron gas (2DEG) density and mobility were $9 \times 10^{12}\,cm^{-2}$ and $1800\,cm^2/V·s$, respectively.

Figure 4 shows the detailed fabrication process for the Ω-shaped GaN nanowire structure. The formation of nanowire structure was achieved by obtaining the $(11\bar{2}0)$ crystal plane with very steep sidewall surface as described in Fig. 1. The initial patterns along the $\langle 11\bar{2}0 \rangle$ direction were defined by e-beam lithography using PMMA, followed by TCP-RIE using a BCl_3/Cl_2 gas mixture, which resulted in trapezoidal shape with sloped sidewalls as shown in Fig. 4(a). Then, TMAH wet etching (5% solution at $90\,°C$) was employed for 30 min to change from a wide trapezoid to a narrow rectangular shape with very steep sidewalls (see Fig. 4(b)). The 20 nm-thick atomic layer-deposited (ALD) HfO_2 spacer mask for subsequent etching process was deposited on the steep sidewall surface as shown in Fig. 4(c). A second dry GaN etch step followed by TMAH lateral wet etching were used to etch the exposed GaN layer for 10 hours to form the Ω-shaped structure in Fig. 4(d) and (e). After removal of the sidewall HfO_2 spacer, a 20 nm-thick ALD Al_2O_3 gate insulator and 30 nm-thick TiN gate metal were deposited (see Fig. 4(f)). Finally, Fig. 4(g) shows the cross-sectional TEM image for the GaN Ω-shaped nanowire structure. The total height and width of this structure were 280 and 55 nm, respectively. The width of the neck between the structure and the buffer layer was around 13 nm.

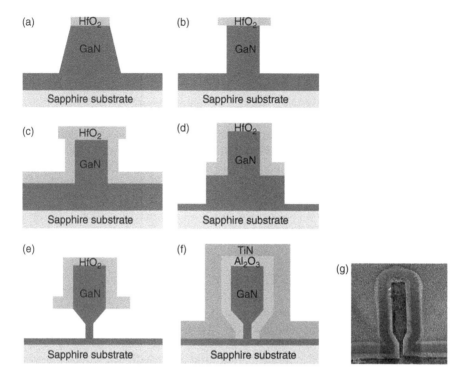

Figure 4. Process flow for GaN-based nanowire Ω-shaped FinFET: vertical (a) and narrow (b) fins structure after TMAH wet etching; (c) 20 nm-thick HfO₂ layer deposition for sidewall spacer formation; fin structure after the second dry (d) and subsequent TMAH wet (e) etching; (f) final Ω-shaped device structure; (g) cross-sectional TEM images of Ω-shaped gate structure.

The GaN nanowire FET fabricated on this Ω-shaped structure exhibits much better performance than the reference device, with extremely low off-state leakage current as low as 10^{-10} mA, the subthreshold swing (SS) value of 59 mV/dec, and high I_{ON}/I_{OFF} ratio ($\sim 10^{10}$) as shown in the I_D–V_G curve of Fig. 5. On the other hand, the reference device had much higher off-state leakage current (10^{-7} mA), an SS of 74 mV/dec, and a lower I_{ON}/I_{OFF} ratio $\sim 10^6$. The difference in the off-state performance between two devices is because active region of the Ω-shaped structure is fully depleted and isolated from the thick GaN buffer region, comparing to the reference tri-gate structure.

4. Gate-all-around vertical GaN FETs

The epitaxial structure of Si-doped GaN/undoped-GaN/Si-doped GaN (70/120/500 nm) stack was grown by MOCVD on sapphire (0001) substrate. The doping densities of Si-doped and undoped GaN layers were 1.0×10^{19} cm^{-3}

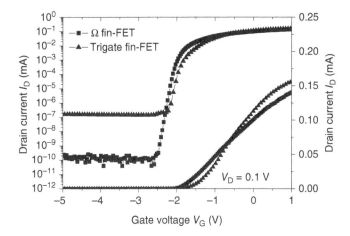

Figure 5. Logarithmic (left) and linear (right) plots of drain current I_D versus gate voltage V_G in Ω-shaped device compare with reference trigate fin-FET.

and 2.0×10^{16} cm^{-3}, respectively. First, a SiO$_2$ mask was patterned by e-beam lithography with both diameter and pitch of 300 nm. The vertical NW pillars were defined by TCP-RIE using a BCl$_3$/Cl$_2$ gas mixture, forming NWs with slanted sidewalls and height of 300 nm. The TMAH wet etch step (5% of solution at 80 °C) was used for 11 min to reduce the diameter of the nanowire as shown in Fig. 6(a), resulting in vertical sidewalls average NW diameter of 100 nm. The TMAH wet etching not only eliminates the plasma damage but also smoothens the etched GaN surface.[17] Then, the SiO$_2$ mask layer was removed prior to the deposition of a 10 nm-thick Al$_2$O$_3$ gate dielectric layer and a 20 nm-thick TiN gate metal by ALD. Then, a 20 nm-thick PECVD-deposited SiO$_2$ mask layer and a photoresist (PR) were deposited, as shown in Fig. 6(b), followed by a blanket etch to remove the PR/SiO$_2$/TiN/Al$_2$O$_3$ layers from the tops of the NWs as in Fig. 6(c). An SC-1 solution (NH$_4$OH:H$_2$O$_2$:H$_2$O = 1:1:5 at 70 °C) was used to define the TiN gate, with the SiO$_2$ mask layer protecting the TiN from being etched in the lateral direction as shown in Fig. 6(d). For the isolation of gate from the drain, an additional 100 nm-thick SiO$_2$ layer was deposited, followed by another blanket etch to enable the deposition of Ohmic Ti/Al contacts (Fig. 6(e)). After rapid thermal annealing at 500 °C for 30 s in N$_2$ ambient, Ni/Au metal layers were deposited for gate pad (Fig. 6(d)).

Figure 7 shows a schematic and the cross-sectional TEM image of GaN-based vertical NW-FET. The number of nanowires in parallel is 12; the height of the TiN gate metal and the separation between the drain end of the gate metal and the drain are 200 and 100 nm, respectively.

Figure 8(a) exhibits the transfer curve at $V_D = 1$ V of the fabricated GaN vertical NW-FET, which demonstrates a normally off operation with a threshold voltage of 0.6 V. The normalized maximum drain current and the maximum

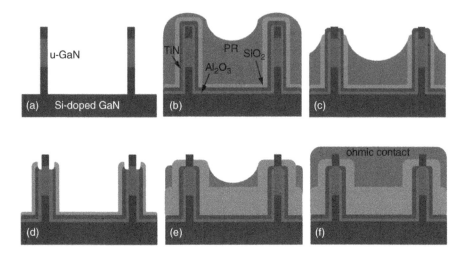

Figure 6. Illustration of the device fabrication processes: (a) GaN vertical nanowire; (b) Al_2O_3/TiN/SiO_2 deposition and PR spin coating; (c) blanket etching of PR and SiO_2; (d) TiN/Al_2O_3 wet etching; (e) SiO_2 deposition, PR spin coating, and PR/SiO_2 dry etching; and (f) Ohmic contact formation.

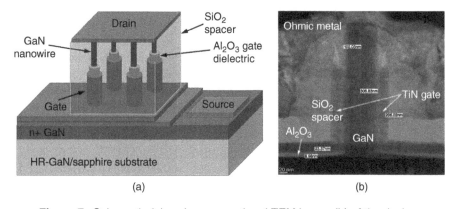

Figure 7. Schematic (a) and cross-sectional TEM image (b) of the device.

transconductance at $V_D = 1$ V, divided by the total gate width (NW diameter multiplied by the number of NWs), are 130 mA/mm and 70 mS/mm, respectively. Figure 8(b) shows the subthreshold characteristics (at $V_D = 0.1$ V) of device with SS of 153–163 mV/dec depending on the V_G sweep direction, good I_{ON}/I_{OFF} ratio as high as 10^9, and excellent off-state leakage current as low as $\sim 10^{-12}$ mA. We attribute the good off-state performance to GAA nanowire structure that eliminates the detrimental effects related to the buffer layer encountered in conventional planar-type FETs. Also, TMAH wet etching effectively smoothens and removes the plasma damage from the etched GaN surface.[3]

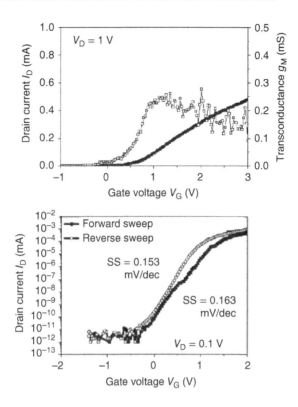

Figure 8. DC characteristics of the fabricated device: (a) transfer curve; (b) sub-threshold characteristics with a double sweep of gate voltage.

5. Conclusion

To summarize, we have proposed and demonstrated GaN-based lateral and vertical nanowire FETs for the first time using top-down approach by combining conventional e-beam lithography and dry etching technique with strong anisotropic TMAH wet etching. The AlGaN-/GaN-based omega-gate NW FETs have been fabricated using TMAH orientation-selective lateral wet etching of ALD-deposited HfO_2 sidewall spacer. The device exhibited excellent off-state performance: an SS of 62 mV/dec, close to the theoretical minimum, and an extremely low leakage current of 10^{-11} mA with a very high I_{ON}/I_{OFF} ratio of 10^{10}. We attribute these remarkable off-state characteristics to the Ω-shaped gate structure, which not only facilitates full depletion of the active region but also eliminates leakage paths between the active region and the thick GaN buffer layer. In addition, vertical GaN-based NW FETs with nanowire diameter of ~100 nm were demonstrated by using top-down fabrication approach involving orientation-selective TMAH wet etching. The device exhibits very low off-state leakage current and high I_{ON}/I_{OFF} ratio due to the GAA layout. The top-down approach provides a viable pathway toward GAA devices for III-nitride semiconductors, which are very promising candidates for steep-switching power device applications.

Acknowledgments

This work was supported by the BK21 Plus funded by the Ministry of Education (21A20131600011), the IT R&D program of MOTIE/KEIT (10048931), and the National Research Foundation of Korea (NRF) grant funded by the Korea Government (MSIP) (No. 2011-0016222, 2013R1A6A3A04057719).

References

1. T. P. Chow and R. Tyagi, "Wide bandgap compound semiconductors for superior high-voltage unipolar power devices," *IEEE Trans. Electron Devices* **41**, 1481–1483 (1994).
2. O. Aktas, Z. F. Fan, S. N. Fan, S. N. Mohammad, A. E. Botchkarev, and H. Morkoc, "High temperature characteristics of AlGaN/GaN modulation doped field-effect transistors," *Appl. Phys. Lett.* **69**, 3872–3874 (1996).
3. K.-W. Kim, S.-D. Jung, D.-S. Kim, *et al.*, "Effects of TMAH treatment on device performance of normally off Al_2O_3/GaN MOSFET,". *IEEE Electron Device Lett.* **32**, 1376–1378 (2011).
4. K. Y. Park, H. I. Cho, H. C. Choi, *et al.*, "Comparative study on AlGaN/GaN HFETs and MIS-HFETs," *J. Korean Phys. Soc.* **45**, S898–S901 (2004).
5. S. Arulkumaran, T. Egawa, H. Ishikawa, T. Jimbo, and Y. Sano, "Surface passivation effects on AlGaN/GaN high-electron-mobility transistors with SiO_2, Si_3N_4, and silicon oxynitride," *Appl. Phys. Lett.* **84**, 613–616 (2004).
6. R. Vetury, N. Q. Zhang, S. Keller, and U. K. Mishra, "The impact of surface states on the DC and RF characteristics of AlGaN/GaN HFETs," *IEEE Trans. Electron Devices* **48**, 560–566 (2001).
7. B. Lu, E. Matioli, and T. Palacios, "Tri-gate normally-off power MISFET," *IEEE Electron Device Lett.* **33**, 360–362 (2012).
8. K.-S. Im, Y.-W. Jo, J.-H. Lee, S. Cristoloveanu, and J.-H. Lee, "Heterojunction-free GaN nanochannel FinFETs with high performance," *IEEE Electron Device Lett.* **34**, 381–383 (2013).
9. K.-S. Im, C.-H. Won, Y.-W. Jo, *et al.*, "High-performance GaN-based nanochannel FinFETs with/without AlGaN/GaN heterostructure," *IEEE Trans. Electron Devices* **60**, 3012–3018 (2013).
10. K.-S. Im, R.-H. Kim, K.-W. Kim, *et al.*, "Normally off single nanoribbon Al_2O_3/GaN MISFET," *IEEE Electron Device Lett.* **34**, 27–29 (2013).

11. K.-S. Im, V. Sindhuri, Y.-W. Jo, *et al.*, "Fabrication of AlGaN/GaN Ω-shaped nanowire fin-shaped FETs by a top-down approach," *Appl. Phys. Express* **8**, 066501-1 (2015).
12. T. Bryllert, L.-E. Wernersson, L. E. Froberg, and L. Samuelson, "Vertical high-mobility wrap-gated InAs nanowire transistor," *IEEE Electron Device Lett.* **27**, 323 (2006).
13. K. Tomioka, M. Yoshimura, and T. Fukui, "A III–V nanowire channel on silicon for high-performance vertical transistors," *Nature* **488**, 189–192 (2012).
14. X. Zhao, J. Lin, C. Heidelberger, E. A. Fitzgerald, and J. A. del Alamo, "Vertical nanowire InGaAs MOSFETs fabricated by top-down approach," *Tech. Dig. IEDM* (2013), pp. 28.4.1–28.4.4.
15. D. Zhuang and J. H. Edgar, "Wet etching of GaN, AlN, and SiC: A review," *Mater. Sci. Eng. Rep.* **48**, 1–46 (2005).
16. S.-Y. Bae, D.-J. Kong, J.-Y. Lee, D.-J. Seo, and D.-S. Lee, "Size-controlled InGaN/GaN nanorod array fabrication and optical characterization," *Opt. Express* **21**, 16854–16862 (2013).
17. Y.-W. Jo, D.-H. Son, C.-H. Won, K.-S. Im, J. H. Seo, I. M. Kang, and J.-H. Lee, "AlGaN/GaN finFET with extremely broad transconductance by side-wall wet etch," *IEEE Electron Device Lett.* **36**, 1008–1010 (2015).
18. W. Chen, J. Lin, G. Hu, *et al.*, "GaN nanowire fabricated by selective wet-etching of GaN micro truncated-pyramid," *J. Crystal Growth* **426**, 168–172 (2015).
19. K. Hiramatsu, K. Nishiyama, A. Motogaito, H. Miyake, Y. Iyechika, and T. Maeda, "Recent progress in selective area growth and epitaxial lateral overgrowth of III-nitrides: Effects of reactor pressure in MOVPE growth," *Phys. Stat. Sol.* **176**, 535–543 (1999).
20. M. Kuroda, T. Ueda, and T. Tanaka, "Nonpolar AlGaN/GaN metal–insulator–semiconductor heterojunction field-effect transistors with a normally off operation," *IEEE Trans. Electron Devices* **57**, 368–372 (2010).

2.8

Scribing Graphene Circuits

N. Rodriguez, R. J. Ruiz, C. Marquez, and F. Gamiz
Department of Electronics, CITIC-UGR, University of Granada, 18071 Granada, Spain

1. Introduction

When it comes to functionality, the pace of evolution in the electronics industry is stunning: most of the devices surrounding us would have seemed science-fiction-like a decade ago. The laptop, that once was one of the most advanced electronic products for the general public, has given way to tablets, smartphones, and smartwatches ... that equal its processing power. The objective that science must pursue is to make sure that this insatiable (and growing) technological mass does not become a pathogen for our society: technology must be at the service of human beings instead of the other way around. Clearly, one of the aims of electronics in the coming years must be the creation of a better symbiosis with people and, in particular, more natural integration to support our everyday activities (including communication, health, protection, and work). In that respect, since its discovery, graphene[1] has certainly become one of the most shining materials for the scientific community and a popular candidate for the podium of the so-called Internet of Things (IoT). In particular, its mechanical, optical, and electrical properties make graphene a great fit for applications in flexible electronics. From this prospective, graphene features key properties for its natural integration in our lives: elasticity, deformability, electrical conductivity, and transparency. However, it is also true that actual applications have not arrived yet, since the fabrication of large patterned graphene structures is a complex task. As a result, considerable effort has been invested in the search for graphene-like substitutes that, while falling short of graphene's outstanding properties gathers, can provide some of its benefits with less technological effort, including easy lithography. One such material is known as the poorest class of graphene: reduced graphene oxide or simply rGO.

This chapter presents a panoramic view of the potential and physical characteristics of graphene oxide (GO) and rGO, from fabrication to physical and electrical properties. The versatility of this material is emphasized by the laser

Future Trends in Microelectronics: Journey into the Unknown, First Edition.
Edited by Serge Luryi, Jimmy Xu, and Alexander Zaslavsky.
© 2016 John Wiley & Sons, Inc. Published 2016 by John Wiley & Sons, Inc.

reduction method, which allows the patterning of conductive (electrically and thermally speaking) rGO surrounded by insulating GO.

2. Graphene oxide from graphite

The term "graphene oxide" (GO) is the name given to graphene samples treated with a strong oxidizer to alter the crystalline nature of its atomic structure. Nowadays, the chemical synthesis of GO from graphite has evolved into a viable technique of producing graphene sheets on a large scale.[2]

In general, graphene oxide is synthesized through the oxidation of graphite, based on Hummers' method,[3] involving oxidizing agents such as concentrated sulfuric acid (H_2SO_4), sodium nitrate ($NaNO_3$), or potassium permanganate ($KMnO_4$). The reaction takes about 2 hours to complete. During its critical steps (involving highly reactive compounds), the temperature should not rise above 25 °C to minimize the risk of explosion during the process. Compared with pristine graphite, GO is strongly oxygenated through functional hydroxyl (—OH) and epoxide (C—O—C) groups in the basal plane of the atoms with sp^3 hybridization, apart from carboxyl groups (COOH) located at the border of the sp^2 hybridization planes.[4] These functional groups modify the surface, leading to compounds based on functionalized graphene oxide. Since the electronic structure is strongly altered by the inclusion of functional groups, GO contains defects and irreversible dislocations. However, its intrinsic conductivity can be partially reestablished after its reduction[5] to values one order of magnitude below pristine graphene. This fact makes the method unattractive when compared to direct exfoliation, for example, for applications where a high crystalline quality is mandatory.

Once the graphite has been oxidized, it can be exfoliated through sonication or continuous stirring in water, leading to a stable homogeneous dispersion, containing primarily monolayer sheets of graphene oxide. The hydrophilic nature of graphene oxide permits easy penetration by water molecules, causing layer splitting. Furthermore, surface charges on the graphene oxide are highly negative when dispersed in water due to carboxylic acid ionization and phenolic-hydroxyl groups,[6] leading to an electrostatic repulsion that avoids the aggregation of the colloidal GO. Finally, GO can be deposited as a thin film over almost any surface and turned into low-quality graphene by chemical methods (using reduction agents such as hydrazine, N_2H_4) or through thermal treatments.[7, 8] One important drawback of the reduction of GO by chemical agents is that most of them are toxic and corrosive. This is a decisive factor for the search for more environmentally friendly methods for the reduction of GO. After the thermal reduction process, GO experiences a ~30% reduction in mass, due to the suppression of functional groups, while mitigating some of the structural defects affecting mechanical and electrical properties. The rGO sheets obtained by this method are electrically conductive, indicating that the resulting product is not

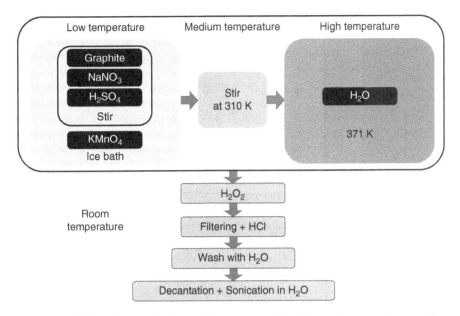

Figure 1. Schematic description of Hummers' method to produce graphene oxide.

pure GO, which is an insulator. This suggests an efficient reduction and a partial restoration of the electronic structure.[9]

As stated before, the common procedure for the oxidation of graphite is based on the method reported by Hummers and Offeman,[3] which is quite efficient and stable compared to other methods.[9] Once the procedure is complete, the resulting solution contains a large quantity of acid, as well as metallic ions and inorganic impurities, accumulated when the sodium- and potassium-based compounds are incorporated. The resulting electrolytes could neutralize the charges in the films, destabilizing the dispersion. As example, in our experiments, we obtained 1 L of solution, where 115 mL correspond to sulfuric acid added at the first stage of the process. The solution is diluted several times to decrease the acid content (Fig. 1).

In the next step, the compound is dried at room temperature and ground. The resulting GO powder is cleaned from metallic ions by a 1 mol HCl wash. The final GO powder can be diluted in deionized water, leading a GO colloid (in our case with a concentration of 4 mg/mL).

3. GO exfoliation

The exfoliation process consists of breaking the interlayer bonds existing in between the layers of graphene oxide, whether by sonication or mechanical stirring, or by chemical compounds with surfactant properties. As mentioned previously, graphene oxide structure is characterized by a large amount of oxygen-containing groups, so GO films can disperse in water. The hydroxyl phenolic groups (—OH) together with the carboxylic groups (—COOH) are responsible for the negative charge of the GO sheets in the aqueous suspension.[10] It has been demonstrated that in very diluted aqueous suspensions, the separation between GO sheets is large and therefore the interaction due to van der Waals forces is weak enough to allow the exfoliation in monolayers.[11] This fact, combined with the electrostatic repulsion mechanism due to the negative charges on the surface of the sheets, leads to the formation of stable well-dispersed GO colloids without the need of polymeric stabilizers or surfactants (which would introduce impurities).

In our case, to obtain a homogeneous suspension of GO with a concentration of 4 mg/mL, 6 g of graphite oxide powder are dispersed in 1.5 L of deionized water, and subsequently sonicated for a long period of time (VMR Ultrasonic Cleaner USC 500 TH) at low temperature. In this way, graphite oxide is exfoliated leading to a stable solution of individual GO layers, without sedimentation or aggregation.

4. Selective reduction of graphene oxide

Both the graphite oxide and the graphene oxide are electrical insulators due to the disruption of the crystallographic network of carbon atoms during the oxidation process. The electrical conductivity can be easily restored by removing the functional groups, thereby also partially restoring the original sp^2 electronic structure. During the reduction process, there is a significant decrease of mass as well as a removal of carbon atoms in CO and CO_2 form; therefore, the reduction process leaves vacancies and topological defects in the films.

The search for ecologically cleaner methods of producing rGO points to laser processing as one of the most efficient and safest procedures: the photo-thermal energy of the laser radiation causes intense vibration in the GO structure generating high localized temperatures (>2000 K). This process has the additional advantage of preserving the insulating nature of the GO in the regions not exposed to the laser. The high temperature induced by the laser beam breaks the C—O and C═O bounds of the carbon sheets; the carbon atoms are reorganized forming graphitic structures, while the fast expansion of the sheets due to the deoxygenation reactions and the pressure generated by the emission of gases prevents the stacking of GO layers.

The use of a controlled laser, illustrated schematically in Fig. 2, is extremely attractive because it permits one to scribe precisely (conductive) rGO patterns surrounded by (insulating) GO regions. Compared to other lithographic techniques, the laser-scribing technology is simple, low-cost, scalable, fast, and fully compatible with flexible substrates.

The samples developed for the studies presented in this chapter were reduced by a 788 nm/5 mW laser mounted in a lightscribe® compatible DVD unit or by

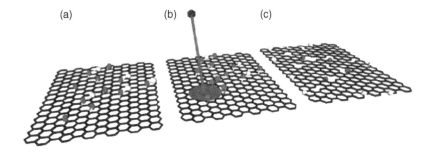

Figure 2. Illustration of (a) GO as described by the Lerf–Klinowski model;[12] (b) GO after laser irradiation leading to the final rGO substrate (c).

a 405 nm/300 mW (power-adjustable) laser controlled by a computer numerical control unit. Any part of the GO surface that is irradiated, even by the infrared laser, turns effectively into rGO. The quality of the reduction can be controlled by the number of passes of the beam over the surface (or by the power intensity). The spatial resolution of the patterns with this system was 20 μm.

The scribing process starts by pouring 8 mL of 4 mg/mL GO solution on a flexible polyethylene terephthalate substrate (PET-3M). After a couple of days, the solvent (water) is completely evaporated, leaving behind an ultrathin layer of GO (characterized by a reddish brown tone). A better uniformity can be achieved if the drying process is carried out in a 3D shaker. After drying, the flexible PET film is laser-printed with the rGO features, which appear black as a consequence of the reduction. The expansion of the graphitic layers can be easily observed by optical microscope.

Since the patterns are directly scribed on the flexible substrate, there is no need for a postprocessing. Despite the simplicity of the method, the dimensions of the pattern are precisely controlled; the number of beam passes (or laser power) also provides some control over the electrical properties of the rGO. Therefore, this technique makes it possible to easily fabricate planar rGO structures surrounded by highly isolating graphene oxide.

5.   Raman spectroscopy

Raman spectroscopy is widely used to characterize graphene and related materials. A large amount of information about the crystal structure, as well as disorder, defects, and thickness can be obtained by this powerful and noninvasive technique.[13–15] Raman spectroscopy uses a monochromatic laser to interact with the molecular vibrational modes in a sample. In the process, the sample absorbs photons and, subsequently, reemits them after losing a sample-dependent fraction of the excitation energy. Most of the dispersed light has the same frequency as the (intense) incident beam. This phenomenon is known as Rayleigh elastic scattering. The second dispersion process is inelastic and is known as Raman scattering: the dispersed light is characterized by frequency shifts down (Stokes) or up (anti-Stokes) compared to the incident beam.[15]

As is true for all carbon-based materials, the most significant resonant modes are located in the spectral region from 1000 to 3000 cm^{-1}. In the case of graphene, the variation observed in the Raman signal for samples with different thickness reflects changes in the electronic band structure, making it possible to distinguish monolayer, bilayer, or few-layer graphene samples.[16]

The most important modes in the Raman spectrum of graphene are the so-called G and 2D modes: the G mode is located around 1580 cm^{-1}, whereas the 2D peak is situated at around 2700 cm^{-1}. Two additional peaks may be also observed: the D peak at approximately 1350 cm^{-1} and the G* peak at 2450 cm^{-1}. The D peak of the spectrum is not visible in pristine graphene because its activation implies that a carrier must be excited and inelastically scattered by a phonon and subsequently a second inelastic scattering by a defect or a zone boundary is required for recombination.

While graphite or graphene yields a very sharp G peak (characteristic of the sp^2 hybridization), in GO this peak is wider and shifted toward longer wavelengths (up to ~1590 cm^{-1}). The D mode is significantly stronger (sometimes even exceeding the G mode) due to the distortion induced in the sp^2 structure by the oxidative synthesis together with the hydroxyl and epoxy groups.[5, 8, 17]

Generally, HRTEM studies have shown that GO preserves graphene-like regions (about 1 nm diameter) and distorted regions (several nanometers) beneath defects (vacancies or additional atoms) arising during synthesis.[13] The ratio between the intensity of the D and G peaks, I_D/I_G, can be used to evaluate the distance between defects in graphene. For GO this ratio is about 1 (532 nm laser excitation) and increases with increasing mean distance between two defects from 1 to about 3 nm (stage 2) followed by a decrease (stage 1, larger than 3 nm).[14-16]

The reduction process of GO can also manifest itself in Raman spectra by the changes in relative intensity of the D and G peaks.[8] Figure 3 shows the Raman spectra of GO and reduced GO. The D peak, located at 1352 cm^{-1} in GO and 1350 cm^{-1} in rGO, originates from a defect-induced breathing mode of sp^2 rings. It is common to all sp^2 carbon lattices and arises from the stretching of C—C bond. The G peak is located around 1600 cm^{-1} for GO; the 2D peak is almost nonexistent. After the reduction process, as shown in Fig. 3(b), the D peak is attenuated and a strong 2D peak emerges, as a result of the partial restoration of the crystallographic structure and the reduction of the number of defects.

6. Electrical properties of graphene oxide and reduced graphene oxide

The linear energy dispersion of carriers in pristine graphene makes the material very promising for solid-state electronic devices, especially if the problematic

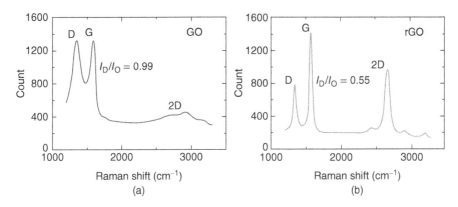

Figure 3. Raman spectra of GO (a) and reduced GO (b). Both Raman spectra were recorded with a low-power (4.7 mW) 532.04 nm laser line.

issue of low I_{ON}/I_{OFF} current ratio due to nonexistent graphene bandgap can be overcome. Extremely high carrier mobility has been predicted by theorists and demonstrated (with some limitations) by experimentalists.[18] The vast majority of the experiments to measure conductivity have been carried out by depositing four or two metallic contacts on 2D graphene layers. These experiments have demonstrated the impact of the substrate (charge transfer) and ambient conditions (absorption of ambient molecules) on the electrical characteristics of graphene.[19, 20] They have also shown that the choice of metal contact plays a fundamental role when evaluating the conductivity in the two-point geometry.[21, 22] In principle, an Ohmic contact should be obtained due to the lack of bandgap in monolayer graphene, but the small density of states (DOS) near the Dirac point could suppress the current injection from the metal to graphene.

When dealing with macroscopic samples (such as chemical vapor deposition (CVD)-graphene or rGO) that may exceed 1 cm^2, the use of deposited contacts becomes laborious (and destructive for the sample). In those cases, four-point contact measurements can be performed as a quick procedure to determine fundamental electrical properties of the sheet avoiding the contact resistance issue. This technique is a well-known characterization method to evaluate the conductivity of bare materials. Even though the experimental procedure is more complex, four-point measurements are preferable to two-point measurements in graphene (and graphene-like) samples, since the nature of the contact may mask the intrinsic properties of material.

Figure 4(a) shows an example of the rGO structure used to test the electrical properties. In this case, the sample corresponds to a matrix of 1 × 1 cm rGO squares reduced at different laser power (from 50 to 130 mW) at a rate of 1 mm/s. The reduction becomes very effective above 70 mW power (for the selected scanning rate). Higher laser power does not benefit significantly the electrical properties. Figure 4(b) shows the sheet resistance as a function of the initial

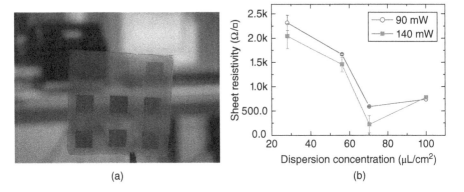

(a)   (b)

Figure 4. Samples of 1 × 1 cm rGO squares reduced at increasing laser intensity from 50 up to 130 mW (a). Sheet resistivity of rGO on PET substrate extracted from four-point contact measurements versus the initial dispersion concentration at two different laser intensities (b).

concentration of the dispersion on the surface, that is, before the water evaporation (the volume concentration is 4 mg/mL). The resistivity always falls as the concentration increases and then saturates (around 70 µL/cm^2 in this case).

For very low concentrations, the inhomogeneity of the graphitic structures hinders the conduction. As the concentration increases, the connection between the graphitic domains improves, resulting in a prominent decrease in resistivity. Further increase in the concentration does not produce any significant decrease in resistance, since the photo-thermal treatment only reduces a superficial layer of the deposited GO film (for a given laser power).

The lowest value of the sheet resistance shown in the right panel of Fig. 4 is 215 Ω/□, lower than the ~450 Ω/□ values that characterize the more refined polycrystalline CVD-graphene available from commercial vendors. This good conductivity, accompanied by the easy selective lithography and cheap production process, makes rGO a promising material for flexible electronics and for the wearable electronics territory.

7. Future perspectives

rGO has been nicknamed the poorest flavor of graphene. Nevertheless, poor does not mean lacking in scientific and industrial interest. This chapter tries to envisage the potential of rGO for electronic applications: a material that, in spite of not achieving the exceptional properties of pristine graphene, retains some of its advantages and can be synthesized and processed in a much more facile, flexible, and cheap way.

The laser-assisted selective photo-thermal reduction of graphene oxide into rGO makes it possible to generate accurate conductive patterns isolated by the graphene oxide on virtually any stable insulating substrate. Flexible electronics is the natural beneficiary of this lithographic technique, allowing the fabrication of large circuits without compromising costs.

However, the above-discussed application constitutes just one of the possible uses of laser-rGO. The range of applications can be further extended to innovations in biology or material science, since interesting sensing capabilities, as well as thermal and electrochemical properties of rGO have been already demonstrated.[23-25]

Acknowledgments

The authors would like to thank BBVA Foundation for the *Ayudas para Investigadores y Creadores Culturales* program and CEMIX-UGR for supporting this work.

References

1. K. S. Novoselov, A. K. Geim, S. V. Morozov, *et al.*, "Electric field effect in atomically thin carbon films," *Science* **306**, 666–669 (2004).
2. S. Stankovich, D. A. Dikin, R. D. Piner, *et al.*, "Synthesis of graphene-based nanosheets via chemical reduction of exfoliated graphite oxide," *Carbon* **45**, 1558–1565 (2007).
3. W. S. Hummers, Jr. and R. E. Offeman, "Preparation of graphitic oxide," *J. Am. Chem. Soc.* **80**, 1339 (1958).
4. D. R. Dreyer, S. Park, C. W. Bielawski, and R. S. Ruoff, "The chemistry of graphene oxide," *Chem. Soc. Rev.* **39**, 228–240 (2010).
5. K. Erickson, R. Erni, Z. Lee, N. Alem, W. Gannett, and A. Zettl, "Determination of the local chemical structure of graphene oxide and reduced graphene oxide," *Adv. Mater.* **22**, 4467–4472 (2010).
6. D. Li, M. B. Müller, S. Gilje, R. B. Kaner, and G. G. Wallace, "Processable aqueous dispersions of graphene nanosheets," *Nature Nanotechnol.* **3**, 101–105 (2008).
7. R. J. Seresht, M. Jahanshahi, A. Rashidi, and A. A. Ghoreyshi, "Synthesis and characterization of graphene nanosheets with high surface area and nanoporous structure," *Appl. Surf. Sci.* **276**, 672–681 (2013).
8. D. Yang, A. Velamakanni, G. Bozoklu, *et al.*, "Chemical analysis of graphene oxide films after heat and chemical treatments by X-ray photoelectron and micro-Raman spectroscopy," *Carbon* **47**, 145–152 (2009).
9. S. Pei and H.-M. Cheng, "The reduction of graphene oxide," *Carbon* **50**, 3210–3228 (2012).
10. H. C. Schniepp, J. L. Li, M. J. McAllister, *et al.*, "Functionalized single graphene sheets derived from splitting graphite oxide," *J. Phys. Chem. B* **110**, 8535–8539 (2006).

11. N. I. Kovtyukhova, P. J. Ollivier, B. R. Martin, *et al.*, "Layer-by-layer assembly of ultrathin composite films from micron-sized graphite oxide sheets and polycations," *Chem. Mater.* **11**, 771–778 (1999).

12. A. Lerf, H. He, M. Forster, and J. Klinowski, "Structure of graphite oxide revisited," *J. Phys. Chem. B* **102**, 4477–4482 (1998).

13. A. C. Ferrari and D. M. Basko, "Raman spectroscopy as a versatile tool for studying the properties of graphene," *Nature Nanotechnol.* **8**, 235–246 (2013).

14. I. Childres, L. A. Jauregui, W. Park, H. Cao, and Y. P. Chen, "Raman spectroscopy of graphene and related materials," chapter 19 in: J. I. Jang, ed., *New Developments in Photon and Materials Research*, Hauppauge, NY: Nova Science Publishers (2013).

15. D. J. Gardiner, *Practical Raman Spectroscopy*, New York: Springer-Verlag (1989).

16. L. M. Malard, M. A. Pimenta, G. Dresselhaus, and M. S. Dresselhaus "Raman spectroscopy in graphene," *Phys. Rep.* **473**, 51–87 (2009).

17. S. Eigler, C. Dotzer, and A. Hirsh, "Visualization of defect densities in reduced graphene oxide," *Carbon* **50**, 3666–3673 (2012).

18. K. I. Bolotin, K. J. Sikes, Z. Jiang, *et al.*, "Ultrahigh electron mobility in suspended graphene," *Solid State Commun.* **146**, 351–355 (2008).

19. D. M. Sedlovets and A. N. Redkin, "The influence of the ambient conditions on the electrical resistance of graphene-like films," *Nanosystems: Phys. Chem. Math.* **5**, 130–133 (2014).

20. C. Gómez-Navarro, R. Thomas Weitz, A. M. Bittner, M. Scolari, A. Mews, M. Burghard, and K. Kern, "Electronic transport properties of individual chemical reduced graphene oxide sheets," *Nano Lett.* **7**, 3499–3503 (2007).

21. S. M. Song and B. J. Cho, "Contact resistance in graphene channel transistor," *Carbon Lett.* **14**, 162–170 (2013).

22. K. Nagashio, T. Nishimura, K. Kita, and A. Toriumi, "Contact resistivity and current flow at metal/graphene contact," *Appl. Phys. Lett.* **97**, 143514 (2010).

23. X. Mu, X. Wu, T. Zhang, D. B. Go, and T. Luo, "Thermal transport in graphene oxide – From ballistic extreme to amorphous limit," *Nature Sci. Rep.* **4**, article no. 3909 (2014).

24. D. Zhang, J. Tong, and B. Xia, "Humidity-sensing properties of chemical reduced graphene oxide/polymer nanocomposite film sensor based on layer-by-layer nano self-assembly," *Sens. Actuators B* **197**, 66–72 (2014).

25. S. Basu and P. Bhattacharyya, "Recent developments on graphene and graphene oxide based solid state gas sensors," *Sens. Actuators B* **173**, 1–21 (2012).

2.9

Structure and Electron Transport in Irradiated Monolayer Graphene

I. Shlimak, A.V. Butenko, E. Zion, V. Richter, Yu. Kaganovskii, L. Wolfson, A. Sharoni, A. Haran, D. Naveh, E. Kogan, and M. Kaveh
Department of Physics and Faculty of Engineering, Jack and Pearl Institute of Advanced Technology, Bar Ilan University, 52900 Ramat Gan, Israel

1. Introduction

Disordered graphene has attracted the attention of many researchers.[1–3] Mainly, this is due to the possibility of obtaining a high-resistance state of graphene films, which is of interest for application in electronics. Experimentally, the disorder is achieved in various ways: by oxidation,[4] hydrogenation,[5] chemical doping,[6] and irradiation by different ions with different energies.[7–11] In this chapter, we discuss ion bombardment as a way of gradually inducing disorder in monolayer graphene.

2. Samples

The initial specimens were purchased from the Graphenea company, who grew monolayer graphene by chemical vapor deposition (CVD) on a copper catalyst and used a wet transfer process to move it to a 300 nm SiO_2/Si substrate. The typical specimen size was 5×5 mm. Graphene films of such a large area are not monocrystalline but rather polycrystalline with an average microcrystal size of up to 10 μm.

On one of these specimens, gold electrical contacts were deposited directly on the graphene surface through a metallic mask (reference sample 0). On the other 5×5 mm specimens, many small devices of 200×200 μm area were prepared by e-beam lithography (EBL), with Ti/Pd 5:45 nm metal electrical contacts. All small samples on the surface of the 5×5 mm specimen were divided into six groups: one group was not irradiated (sample 1), whereas the other groups were irradiated with 35 keV C^+ ions at five doses: 5×10^{13}, 1×10^{14}, 2×10^{14}, 4×10^{14}, and 1×10^{15} cm^{-2} (samples 2–6, respectively).

Future Trends in Microelectronics: Journey into the Unknown, First Edition.
Edited by Serge Luryi, Jimmy Xu, and Alexander Zaslavsky.

3. Raman scattering (RS) spectra

In the Raman spectra measurements, the samples were excited with a $\lambda = 532$ nm laser beam at low power (<2 mW) to avoid heating and film damage. Figure 1(a) compares the Raman spectra of the unirradiated reference sample 0 and sample 1, which was also not irradiated but had gone through EBL. Reference sample 0 has a typical Raman spectrum for monolayer graphene film,[7] consisting of three main lines: a weak D-line at 1350 cm^{-1} related to the intervalley double resonant process in graphene with defects (edges, vacancies, etc.); a 2D-line at 2700 cm^{-1} related to an intervalley two-phonon mode, characteristic of the

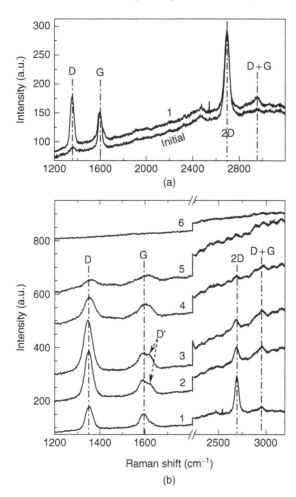

Figure 1. Raman spectra of graphene monolayers. (a) Reference sample 0 versus sample 1 (nonirradiated, but after EBL); (b) samples 1–6 versus irradiation dose in cm^{-2}: 0 (sample 1), 5×10^{13} (2), 10^{14} (3), 2×10^{14} (4), 4×10^{14} (5), and 10^{15} (6). The lines are shifted for clarity.

perfect crystalline honeycomb structure; and a graphite-like G-line at $1600\,\text{cm}^{-1}$ that is common for different carbon-based films. Usually, the intensity ratio I_D/I_G between the D- and G-lines is used as a measure of disorder in graphene films. This I_D/I_G ratio was ~0.15 in the reference sample 0, indicating that the initial large specimen was of reasonably good quality, albeit not perfect.

Analogous measurements on sample 1 showed that I_D/I_G increased from ~0.15 to ~1.8 (see Fig. 1(a)). It means that EBL introduces disorder even without ion irradiation. The damage in our case could be due to the lift-off process during EBL near the edges of the polycrystalline film.

Ion bombardment leads to further increase in disorder. Figure 1(b) shows the transformation of the Raman spectra in samples 2–6 with increasing irradiation dose Φ. As the dose increases to $\Phi = 1 \times 10^{14}\,\text{cm}^{-2}$, the amplitude of the defect-associated D-line increases, while the crystalline 2D-line quickly disappears. Furthermore, new defect lines appear: the D' line at $1620\,\text{cm}^{-1}$ and (D+G)-line at $2950\,\text{cm}^{-1}$. As for the G-line, it remains approximately constant, broadening slightly because of the appearance of the nearby D'-line. At $\Phi = 1 \times 10^{14}\,\text{cm}^{-2}$, the increasing intensity of the D-line enhances the I_D/I_G ratio to ~3.2. Further increase in the dose leads to decrease and broadening of the D-line, so I_D/I_G decreases. At higher doses, the G-line also broadens and becomes weaker; eventually, at the maximum $1 \times 10^{15}\,\text{cm}^{-2}$ dose all Raman scattering structure disappears (sample 6). We are not aware of any reports of all RS lines disappearing in disordered graphene. Usually, dependence of I_D/I_G on irradiation-induced disorder displays two different behaviors. In the "low defect density" regime, I_D/I_G increases with irradiation dose Φ, whereas in the "high defect density" regime I_D/I_G falls with as Φ increases because the amorphization of the graphene structure attenuates all Raman peaks. However, the complete disappearance of all RS lines cannot be explained by amorphization because G-line is observed even in amorphous carbon films.[12] We assume that at the maximum level of disorder achieved in our experiments, the graphene film ceases to be continuous and splits into separate spots of small size (quantum dots). The small size of the quantum dots makes it impossible to form phonons responsible for the structural line in the RS.

The degree of disorder can be characterized by the concentration of defects N_D or by the mean distance between defects $L_D = N_D^{-1/2}$. Our irradiation conditions (35 keV C^+ ions) were the same as used by Buchowicz et al.,[8] chosen such that the end-of-range damage would be away from the graphene film. In this case, concentration of induced defects N_D is much less than the dose Φ: $N_D = k\Phi$, where $k \ll 1$. Simulations showed that for our case, $k = 6$–8%.[13]

We will use the empirical model that describes the dependence of I_D/I_G versus L_D in both low and high defect density regimes developed by Lucchese et al.[14] In this model, a single defect causes modification of two different length scales, r_A and r_S ($r_A > r_S$). In the immediate vicinity of a defect, the area $S = \pi r_S^2$ is structurally disordered, but at $r_S < r < r_A$, the lattice structure is saved, though the proximity to a defect leads to breaking of selection rules and contribution to the D-line. The activated area responsible for the D-line is $A = \pi(r_A^2 - r_S^2)$. In the

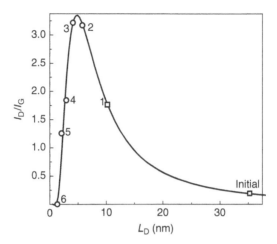

Figure 2. Ratio I_D/I_G versus mean distance between defects L_D. Solid line represents Eq. (1) with $C_A = 5.3$, $C_S = 0$, $r_S = 1.8$ nm, and $r_A = 5.5$ nm. Numbers near the points indicate sample number.

low defect density regime, the D-line intensity linearly increases with N_D, which means that $I_D/I_G \sim L_D^{-2}$. The maximum value of I_D is achieved when L_D falls down to r_A. Further decrease of L_D leads to overlap between A and S areas, at which point the I_D/I_G ratio begins to fall. The final equation for the dependence I_D/I_G on L_D has the form[15]

$$\frac{I_D}{I_G} = C_A e^{-\pi r_S^2 / L_D^2} \left[1 - e^{-\pi \left(r_A^2 - r_S^2 \right) / L_D^2} \right] + C_S \left[1 - e^{-\pi r_S^2 / L_D^2} \right]. \tag{1}$$

Figure 2 shows the result of fitting the theoretical curve (1) with experimental data. First, the experimental points for irradiated samples 2–6 were plotted. We choose $k = 6\%$; L_D for these samples was defined as $L_D = (0.06\Phi)^{-1/2}$. Then, the curve (1) was calculated to fit the experimental points. The best fit shown in Fig. 2 was obtained for $C_A = 5.3$, $C_S = 0$, $r_S = 1.8$ nm, and $r_A = 5.5$ nm. Finally, the values of I_D/I_G for two nonirradiated samples – the reference sample 0 and the post-EBL sample 1 – were placed on the curve for the purposes of L_D (and hence N_D) estimation. We obtain $N_D \sim 8 \times 10^{10}$ cm^{-2} in the reference sample 0 and $N_D \sim 10^{12}$ cm^{-2} in sample 1.

4. Sample resistance

Room-temperature measurements of the current–voltage (I–V) characteristics for all samples showed that for highly irradiated samples 5 and 6, the I–V

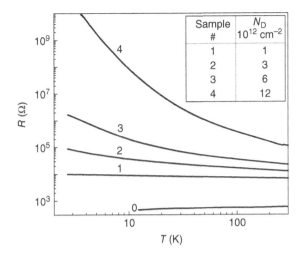

Figure 3. Resistivity of disordered monolayer graphene samples versus temperature T, with sample numbers indicated. Inset shows the density of structural defects N_D in samples.[15]

is strongly nonlinear even at very small currents. For this reason, we have focused our studies of the temperature dependence of sample resistance $R(T)$ to samples 0–4. (We note that the resistance was equal to resistivity due to the square shape of samples.) The $R(T)$ was measured by two-probe method in a helium cryostat down to $T = 1.8$ K in magnetic fields B up to 4 T. Figure 3 shows $R(T)$ curves on a log–log scale for samples 0–4 in zero B field. Sample 0 shows typical metallic behavior, when R slightly decreases with decrease of T. For sample 1, R slightly increases with decreasing T, characteristic for a "dirty" metal. For other samples, R changes with T exponentially, which is characteristic for strongly localized carriers.

• *Weak localization (WL)*

Plot of the temperature dependence of conductivity on the scale σ versus $\ln(T)$ in Fig. 4 shows logarithmic behavior of σ at low T, characteristic of weak localization (WL).[16] The WL regime in monolayer graphene is interesting due to the fact that charge carriers are chiral Dirac fermions that reside in two nonequivalent valleys at the K and K' points of the Brillouin zone. Due to chirality, Dirac fermions acquire a phase of π upon intravalley scattering, which leads to destructive interference with its time-reversed counterpart and weak antilocalization (WAL). Intervalley scattering leads to restoration of WL because fermions in K and K' valleys have opposite chiralities.

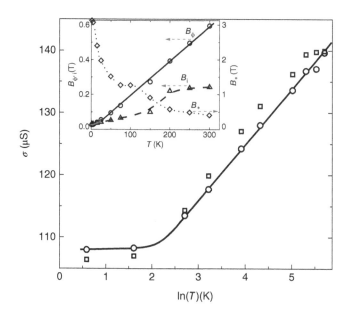

Figure 4. Conductivity of sample 1 versus $\ln(T)$. Circles and squares present experimental data, and theoretical calculation with parameters determined from fitting the magnetoconductance in Fig. 5. These parameters are shown in inset: B_ϕ and B_i (solid and dashed lines, left axis) and B. (right axis).

Quantum corrections to the conductivity of graphene have been studied theoretically.[17, 18] It was predicted that WAL corrections will dominate at relatively high T, while WL corrections will dominate at low T. For the magnetoconductance (MC) $\Delta\sigma(B, T)$, the theory predicts[19]

$$\Delta\sigma(B, T) = \left(\frac{e^2}{\pi h}\right)\left[F\left(\frac{B}{B_\phi}\right) - F\left(\frac{B}{(B_\phi + 2B_i)}\right) - 2F\left(\frac{B}{(B_\phi + 2B_*)}\right)\right], \quad (2)$$

where $F(z) = \ln(z) + \Psi(1/z + 1/2)$ and $B_{\phi i*} = (\hbar c/4De)\tau_{\phi i*}^{-1}$, where Ψ is the digamma function, τ_ϕ is the coherence time, τ_i^{-1} is the intervalley scattering rate, and τ_*^{-1} is the combined scattering rate of intravalley and intervalley scattering and of trigonal warping.

Fitting Eq. (2) to experimental data for MC of sample 1 at different temperatures is illustrated in Fig. 5. In the process of fitting, we were able to extract all three parameters $B_{\phi i*}$ entering the equation. These parameters are shown in inset in Fig. 4. One can see that, in agreement with theoretical prediction, $B_\phi \sim 1/\tau_\phi \sim T$. The saturation of τ_ϕ at low temperatures is well known in classical 2D systems and may be connected with existence of dephasing centers (e.g., magnetic impurities).[20, 21]

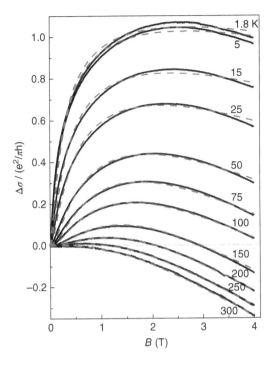

Figure 5. Magnetoconductance of sample 1 versus perpendicular magnetic field B at various T: solid lines – experiment, dashed lines – Eq. (2) with fitting parameters shown in the inset of Fig. 4.

• *Strong localization*

Resistance of samples 2–4 exhibited pronounced insulating behavior, as shown in Fig. 3. Plotting the data on the Arrhenius scale, as $\ln(R)$ versus $1/T$ showed that the energy of activation continuously decreases with decreasing T, which is characteristic for the variable-range-hopping (VRH) conductivity.[22] There are two kinds of VRH depending on the structure of the density of states (DOS) $g(E)$ in the vicinity of the Fermi level E_F. When $g(E) = g(E_F)$ is constant, $R(T)$ is described by the Mott Law, also known as the "$T^{-1/3}$"-law in the case of 2D conductivity:

$$R(T) = R_0 \exp\left(T_M/T\right)^{1/3}, \quad T_M = C_M\left[g\left(E_F\right)a^2\right]^{-1}. \tag{3}$$

Here $C_M = 13.8$ is the numerical coefficient[22] and a is the radius of localization.

The Coulomb interaction between localized carriers leads to appearance of a soft Coulomb gap in the vicinity of E_F. In the 2D case, this gap has a linear form:

$$g(E) \sim |E - E_F|\left(\frac{e^2}{\kappa}\right)^{-2}, \tag{4}$$

where κ is the dielectric constant of the material, leading to the Efros–Shklovskii (ES) VRH or "$T^{-1/2}$"-law:

$$R(T) = R_0 \exp\left(\frac{T_{ES}}{T}\right)^{1/2}, \quad T_{ES} = C_{ES}\left(\frac{e^2}{\kappa a}\right), \tag{5}$$

where $C_{ES} = 2.8$.[22]

Coulomb interaction can alter the DOS only near E_F. Far from E_F, the DOS is restored to its initial value, which is approximately equal to $g(E_F)$, as illustrated in the inset in Fig. 6. Denoting the half-width of the Coulomb gap as Δ one can conclude, therefore, that the ES dependence should be observed when $kT < \Delta$, whereas in the opposite $kT > \Delta$ regime one should observe the Mott Law.

In Fig. 6, $\log(R)$ is plotted versus $T^{-1/3}$. At high T, $R(T)$ follows the $T^{-1/3}$ dependence for all samples, but deviations to a stronger power-law dependence are observed with decreasing T for samples 3 and 4, as the samples approach the ES-VRH behavior, which should dominate at lowest temperatures. The slopes of the straight lines on $T^{-1/3}$ and $T^{-1/2}$ scale give the values of T_M and T_{ES}. We get correspondingly 300 and 50 K for sample 3 and 6000 and 500 K for sample 4.

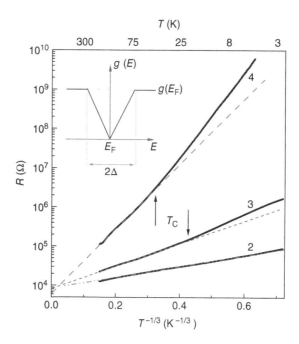

Figure 6. Plot of $\log(R)$ versus $T^{-1/3}$ for samples 2–4 (sample numbers next to the curves). Inset shows a schematic illustration of the 2D Coulomb gap in the DOS near E_F. Arrows indicate the crossover temperature T_C between Mott- and ES-VRH.

In VRH, only localized states within an optimal energy band $\Delta E(T)$ near the Fermi level are involved in the hopping process. This band becomes continuously narrower with decreasing T. The crossover temperature T_C can be determined from the equality $\Delta E(T_C) = \Delta$, which gives[23]

$$T_C = \left(\frac{C_M^2}{C_{ES}^3} \right) \left(\frac{T_{ES}^3}{T_M^2} \right) \approx 8.6 \left(\frac{T_{ES}^3}{T_M^2} \right). \tag{6}$$

The values of T_C calculated from Eq. (6) for samples 3 and 4 are shown as arrows in Fig. 6. One can see a good agreement with experiment.

5. Hopping magnetoresistance

The magnetoresistance (MR) of samples 2–4 was measured at temperatures down to 1.8 K in perpendicular $B_\perp$ and in-plane (parallel) $B_\parallel$ magnetic fields up to $B = 8$ T. It turns out that $B_\perp$ leads to the increase of conductivity, or negative MR, while $B_\parallel$ results in positive MR at low temperatures.[24] This strong anisotropy indicates unambiguously different mechanisms of MR: negative MR in perpendicular fields is connected with orbital effects, while positive MR in parallel fields is determined by the spin polarization.

•   *Negative MR in perpendicular fields*

Figure 7 shows the MR curves $\Delta R(B)/R(0) \equiv [R(B) - R(0)]/R(0)$ at different T for all three samples on a log–log scale. On this scale, the slope of the curve is equal to the power α in $\Delta R/R \sim B^\alpha$. Quadratic dependence ($\alpha = 2$) is observed at low fields up to some value B^*.

Theoretically, the effect of orbital negative MR in the VRH regime has been discussed earlier.[25-27] The idea suggested in Ref. 25 is based on the following considerations. In VRH, the hopping distance r_h increases with decreasing T and becomes much larger than the mean distance between localized centers. As a result, the probability of a long-distance hop is determined by the interference of many tunneling paths via intermediate sites that may include scattering processes – see Fig. 8 for a schematic illustration. All these scattered waves together with the nonscattered direct wave contribute additively to the amplitude of the wave function ψ_{12} that reflects the probability for a charge carrier localized on site 1 to appear on site 2.

An important feature is that there is no backscattering, and scattered waves decay rapidly with increasing distance as $\exp(-2r/\xi)$, where ξ is the localization radius. Therefore, only the shortest paths contribute to ψ_{12}. All these paths are concentrated in a cigar-shaped domain of the length r_h, width $D \approx (r_h \xi)^{1/2}$, and area $A \approx r_h^{3/2} \xi^{1/2}$.

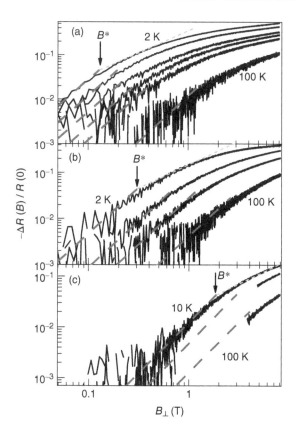

Figure 7. Negative MR of samples 2 (a), 3 (b), and 4 (c) at different T plotted on a log–log scale. Long-dashed lines correspond to the quadratic dependence $\Delta R/R \sim B^2$, whereas short-dashed lines correspond to linear dependence.

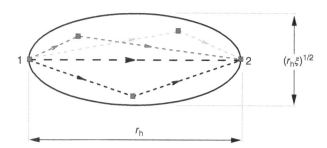

Figure 8. Schematics of the cigar-shaped region with localized states contributing to the probability of an electron tunneling from center 1 to center 2.

As a result of averaging over different configurations, the contribution of the scattering sites to the total hopping probability vanishes due to destructive interference. The perpendicular magnetic field suppresses the interference, which leads to an increase of the hopping probability and, therefore, to the negative MR. In accordance with theoretical considerations, negative MR as a function B is linear at moderate fields and quadratic at very low fields. We assume that quadratic dependence ends at a magnetic field B^* when the magnetic flux through the average cigar-shaped area $\Phi_B = B^*A$ will be equal to the magnetic flux quantum $\Phi_0 = h/2e \approx 2.07 \times 10^{-15}$ Wb. This gives $B^* = \Phi_0/A \sim r_h^{-3/2}\xi^{-1/2} \sim T^{1/2}$. The square-root dependence $B^* \sim T^{1/2}$ has been indeed observed in experiment – see Fig. 9.[24]

We also use the values of B^* in an attempt to normalize the negative MR data for all samples and temperatures below 25 K. In Fig. 10, MR curves on a linear scale are plotted as a function of dimensionless parameter B/B^*. One can see that all curves are fully described by a single common magnetic field dependence.

- *Positive MR in parallel fields*

In our samples, positive MR in parallel fields was observed at low temperatures (see Fig. 11). Earlier, positive MR in VRH regime had been observed in different 2D systems, particularly in a 2D electron gas formed in a GaAs/Al$_x$Ga$_{1-x}$ heterostructure.[28] The parallel magnetic field couples only to the electron spin, meaning that the spin state of localized electrons influences the hopping conductivity despite the fact that it is not included explicitly in Eqs (3) and (5).

A possible mechanism of positive MR was suggested by Kurobe and Kamimura[29] and studied in more detail in Ref. 30. In this model, it is recognized

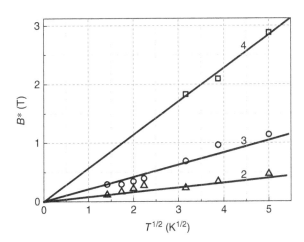

Figure 9. The values of B^* as a function of $T^{1/2}$. The sample numbers are indicated near the straight lines.

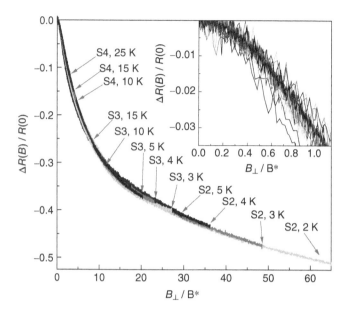

Figure 10. Negative MR curves for different samples and temperatures plotted versus $B_\perp/B*$. The arrows and numbers show the end of each curve and indicate samples (2, 3, 4) and T. The inset shows negative MR data for small values $B_\perp/B* < 1$.

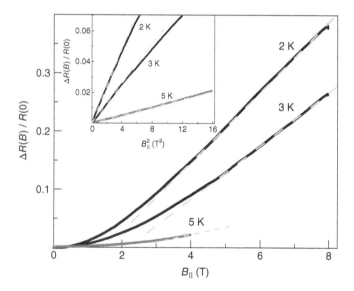

Figure 11. Positive MR of sample 3 plotted as a function of $B_\|$. Inset shows the MR data for low fields on a quadratic scale.

that a certain fraction of the states can accommodate two electrons. Double occupancy is possible if the on-site Coulomb repulsion U between the electrons is smaller than the energy distribution occupied by localized states. It was already mentioned that in VRH, only localized states with energy level within the narrow optimal band of width $\Delta E(T)$ around E_F are involved in the hopping process. However, for some states, which cannot participate in VRH at a given temperature because the energy of the first electron $E^{(1)}$ is well below E_F, the energy of the second electron $E^{(2)} = E^{(1)} + U$ may be located just within the optimal band. This allows those states to participate in the VRH at zero magnetic field. In the strong field limit, all spins are polarized and, therefore, transitions through the double occupied states are suppressed, resulting in positive MR. In this mechanism, contribution of the doubly occupied states in zero-field VRH is significant only when the width of the optimal band $\Delta E(T)$, which decreases with T, falls below U. In the opposite limit, $U \ll \Delta E(T)$, the localized states will either participate or not participate in VRH independently of the existence of doubly occupied states. This explains why positive MR is observable only at low T. At moderate fields, theory predicts the linear dependence $\Delta R/R \sim B$, while at weak fields one expects[30] a quadratic dependence, $\Delta R/R \sim B^2$. This prediction agrees with experiment, as shown in Fig. 11. Theory also predicts that positive MR should saturate at strong fields when all electron spins are polarized. In our samples, no saturation was observed in magnetic fields up to 8 T.

References

1. M. I. Katsnelson, *Graphene: Carbon in Two Dimensions,* Cambridge, UK: Cambridge University Press, 2012.

2. L. E. F. Foa Torres, S. Roche, and J.-C. Charlier, *Introduction to Graphene-Based Nanomaterials: From Electronic Structure to Quantum Transport,* Cambridge, UK: Cambridge University Press, 2014.

3. J. H. Warner, F. Schaffel, M. Rummeli, and A. Bachmatiuk, *Graphene: Fundamentals and Emergent Applications,* Amsterdam: Elsevier, 2012.

4. L. Liu, S. Ryu, M. R. Tomasik, *et al.,* "Graphene oxidation: Thickness-dependent etching and strong chemical doping," *Nano Lett.* **8**, 1965–1970 (2008).

5. D. C. Elias, R. R. Nair, T. M. G. Mohiuddin, *et al.,* "Control of graphene's properties by reversible hydrogenation: Evidence for graphane," *Science* **323**, 610–613 (2009).

6. H. Liu, Y. Liu, and D. Zhu, "Chemical doping of graphene," *J. Mater. Chem.* **21**, 3335–3345 (2011).

7. R. Saito, M. Hofmann, G. Dresselhaus, A. Jorio, M. S. Dresselhaus, "Raman spectroscopy of graphene and carbon nanotubes," *Adv. Phys.* **30**, 413–550 (2011).

8. G. Buchowicz, P. R. Stone, J. T. Robinson, C. D. Cress, J. W. Beeman, and O. D. Dubon, "Correlation between structure and electrical transport in

ion-irradiated graphene grown on Cu foils," *Appl. Phys. Lett.* **98**, 032102 (2011).

9. B. Guo, Q. Liu, E. Chen, H. Zhu, L. Fang, and J. R. Gong, "Controllable N-doping of graphene," *Nano Lett.* **10**, 4975–4980 (2010).

10. Q. Wang, W. Mao, D. Ge, Y. Zhang, Y. Shao, and N. Ren, "Effects of Ga ion-beam irradiation on monolayer graphene," *Appl. Phys. Lett.* **103**, 073501 (2013).

11. A. C. Ferrari, J. C. Meyer, V. Scardaci, *et al.*, "Raman spectrum of graphene and graphene layers," *Phys. Rev. Lett.* **97**, 187401 (2006).

12. A. C. Ferrari and J. Robertson, "Interpretation of Raman spectra of disordered and amorphous carbon," *Phys. Rev. B* **61**, 14095–14107 (2000).

13. O. Lehtinen, J. Kotakoski, A. V. Krasheninnikov, A. Tolvanen, K. Nordlund, and J. Keinonen, "Effects of ion bombardment on a two-dimensional target: Atomistic simulations of graphene irradiation," *Phys. Rev. B* **81**, 153401 (2010).

14. M. M. Lucchese, F. Stavale, E. H. Ferreira, *et al.*, "Quantifying ion-induced defects and Raman relaxation length in graphene," *Carbon* **48**, 1592–1597 (2010).

15. I. Shlimak, A. Haran, E. Zion, *et al.*, "Raman scattering and electrical resistance of highly disordered graphene," *Phys. Rev. B* **91**, 045414 (2015).

16. B. L. Altshuler, A. G. Aronov, and D. E. Khmelnitsky, "Effects of electron–electron collisions with small energy transfers on quantum localization," *J. Phys. C: Solid State Phys.* **15**, 7367–7386 (1982).

17. I. L. Aleiner and K. B. Efetov, "Effect of disorder on transport in graphene," *Phys. Rev. Lett.* **97**, 236801 (2006).

18. P. M. Ostrovsky, I. V. Gornyi, and A. D. Mirlin, "Electron transport in disordered graphene," *Phys. Rev. B* **74**, 235443 (2006).

19. E. McCann, K. Kechedzhi, V. I. Fal'ko, H. Suzuura, T. Ando, and B. L. Altshuler, "Weak-localization magnetoresistance and valley symmetry in graphene," *Phys. Rev. Lett.* **97**, 146805 (2006).

20. D. J. Bishop, D. C. Tsui, and R. C. Dines, "Non-metallic conduction in electron-inversion layers at low temperatures," *Phys. Rev. Lett.* **44**, 1153–1156 (1980).

21. V. F. Gantmakher, *Electrons and Disorder in Solids*, New York: Oxford University Press, 2005.

22. B. I. Shklovskii and A. L. Efros, *Electronic Properties of Doped Semiconductors*, Berlin: Springer-Verlag, 1984.

23. E. Zion, A. Haran, A. V. Butenko, *et al.*, "Localization of charge carriers in monolayer graphene gradually disordered by ion irradiation," *Graphene* **4**, 45–53 (2015).

24. I. Shlimak, E. Zion, A. V. Butenko, *et al.*, "Hopping magnetoresistance in ion irradiated monolayer graphene," *Physica E* **76**, 158–163 (2016).

25. V. L. Nguen, B. Z. Spivak and B. I. Shklovskii, "Tunnel hopping in disordered systems," *Sov. Phys. JETP* **62**, 1021–1029 (1985).

26. B. I. Shklovskii and B. Z. Spivak, "Scattering and interference effects in variable range hopping conduction," chapter in: M. Pollak and B. Shklovskii, eds. *Hopping Transport in Solids*, Amsterdam: Elsevier, 1991, p. 217.
27. W. Schirmacher, "Quantum-interference magnetoconductivity in the variable-range hopping-regime," *Phys. Rev. B* **41**, 2461–2468 (1990).
28. I. Shlimak, S. I. Khondaker, M. Pepper, and D. A. Ritchie, "Influence of parallel magnetic fields on a single-layer two-dimensional electron system with a hopping mechanism of conductivity," *Phys. Rev. B* **61**, 7253–7256 (2000).
29. A. Kurobe and H. Kamimura, "Correlation-effects on variable-range-hopping conduction and the magnetoresistance," *J. Phys. Soc. Japan* **51**, 1904–1913 (1982).
30. K. A. Matveev, L. I. Glazman, P. Clark, D. Ephron, and M. R. Beasley, "Theory of hopping magnetoresistance induced by Zeeman splitting," *Phys. Rev. B* **52**, 5289–5297 (1995).

2.10

Interplay of Coulomb Blockade and Luttinger-Liquid Physics in Disordered 1D InAs Nanowires with Strong Spin–Orbit Coupling

R. Hevroni, V. Shelukhin, M. Karpovski, M. Goldstein, E. Sela, and A. Palevski
Raymond and Beverly Sackler School of Physics and Astronomy, Tel-Aviv University, Tel Aviv 69978, Israel

Hadas Shtrikman
Department of Condensed Matter Physics, Weizmann Institute of Science, Rehovot 76100, Israel

1. Introduction

Ballistic 1D nanowires (NWs) with strong spin–orbit coupling are theoretically predicted[1] to exhibit nonmonotonic (up and down) conductance steps of size $G_0 = e^2/h$ as the electron density is varied by the gate voltage V_G. Although many attempts have been made to measure these conductance steps, they have not been observed yet in either InAs or InSb NWs. This indicates that disorder plays an essential role, preventing the motion of the electrons between the contacts from being ballistic. It is well known that in 1D systems, electron–electron interactions, described by the Luttinger-liquid (LL) model, amplify the role of disorder significantly, causing the conductance to vanish at zero temperature even for very weak disorder.[2] Experimentally, however, the effects of the interactions in NWs with strong spin–orbital scattering have not yet been reported.

In this chapter we report on experimental studies of the Coulomb blockade in disordered InAs NW at low temperatures. We demonstrate that sequential tunneling is strongly affected by electron–electron interactions. The analysis of the temperature dependence of the conductance and of the lineshape of the sequential tunneling in the Coulomb blockade regime within the framework of the existing theories allows us to deduce the corresponding LL parameter g. We show that in our NWs the effective LL parameter reaches a value less than 1/2, leading to a decrease in the Coulomb blockade peak-to-valley difference as the temperature is reduced. To the best of our knowledge, this phenomenon, predicted by the LL model, has never been experimentally observed before. While

Future Trends in Microelectronics: Journey into the Unknown, First Edition.
Edited by Serge Luryi, Jimmy Xu, and Alexander Zaslavsky.
© 2016 John Wiley & Sons, Inc. Published 2016 by John Wiley & Sons, Inc.

there were a number of experimental papers[3, 4] in which the Coulomb peaks decreased with decreasing temperature, this behavior was sporadic, that is, did not occur for consecutive peaks. Thus, these previous results do not follow the predictions of the LL theory but are rather consistent with stochastic Coulomb blockade,[5] while the opposite is true for our results, as we discuss below.

2. Sample preparation and the experimental setup

Our InAs NWs, approximately 2 μm long and 50 nm in diameter, were grown by Au-assisted vapor–liquid–solid MBE on a 2 in. SiO_2/Si substrate. A ~1 nm gold layer was evaporated *in situ* in a chamber attached to the MBE growth chamber after degassing the substrate at 600 °C. The substrate was heated to 550 °C after being transferred to the growth chamber to form gold droplets, then cooled down to the growth temperature of 450 °C. Indium and As_4 were evaporated at a V/III ratio of 100. The NWs were studied by SEM and TEM and were found to have a uniform morphology with no tapering and a pure wurtzite structure with a negligible number of stacking faults.[6] The NWs were deposited randomly from an ethanol suspension onto 300 nm-thick SiO_2 thermally grown on a p^+-Si substrate, to be used as a back gate. The NWs were then mapped with respect to alignment marks using optical microscopy and e-beam lithography and evaporation were used to deposit Ti/Al (5/100 nm) contact leads. A short dip in an ammonium polysulfide solution was used for removing the oxide from the InAs NWs surface prior to contact deposition.[7] The leads were separated by ~650 nm (see Fig. 1).

Generally, InAs NWs are highly sensitive to surface impurities and other imperfections (such as surface steps and dangling bonds) since the conductance electrons are near the surface. Hence, impurities resulting from sample fabrication and the external environment, as well as the substrate on which the sample is placed, induce disorder potential barriers.

Conductance was measured by a four-terminal method using a low-noise analog lock-in amplifier. The current was passed between two probes (I_+ and I_- in Fig. 1), while the voltage was measured by two different probes (V_+ and V_- in Fig. 1). It should be noted that I and V probes are connected to the NW at the same point, so that the contact resistance is always included in the conductance measured. The measurements were done in a cryogenic system in the 1.7–4.2 K temperature range.

3. Experimental results

Figure 2 shows the measured conductance G as a function of gate voltage V_G of an InAs NW. The as-grown NWs are conducting and the gate voltage bias required to pinch off the conductance is $V_G = -0.35$ V. A series of distinct

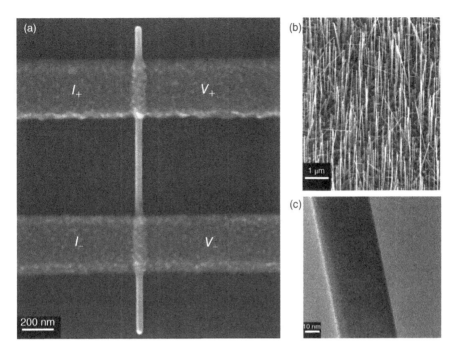

Figure 1. (a) SEM image of a typical sample showing the four-point conductance measurement geometry: current passed between I_+ and I_- probes, voltage measured between V_+ and V_- probes; (b) SEM image of as-grown InAs NWs; (c) TEM image of an InAs NW.

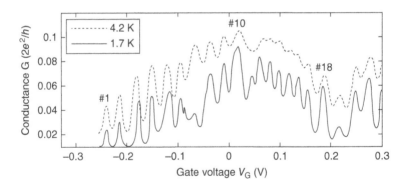

Figure 2. Conductance of an InAs NW (diameter: 50 nm; length: 650 nm) versus gate voltage, at 4.2 K (dashed line) and 1.7 K (solid line). The first, tenth, and eighteenth conductance peaks are labeled accordingly.

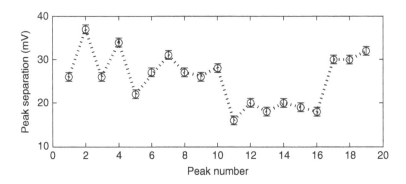

Figure 3. Peak separation versus peak number at $T = 1.7\,\text{K}$. Separation at peak n indicates the distance in millivolts between peak n and peak $(n+1)$.

conductance peaks is clearly observed, with a typical spacing of $\delta V_G \sim 25\,\text{mV}$. At lower temperatures, the conductance peaks become sharper, but the peak conductance values are reduced. This behavior indicates the occurrence of a Coulomb blockade, similar to quantum dots.

The peak spacing in our device is shown in Fig. 3. It is well known that the conductance peaks in quantum dots are equally spaced if the energy level spacing Δ between single electron states in the dot is negligible relative to the charging energy E_C. In the opposite limit, the conductance peaks are irregularly spaced,[8] which is the case in our NW sample since the distance between the peaks varies by over 50%. In this regime every peak corresponding to an odd number of electrons in the dot should be separated from the previous one by a roughly constant value proportional to E_C, whereas the next peak should be separated by value proportional to $(E_C + \Delta)$ that varies from level to level. Indeed, we see that every second peak of the first 10 peaks has a gate voltage spacing of $\delta V_G = 25\,\text{mV}$, with the exception of the distance between the fifth and sixth peaks that is slightly lower ($\sim 21\,\text{mV}$). This δV_G value should thus correspond to the charging energy, and yields a value of $C_G = 6.4 \times 10^{-18}\,\text{F}$ for the gate capacitance. Since the geometry of the sample and its dimensions are known, we can estimate the size L_{QD} of the quantum dot from the expressions for the capacitance of a cylinder in the vicinity of a conducting plate, $C_G = 2\pi\varepsilon L_{QD}/\ln(4d/D_{NW})$, where ε is the dielectric constant, D_{NW} is the NW diameter, and d is the SiO_2 thickness. Substituting the values of the sample dimensions, the average dielectric constant of ^4He and SiO_2 ($\varepsilon = 2.5\varepsilon_0$) and the estimated value of the capacitance give $L_{QD} \approx 200\,\text{nm}$.

We see that L_{QD} is smaller than the NW length ($L = 2\,\mu\text{m}$) by an order of magnitude, and smaller by more than a factor of 3 compared to the 650 nm separation between the voltage leads. Thus it is legitimate to assume that the QD is formed as a puddle of 1D electrons separated by two barriers on both sides, lying somewhere in the NW segment between the leads. In addition, we see that the segments of the NW to the right and left of the dot are long enough so that their single-particle level spacing and charging energies are well below the temperatures reached in our experiment, allowing us to describe them as infinite 1D leads. In such a case it is reasonable to carry out our data analysis in the framework of the theory of tunneling between two 1D NWs through a quantum dot.

Resonant and sequential tunneling were well studied theoretically and experimentally in the past for both interacting and noninteracting 1D electrons, see, for example, Ref. 2 and references therein. In our system the peak widths are found to scale linearly with the temperature T. We thus try to explain our results using Furusaki's expression[9] for the conductance due to sequential electron tunneling in a QD connected to LL leads. The lineshape of a single conductance peak as a function of the energy E (distance from the peak) is then

$$G(E, T) = A[G_0\gamma(T)/T \cosh(E/2k_BT)][\Gamma(1/2g + iE/2\pi k_BT)]^2, \qquad (1)$$

where A is a constant related to the asymmetry and height of the barriers defining the dot, the factor $\gamma(T) \sim T^{1/g-1}$ accounts for the renormalization of the tunneling rates by the LL effects, and $\Gamma(z)$ is the gamma function. Note that the temperature variation of G at the peak then becomes

$$G_{MAX}(T) \sim T^{1/g-2}. \qquad (2)$$

In all the above expressions, g is the effective LL interaction parameter; $g = 1$ for a noninteracting NW and decreases ($g < 1$) with increasing repulsive interactions. It is a combination of the charge and spin interaction parameters, as we discuss below.

The experimental data in Fig. 2 shows that both the height and the width of the conductance peaks decrease as temperature is reduced. Thus, the interaction parameter g should be smaller than 1/2. Our experimental data in Fig. 4 shows that indeed the temperature dependence of the height of each peak can be well described by the power law, Eq. (2), from which we can deduce the value of g for each peak. For the first two peaks, we find $g = 0.38 \pm 0.03$.

In order to verify that the lineshape of the Coulomb blockade peaks as a function of V_G can be described by Eq. (1), we fitted the gate voltage dependence of our data to a sum of terms (one for each peak) of the form of Eq. (1), with $E = \alpha(V_G - V_0)$, where V_0 is the gate voltage value at the peak and α is the ratio between the gate capacitance and the total capacitance of the dot. Parameters α and A are used as fitting parameters, and for g we plug in the value extracted from the data analysis presented in Fig. 4. The result for the first three peaks is shown in Fig. 5.

We find that, as expected, $\alpha = 0.1 \pm 0.005$ does not vary between the peaks and/or as function of temperature, indicating that Eq. (1) indeed gives a consistent description of our data set.

We have performed a similar fitting procedure for peaks #17–19 (see Figs 6 and 7). We find $g \sim 0.5$. It is indeed expected that the interaction constant should

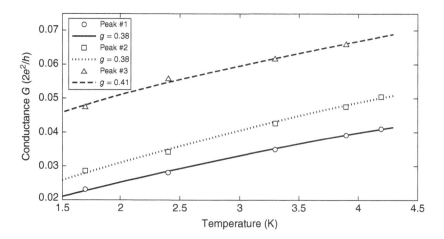

Figure 4. Experimental conductance peak heights versus temperature and fits to Eq. (2), for the first three peaks of Fig. 2.

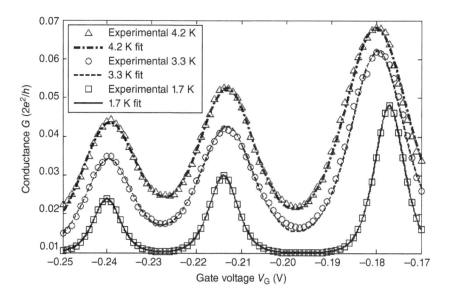

Figure 5. Fit of the first three peaks in Fig. 2 to Eq. (1) for $T = 4.2$, 3.3, and 1.7 K.

increase as the Fermi energy E_F increases, since g depends on the ratio between the Coulomb energy U and the Fermi energy in the NW.

As we pointed out earlier, the reduction of the conductance peaks at low temperatures has been observed in previous experiments[3, 4] but with markedly different results. In previous experiments, the decrease was sporadic, occurring only for nonconsecutive peaks, and thus cannot be accounted for by the LL

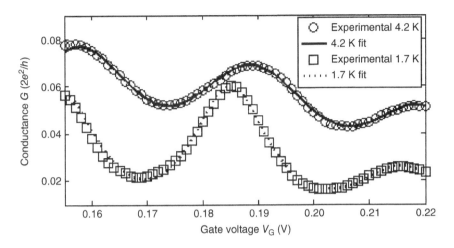

Figure 6. Fit of peaks #17–19 in Fig. 2 to Eq. (1) with $\alpha = 0.08$ and $g = 0.5$ at $T = 4.2$ and 1.7 K.

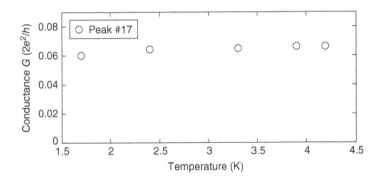

Figure 7. The maximum conductance of peak #17 versus T, extracted from the fitting the data of Fig. 6 to Eq. (1). The peak height is almost constant as function of temperature, indicating $g \sim 0.5$ according to Eq. (2).

picture, but rather indicates a stochastic Coulomb blockade.[5] In contrast, in our system the peak reduction occurs in a similar way for several consecutive peaks.

Now we address the question of why in our InAs NWs the effective LL parameter g is smaller than 1/2 at low filling (so that G_{MAX} decreases with decreasing temperature), while other experimental studies of 1D quantum wires, for example, carbon nanotubes,[10] GaAs wires formed at the cleaved edge overgrowth of a GaAs/AlGaAs heterostructure,[11] or V-groove GaAs NWs,[12] all exhibit effective LL parameters higher than 1/2 (so that G_{MAX} increases with decreasing temperature). We believe that there are two main reasons that contribute to the lower value of the LL parameter in our InAs NWs.

The first is related to the environment of the quantum wires. Both types of GaAs wires reported in the literature[11, 12] were created within 2DEG structures embedded well inside a semiconductor material with a large dielectric constant (AlGaAs and GaAs). In contrast, while the InAs NWs have a similar dielectric constant to GaAs, they are placed on SiO_2 surface, so the surrounding materials, namely liquid ^4He and SiO_2, possess much smaller dielectric constants. These reduced dielectric constants enhance the effect of Coulomb interaction in our system as compared to the GaAs wires reported before.

The second reason for observing a smaller LL parameter in our system is related to an inherent property of InAs – the strong spin–orbit coupling that breaks the spin rotation symmetry. The effective LL parameter g is related to the interaction parameters in the charge and spin channels, g_C and g_S, respectively, by[9]

$$\frac{1}{g} = \frac{1}{(2g_C)} + \frac{1}{(2g_S)}. \tag{3}$$

In GaAs, the spin–orbit coupling is very small; therefore, spin-rotation symmetry dictates that $g_S = 1$. Thus, $g < 1/2$ can only be obtained if the interaction in the charge sector is extremely strong, $g_C < 1/4$. In carbon nanotubes, the additional valley degeneracy results in $1/g = 1/(4g_C) + 3/4$, so reaching $g < 1/2$ requires an even stricter condition $g_C < 1/5$. On the other hand, in InAs spin rotation symmetry is broken, allowing for $g_S < 1$ and making it easier to reach $g < 1/2$.

4. Conclusion

We believe that the combination of a lack of orbital degeneracy due to strong spin–orbit coupling and of the low effective dielectric constant makes our InAs NW a unique system where strong effective interactions, $g < 1/2$, can be achieved, and thus a decrease in Coulomb blockade peak heights with decreasing temperature can be observed.

Acknowledgments

We are thankful to Ronit Popovitz-Biro for professional TEM study of the InAs NWs. We gratefully acknowledge support by the ISF BIKURA program and GIF (MG); ISF and Marie Curie CIG grants (ES); as well as ISF grant #532/12 and IMOST grants #3-11173 (AP and HS) & #3-8668 (HS).

References

1. Y. V. Pershin, J. Nesteroff, and V. Privman, "Effect of spin–orbit interaction and in-plane magnetic field on the conductance of a quasi-one-dimensional system," *Phys. Rev. B* **69**, 121306 (2004).

2. I. Krive, A. Palevski, R. Shekhter, and M. Jonson, "Resonant tunneling of electrons in quantum wires," *Fiz. Nizk. Temp.* **36**, 155–180 (2010).

3. S. B. Field, M. A. Kastner, U. Meirav, *et al.*, "Conductance oscillations periodic in the density of one-dimensional electron gases," *Phys. Rev. B* **42**, 3523–3536 (1990).

4. J. Moser, S. Roddaro, D. Schuh, M. Bichler, V. Pellegrini, and M. Grayson, "Disordered AlAs wires: Temperature-dependent resonance areas within the Fermi-liquid paradigm," *Phys. Rev. B* **74**, 193307 (2006).

5. I. M. Ruzin, V. Chandrasekhar, E. I. Levin, and L. I. Glazman, "Stochastic Coulomb blockade in a double-dot system," *Phys. Rev. B* **45**, 13469–13478 (1992).

6. H. Shtrikman, R. Popovitz-Biro, A. Kretinin, and P. Kacman, "GaAs and InAs nanowires for ballistic transport," *IEEE J. Sel. Top. Quantum Electron.* **17**, 992–934 (2011).

7. D. B. Suyatin, C. Thelander, M. T. Björk, I. Maximov, and L. Samuelson, "Sulfur passivation for Ohmic contact formation to InAs nanowires," *Nanotechnology* **18**, 105307 (2007).

8. M. Pustilnik and L. Glazman, "Kondo effect in quantum dots," *J. Phys. Condens. Matter* **16**, R513–R537 (2004).

9. A. Furusaki, "Resonant tunneling through a quantum dot weakly coupled to quantum wires or quantum Hall edge states," *Phys. Rev. B* **57**, 7141–7148 (1998).

10. M. Bockrath, D. H. Cobden, J. Lu, *et al.*, "Luttinger-liquid behavior in carbon nanotubes," *Nature* **397**, 598 (1999).

11. O. M. Auslaender, A. Yacoby, R. de Picciotto, K. W. Baldwin, L. N. Pfeiffer, and K. W. West, "Experimental evidence for resonant tunneling in a Luttinger liquid," *Phys. Rev. Lett.* **84**, 1764–1767 (2000).

12. E. Levy, I. Sternfeld, M. Eshkol, *et al.*, "Experimental evidence for Luttinger liquid behavior in sufficiently long GaAs V-groove quantum wires," *Phys. Rev. B* **85**, 045315 (2012).

Part III

Microelectronics in Health, Energy Harvesting, and Communications

While the drivers of the "next economy" are a subject of debate, there is no doubt that microelectronics will continue to be a primary enabler. But to many, it is not as clear that the next major innovations will be enabled by the famed "roadmap" advances and breakthroughs in microelectronics or by creation of new transistors or even new semiconductors. It is not even certain that those major innovations will be in the realm of electronics. Instead, they could be in healthcare, or on and even in our body, or in distributed energy sources and manufacturing. It is just as interesting to raise the questions of what the impact will be if the semiconductor "roadmap" evolution truly comes to an end. Which industries would survive? Is there anything on the horizon that will take over the role of the current information economy? In this chapter, one may glimpse some signs of such future industries and innovations. One thing seems clearer than others: the continuation of the exponential progress in data transmission and storage calls for greater bandwidth, likely meaning photonics, as also highlighted here.

Contributors

3.1 B. H. W. Hendriks, D. Mioni, W. Crooijmans, and H. van Houten

3.2 D. A. Borton

3.3 A. Romani, M. Dini, M. Filippi, M. Tartagni, and E. Sangiorgi

3.4 R. Tao, G. Ardila, R. Hinchet, A. Michard, L. Montès, and M. Mouis

3.5 N. Ledentsov Jr., V. A. Shchukin, N. N. Ledentsov, J.-R. Kropp, S. Burger, and F. Schmidt

3.6 X. Zhang, V. Mitin, G. Thomain, T. Yore, Y. Li, J. K. Choi, K. Sablon, and A. Sergeev

3.7 M. C. M. M. Souza, G. F. M. Rezende, A. A. G. von Zuben, G. S. Wieder-hecker, N. C. Frateschi, and L. A. M. Barea

3.1

Image-Guided Intervention and Therapy: The First Time Right

B. H. W. Hendriks, D. Mioni, W. Crooijmans, and H. van Houten
Philips Research, High Tech Campus 34, Eindhoven, The Netherlands

1. Introduction

In 2006, Dr Elias Zerhouni, director of the U.S. National Institutes of Health, outlined the vision that "medicine in the future has to be predictive, personalized, and very precise to the individual, and it has to be pre-emptive."[1] He stressed the importance of imaging in understanding complex biological systems, a topic we reviewed in FTM-2006.[2] He also articulated a second and perhaps even bolder vision: "Twenty-five years from now, I hope that we won't perform any more open surgery. There would be no need to essentially take the risk of full exposure of the human body to go to a targeted region that needs to be affected."[1] How this might be achieved is the topic we address in this chapter.

In cardiology, one of the main causes of mortality today, major progress has already been made through minimally invasive interventions, such as placing a stent. Heart rhythm disorders can also be treated using catheters, by first mapping the disturbed pattern of electrical activity around the heart chambers, followed by selectively altering the current paths through local tissue ablation. Valve replacement is also rapidly gaining ground. These procedures require careful navigation and steering of the various catheters, which can be optimized using 3D X-ray and ultrasound imaging methods, combined in real time with physiological models and image processing – thereby enabling proper eye–hand coordination.

However, a catheter today is a purely mechanical device, controlled via external imaging and manual steering. Our vision for the future is that the efficiency and positive outcome of catheter-based cardiovascular procedures can be drastically improved by adding in-body sensing and imaging to catheters. Advances in miniaturization technology allow us to build intelligence into the catheter, to provide local imaging, localization, and control capabilities. Examples are MEMS-based ultrasound transducers[3] that can be mounted onto the tip of a catheter, and real-time 3D optical shape sensing along its length. We expect that

Future Trends in Microelectronics: Journey into the Unknown, First Edition.
Edited by Serge Luryi, Jimmy Xu, and Alexander Zaslavsky.
© 2016 John Wiley & Sons, Inc. Published 2016 by John Wiley & Sons, Inc.

such smart catheter devices will disrupt the field of minimally invasive procedures. The ultimate solution will have a control loop between in-body sensing information and external imaging and therapy planning and delivery – enabled by adaptive therapy planning software.

In the field of cancer, the other major cause of death, we believe that a lot can be gained, thanks to advances in imaging and genomics that are enabling precision diagnostics. Precision diagnostics can be realized through image-guided focal biopsies, followed by molecular pathology. Image-guided targeted therapy delivery is possible through local administration of drugs, or by precise local delivery of energy to the tumor. These procedures require again smaller and smarter interventional instruments that can be controlled with a high level of precision, such as a photonically enhanced biopsy needles that can discriminate between tissue types.

Apart from the advances in disease management, the current healthcare system is also rapidly changing due to various economic realities the world is facing. These realities require a more integrated approach along the care continuum. Informatics solutions and healthcare transformation services play an important role in enabling this evolution.

It is our firm belief that the predictive, personalized, precise, and preemptive medicine of the future requires a care continuum approach, where image-guided therapy will play an important role.

2. Societal challenge: Rapid rise of cardiovascular diseases

Today, almost 30% of all global deaths are caused by cardiovascular disease (CVD) and by 2030 about 24 million people will die every year from it. This creates a burden on the healthcare system with costs up to 500 billion euros.[4] Major contributors to heart disease include (i) electrical signal disorders (atrial fibrillation, sudden cardiac arrest); (ii) blockage of coronaries (insufficient power for contraction); (iii) valve problems (inefficient blood circulation); and (iv) damaged myocardial tissue (insufficient contraction, insufficient output).

In the recent past, treating heart disease meant open-heart surgery, where the patient's chest was exposed for direct access to the heart. This kind of procedure was expensive, due to long recovery stays in the hospital and a high risk of complications. Furthermore, the patient had to be in a reasonable health to be eligible for this kind of procedure.

The introduction of minimally invasive image-guided procedures based on X-ray guidance in combination with catheters that can be inserted into the body via a small incision has completely revolutionized this treatment. This represents a significant improvement for the patients, due to fewer restrictions on patient eligibility and lower risks. Furthermore, the faster recovery and shorter hospital stays has helped to reduce the burden of CVD on the healthcare system.

• *Innovative products using X-rays*

X-rays, discovered by the German physicist Wilhelm Conrad Röntgen, led to the development of X-ray systems for diagnostic and interventional procedures. One of the first diagnostic X-ray system was the Metallix machine, introduced in 1928 (see Fig. 1).[5] Subsequent versions played an important role in tuberculosis testing in the 1950s and later for cardiovascular procedures (see Fig. 1). Over the years, advances in X-ray tubes, imaging intensifiers, and source-detector geometries have improved the imaging capabilities tremendously. Furthermore, the intro-duction of a source-detector geometry (C-arm geometry, angiography system) that could be rotated around the patient became the standard geometry for car-diovascular applications. Rotational cone beam computed tomography (CT), in which a complete series of projections is acquired on a conventional angiogra-phy system while the C-arm made a continuous rotation over an angle greater than 180° around the patient made it possible to make three-dimensional data

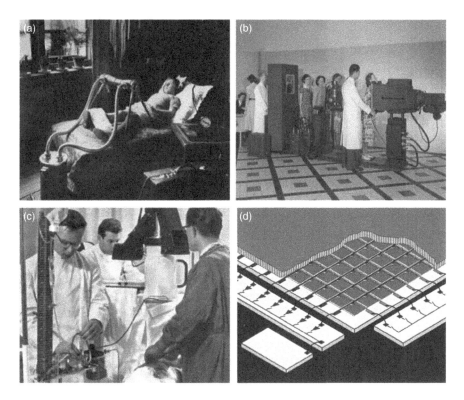

Figure 1. Developments in X-ray imaging at Philips: (a) Metallix X-ray machine (1928); (b) tuberculosis testing (1951); (c) heart catheterization using an image inten-sifier (1956); and (d) digital flat X-ray detector (2002). Courtesy of Philips Research.

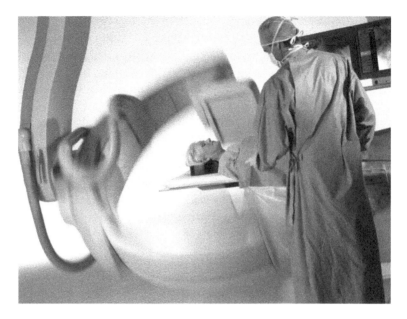

Figure 2. Rotational cone beam CT using X-rays, allowing volumetric imaging during interventions. Courtesy of Philips Research.

sets, as illustrated in Fig. 2. To overcome the limited soft tissue visibility, which is easily seen in computed tomography (CT)[6] and magnetic resonance imaging (MRI),[7] the rotational angiography system was further improved to make CT-like imaging possible on these systems. A further development worth mentioning is the introduction of a more intuitive working environment and data management across the entire procedure, first introduced for electrophysiology procedures (EP).[5] Due to the increasing number of older people, the prevalence of heart attacks resulting from irregular rhythms in the heart's EP system has been increasing. To treat these disorders, an electrophysiologist uses X-ray imaging and many other medical devices, such as electrophysiology recorders, ultrasound scanners, and ablation and navigation equipment. To deal with this complexity, an EP-cockpit was developed that provides an integrated package of equipment that could be controlled from one console. This allows performing the diagnostics, planning the procedure, and executing the treatment with real-time feedback to the physiologist.

Apart from the hardware developments, improvements have been made in image processing due to the increasing computing power and advances in X-ray detectors such as the one shown in Fig. 1(d). The rotational cone beam CT acquisition of Fig. 2 enables three-dimensional (3D) visualization of structures such as the vessel tree, shown in Fig. 3(a). Such data are essential for assessment of brain aneurysms and to perform precise measurements important for therapy

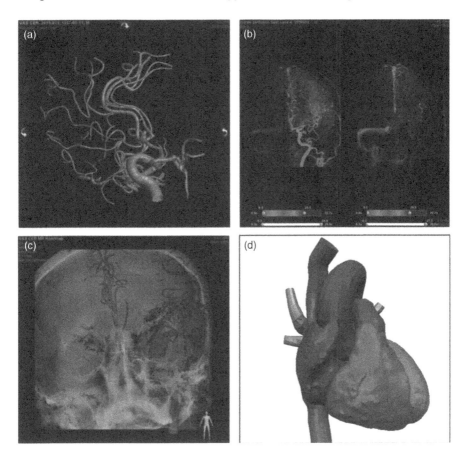

Figure 3. Anatomical and functional imaging: (a) three-dimensional assessment of brain aneurysm; (b) brain perfusion assessment; (c) real-time catheter navigation in three dimensions; and (d) patient-specific electronic organ models. Courtesy of Philips Research.

planning. By using contrast injections, blood flow can be assessed (see Fig. 3(b)), which is important for determining occlusions in the blood vessels. Combining images from different imaging modalities enhances the visualization of tissue structures by combining the best of different worlds – see Fig. 3(c). When one of the imaging modalities is performed in real time, for instance real-time low-dose two-dimensional X-ray fluoroscopy combined with the preoperatively acquired 3D cone beam CT image, applications such as real-time 3D catheter navigation for therapy delivery become possible. Further improvement has been achieved by adding anatomical intelligence through physiological modeling, as shown in Fig. 3(d). Using patient-specific electronic organ models allows for efficient and quantitative inspection of the data, as well as automated anatomical structure recognition. This is key for more precise therapy planning, such as aortic valve annulus diameter measurement, important for selection of the right transcatheter aortic valve implant.

Due to these innovative X-ray products, treating heart disease has become a simpler procedure with a highly positive impact on the patient's quality of life. Image guidance has opened the way for using minimally invasive techniques, where patients are quickly treated and are often discharged from the hospital the same day rather than having to stay for week-long hospital recoveries.

• *Innovative products using ultrasound*

Similar to X-rays, ultrasound has a long history and is widely used in a range of clinical applications. It dates back to 1942, when ultrasound was first used as a diagnostic tool to locate brain tumors and the cerebral ventricles,[8] followed by the diagnosis of gallstones and foreign bodies in 1948. In the field of obstetrics and gynecology, ultrasound was first employed in 1958. Ultrasound advances are strongly linked to the development of piezoelectric materials for transducers, as well as the manufacturing of microbeam-forming arrays that form the heart of the handheld ultrasound transducers for 3D ultrasound imaging. Advances in electronics and computing power have resulted in high-performance ultrasound imaging systems (see Fig. 4(a)) and ultramobile tablet-based ultrasound systems, illustrated in Fig. 4(a) and (b), respectively. The miniaturization of the transducer array and driving electronics made it possible to integrate ultrasound transducers into catheters, enabling ultrasound imaging in cardiovascular and endoscopic procedures (see Fig. 4(c)).

A new technology that has the potential to further advance ultrasound imaging is based on capacitive micromachined ultrasonic transducer (cMUT) technology.[3] This technology, based on silicon IC processing, allows the ultrasound interconnect and the transducer to be made in one process (i.e., both flexible and rigid components made in one step). An arbitrary shape with a highly flexible interconnect can be made, as illustrated in Fig. 5.

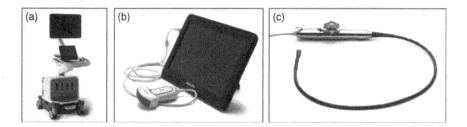

Figure 4. Developments in ultrasound: (a) high-performance ultrasound imaging system; (b) tablet-based ultrasound system; and (c) transesophageal echocardiogram (TEE) probe. Courtesy of Philips Research.

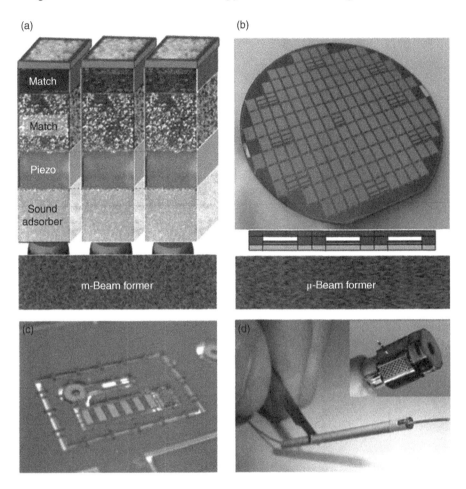

Figure 5. Advances in transducer technology: from piezoelectric to cMUT on a beam-former chip. (a) Conventional piezoelectric beam-former technology; (b) cMUT beam-former chip elements; (c) cMUT transducer and highly flexible interconnects made with standard IC processing; and (d) assembly of ultrasound transducers on a catheter (inserting ultrasound transducer tip). Courtesy of Philips Research.

The current state of the art is 3D real-time ultrasound. In obstetric gynecology, this allows almost realistic imaging of the fetus (see Fig. 6(a) and (b)). In cardiology, the transesophageal echocardiogram (TEE) probe enables 3D visualization of heart structures such as the mitral valve that are important for structural heart disease treatments (see Fig. 6(c) and (d)).[9]

Both external as well as in-body imaging has played an important role in driving cardiovascular procedures to the current state of the art. Today, these cardiovascular procedures are performed in hybrid rooms, shown in Fig. 6(e),

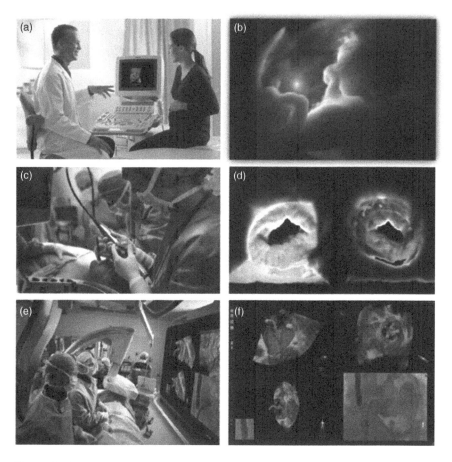

Figure 6. Advances in ultrasound imaging: (a) and (b) three-dimensional live ultra-sound in obstetric gynecology; (c) and (d) mitral valve imaging with TEE probe for cardiac interventions; and (e) and (f) combined X-ray and ultrasound imaging for structural heart disease interventions. Courtesy of Philips Research.

where combined X-ray and ultrasound enables soft tissue anatomical imaging, including functional and blood flow measurements with real-time 3D insight, as shown in Fig. 6(f).

3.   Societal challenge: Rapid rise of cancer

According to the World Health Organization,[10] cancer is one of the leading causes of morbidity and mortality, with approximately 14 million new cases and 8.2 million deaths in 2012. The number of new cancer cases is expected to grow to 24 million in 2030. The most common cancers are lung, breast, prostate, liver, and colorectal. Lung cancer is the most common cancer, with 1.8 million new cases and 1.59 million deaths in 2012. Breast cancer is the most frequent cancer in women, while prostate cancer is the second most common cancer in men. Liver cancer is largely a problem in less developed regions, where it accounts for 83% of the cancer cases reported.

- *Innovative products precise diagnostics*

Various imaging modalities are used in oncology. Apart from the X-ray and ultrasound imaging that we discussed earlier for cardiovascular applications, there are three other commonly used imaging modalities: computer tomography (CT) imaging,[6] MRI,[7] and positron emission tomography (PET).[11] In CT imaging, an X-ray source and detector rotate while the patient is advanced through the bore of the system. This way of imaging provides cross-sectional and tomographic images of the patient with good tissue contrast at the expense of exposure to a fairly large dose of X-ray radiation. MRI uses magnetic fields and radio waves to investigate the anatomy and physiology of the patient's tissue. A strong magnetic field (several Tesla) is used to align the magnetic moments of protons. When a radiofrequency pulse is applied, the magnetization alignment is altered, which in turn causes a change in the magnetic flux that can be detected by the receiver coils of the system. Since this change depends on the local magnetic field near the proton, by applying additional magnetic fields (gradient fields) the distribution of the protons can be determined. MRI provides superb soft tissue contrast at the expense of longer acquisition times. It is also more costly than CT imaging. Finally, PET is a nuclear imaging technique that detects pairs of gamma rays that are emitted by a positron-emitting radionuclide (tracer). Functional processes in the body can then be imaged by administering a tracer conjugated to a biologically active molecule (such as fluorodeoxyglucose) to the patient. The metabolic activity related to the uptake of the tracer is used to investigate the presence of cancer metastasis.

Spectral CT is one of the recent technology advances that have become available in the clinic (see Fig. 7(a)).[12] In addition to the intensity-based contrast typically used in CT, this imaging technique provides the energy levels of the detected X-ray photons. In oncology, it is expected that spectral CT will improve lesion characterization and quantification in the diagnostic stage, and also provide a better understanding of the efficacy of cancer treatments applied.

Another new development is digital MRI, shown in Fig. 7(b), in which digital broadband architectures, available on modern MRI systems, lead to enhanced image acquisition speed, signal-to-noise ratio (SNR), and volumetric imaging. In oncology it results in finer visualization of soft tissue organs, such as the breast, prostate, or liver.

Furthermore, the combination of digital PET and CT, shown in Fig. 7(c), results in the enhancement of resolution, sensitivity, and quantitative accuracy

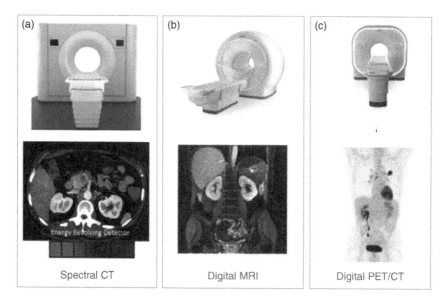

Figure 7. The most commonly used imaging modalities for precision diagnostics in oncology: (a) spectral computed tomography; (b) digital magnetic resonance imaging; and (c) digital positron emission tomography combined with computed tomography. Courtesy of Philips Research.

compared to analog PET systems, making cancerous lesions easier to detect and their metabolic activity easier to measure. Once radiologists using imaging have found a suspicious lesion, the next step is to harvest a tissue sample for further analysis. To obtain these tissue samples, biopsy needles are inserted into the lesion using image guidance. To further enhance the biopsy yield of cancerous tissue, more accurate biopsy taking methods have been developed. For example, in prostate biopsies, preoperative MRI images that show a lesion in the prostate are combined with real-time transrectal ultrasound images. Combining the imaging with electromagnetic sensing, the tip of the needle can be guided to the lesion, as illustrated in Fig. 8(a).[13] Another way to improve the biopsy yield is by equipping the biopsy needle with optical fibers enabling tissue discrimination at the tip of the needle via optical spectroscopy, as shown in Fig. 8(b). In this way, the needle can confirm whether the lesion has been reached before the actual biopsy is taken.[14]

Treatment selection is determined after histopathology and increasingly after additional molecular diagnostics. Oncology dashboards are under development to support treatment selection. For instance in a prostate dashboard, the information from different disciplines dealing with prostate cancer patients (urology, radiology, pathology, and genetics) comes together in an integrated manner to show 36 anatomical zones of the prostate, where the pathological findings are annotated by the respective specialists, allowing for a more

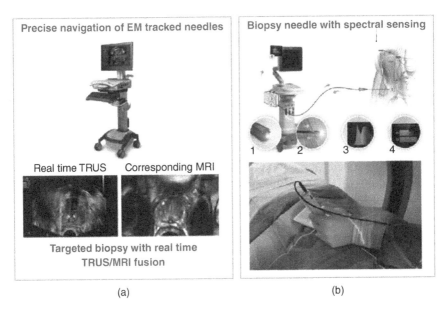

Figure 8. Two ways to improve biopsies: (a) preoperative MRI is overlaid with real-time transrectal ultrasound in combination with electromagnetically tracked needles for precise navigation to the targeted lesion; (b) a biopsy needle with integrated fibers allows optical spectroscopy at the tip of the needle to provide real-time tissue discrimination during image-guided biopsies. Courtesy of Philips Research.

personalized treatment including active surveillance, surgery, radiotherapy, brachytherapy,[15] or focal therapy.

Apart from surgery and chemotherapy, radiotherapy is an important option in cancer treatment.[16] Nowadays, over 50% of patients are treated with radiation therapy. In this treatment, high-energy beams are directed toward the cancerous cells to destroy them. Since the radiation also destroys normal cells, careful planning and delivery are important to minimize the comorbidity. Today's treatment delivery with radiotherapy starts by making a simulation scan for every treatment plan. By applying tumor characterization and organ risk knowledge, the final treatment plan is made. This planning phase can take several days, delaying treatment delivery. Typically, radiotherapy is delivered in 30 sessions over the course of 6 weeks. Since no real-time image guidance or replanning is used, changes due to setup uncertainty, weight loss, tumor response, and anatomy changes impact the radiotherapy delivery and result in suboptimal treatment. To overcome these drawbacks, current advances are aiming toward real-time adaptive planning and personalized treatment delivery. For this, radiotherapy systems are combined with imaging systems, allowing real-time imaging of the anatomy and tumor characteristics. The therapy delivery plan can then be adapted during the treatment.

Another cancer treatment where imaging plays an important role is in embolization therapy.[17] In this procedure, various substances (such as coils, ethanol, polyvinyl alcohol, and microspheres) are injected in the blood vessel feeding the tumor to block or reduce the blood flow. First, an image of the vascular tree is made to determine the main vessel feeding the tumor. A catheter is advanced under image guidance toward this vessel and small particles are injected into the blood vessel to plug it up, a procedure called transarterial embolization (TAE). The procedure can be enhanced by loading the particles with chemotherapy or with radioactive particles.

Local treatment of the tumors can be performed percutaneously by placing a needle tip inside the tumor under ultrasound or CT image guidance and ablating the tissue.[18] Different types of ablation are used. In radiofrequency ablation, high-energy radio waves are applied at the tip of the needle, heating the tumor and destroying cancer cells. A similar technique is to apply microwaves to heat the tissue. Another local therapy technique is freezing with very cold gases that are passed through the needle, a technique known as cryotherapy. A final method used is the injection of concentrated alcohol directly into the tumor to destroy the cancer cells. For all these techniques, imaging plays an essential role for planning and accurate needle placement to make sure that all tumor cells are treated while normal tissue and critical structures are spared as much as possible.

These examples in oncology show that imaging and image guidance play an important role to realize Zerhouni's vision to make the future of medicine predictive, personalized, precise to the individual, and preemptive.

4.   Drivers of change in healthcare

Today, changes in healthcare are driven by three main economic realities. Clinical and economic outcomes are driving provider reimbursement, where compliance with the standard of care results in the "consumerization" of healthcare. Then, there is a move from treating illness to maintaining population wellness, shifting the emphasis to avoidance of injuries, complications, and readmissions. Finally, connecting everyone through common health information will unlock the power in the rich but highly disconnected islands of information in currently existing systems.

Healthcare needs to be seen along the "health continuum." We see professional healthcare and consumer markets converging into a single health continuum, enabled by connected health technology. This continuum starts with a focus on healthy living and prevention, which empowers consumers to take control over their own health and enables countries to increase the health of the overall population. Next, the continuum includes definitive diagnostics and minimally invasive treatments, optimized with respect to both quality and cost. And finally, the continuum encompasses recovery and home care, which means care shifts as soon as possible to settings outside the hospital that are more comfortable and cost-effective. Telecare will enable patients to recapture a healthy life again, and avoid relapsing. Governments, insurers, medical professionals, patients, and caregivers all need to work together to realize the health continuum.

5. Conclusion

In conclusion, medical imaging has dramatically changed medical practice over the last century, enabling specialists to see inside the human body without incision and perform intricate therapeutic procedures without open surgery.

Nowadays, the challenges have increased substantially with today's healthcare industry drivers, but at the same time medical imaging provides new avenues for technology innovation that will play a central role in realizing Zerhouni's vision that "medicine in the future has to be predictive, personalized, and very precise to the individual, and it has to be pre-emptive."

We have discussed the impact of imaging on the diagnostic and therapeutic stages of the world's most common diseases that have an unprecedented impact on our modern societies. In line with the WHO Global Noncommunicable Disease Action Plan, individuals taking ownership of their health and acting on modifiable risk behaviors (physical activity, healthy diet, and nonuse of tobacco) will open up completely new spaces for innovation. The overarching goal is the prevention of a significant amount of premature deaths. This is a topic for a future chapter in trends in microelectronics.

Acknowledgments

The authors thank T. M. Bydlon for attentive reading of the manuscript.

References

1. E. Ridley, "Imaging poised to transform the future of medicine," see www .auntminnie.com/index.aspx?sec=ser&sub=def&pag=dis&ItemID=69642 (accessed on Apr. 22, 2016).
2. H. van Houten and H. Hofstraat, "Towards molecular medicine," chapter in: S. Luryi, J. M. Xu, and A. Zaslavsky, eds., *Future Trends in Microelectronics: Up the Nano Creek*, Hoboken, NJ: Wiley, 2007, pp. 90–100.
3. Ö. Oralkan, A. S. Ergun, J. A. Johnson, *et al.*, "Capacitive micromachined ultrasound transducers: Next-generation arrays for acoustic imaging?" *IEEE Trans. Ultrason. Ferroelectr. Freq. Control* **49**, 1596–1610 (2002).
4. "World Health Organization fact sheet on cardiovascular diseases (CVDs)," see www.who.int/mediacentre/factsheets/fs317/en (Sept. 23, 2015).

5. J. A. M. Hofman, "The art of medical imaging: Philips and the evolution of medical X-ray technology," *Medicamundi* **54**, 5–21 (2010).

6. D. Brenner and E. Phil, "Computed tomography: An increasing source of radiation exposure," *New Eng. J. Med.* **357**, 2277–2284 (2007).

7. D. Plewes and W. Kucharczyk, "Physics of MRI: A primer," *J. Magn. Reson. Imag.* **35**, 1038–1054 (2012).

8. See article by J. Woo at www.ob-ultrasound.net/dussikbio.html (accessed on Apr. 22, 2016).

9. Y. Mehta and S. Dhole, "Trans-esophageal echocardiography – A review," *Ind J. Anaesth.* **46**, 315–322 (2002).

10. "Word Health Organization fact sheets on cancer," see www.who.int/mediacentre/factsheets/fs317/en (Sept. 23, 2015).

11. G. Kelloff, J. Hoffman, H. Scher, *et al.*, "Progress and promise of FDG-PET imaging for cancer patient management and oncologic drug development," *Clin. Cancer Res.* **11**, 2785–2808 (2005).

12. J. Schlomka, E. Roessl, R. Dorscheid, *et al.*, "Experimental feasibility of multi-energy photon-counting K-edge imaging in pre-clinical computed tomography," *Phys. Med. Biol.* **53**, 4031–4047 (2008).

13. J. Krücker, S. Xu, N. Glossop, *et al.*, "Fusion of real-time trans-rectal ultrasound with pre-acquired MRI for multi-modality prostate imaging," *Proc. SPIE* **6509**, 650912 (2007).

14. J. Spliethoff, W. Prevoo, M. Meier, *et al.*, "Real-time *in vivo* tissue characterization with diffuse reflectance spectroscopy during transthoracic lung biopsy: A clinical feasibility study," *Clin. Cancer Res.* **22**, 357–365 (2015); doi: 10.1158/1078-0432.CCR-15-0807.

15. G. Koukourakis, N. Kelelis, V. Armonis, and V. Kouloulias, "Brachytherapy for prostate cancer: A systematic review," *Adv. Urol.* **2009**, article no. 327945 (2009).

16. See www.cancer.gov/about-cancer/treatment/types (Sept. 23, 2015).

17. Y.-S. Guan, Q. He, and M.-Q. Wang, "Transcatheter arterial chemo-embolization: History for more than 30 years," *ISRN Gastroenterol.*, article no. 480650 (2012).

18. See www.cancer.org/cancer/livercancer/detailedguide/liver-cancer-treating-tumor-ablation (Sept. 23, 2015).

3.2

Rewiring the Nervous System, Without Wires

D. A. Borton
*School of Engineering and the Brown Institute for Brain Science,
Brown University, Providence, RI 02912, USA*

1. Introduction

The human nervous system is arguably the most advanced information-processing network in existence. Multiarea, recurrent networks in our brains enable flexible behavior in a dynamic and uncertain world. Amazingly, small-scale networks of the spinal cord can regulate intelligent responses on a millisecond timescale to a poorly placed foot on a stair, initiating an ever-useful correction loop and avoiding catastrophe. Large-scale networks across vast landscapes of the brain regulate everything from the decision to take a step forward to knowing when to take a step backward.

In the healthy nervous system, the development of intention and motor execution is a dynamic and highly distributed process that originates in the brain. The intended action is transmitted along the axon super highway of the spinal cord and down to smart circuits in the spinal cord that transform the descending command into coordinated patterns of muscle activation. A continuous stream of sensory and motor information is actively coalesced in the brain to ensure the accurate and sequenced execution of movement. The output of the motor cortex and brainstem motor regions is under the continuous influence of other structures of the brain, including the cerebellum and basal ganglia, which are essential for producing smooth movements. These contextual information streams form loops interacting with one another at key integration centers, such as the association cortices and the thalamus. Failures of function in one seemingly insignificant processing loop can, and often does, lead to dramatic consequences that induce transient or permanent deficits in cognitive ability or motor control.

For example, in Parkinson's disease (PD), the premature loss of dopaminergic transmission in the *substantia negra* disturbs the basal ganglia loops and motor output circuitry of the brain, causing tremor, gait disturbance, rigidity, and deficits in cognitive function. Demyelination – the phenotype of multiple sclerosis (MS) – reduces the conductivity of axons, which alters their ability to convey information to distant circuits. Stroke, while focal in nature, induces complex

Future Trends in Microelectronics: Journey into the Unknown, First Edition.
Edited by Serge Luryi, Jimmy Xu, and Alexander Zaslavsky.
© 2016 John Wiley & Sons, Inc. Published 2016 by John Wiley & Sons, Inc.

and far-reaching effects on the central nervous system (CNS). Less subtle, but particularly palpable is the dramatic consequence resulting from severe spinal cord injury (SCI), which, in extreme cases, can make a person completely unable to control or interact with the world around them. Nervous system disorders have long-term health, economic, and social consequences. Despite the best available medical treatments, millions of individuals endure a long life with sensorimotor and cognitive deficits that dramatically affect their quality of life.

The goal of this chapter is to discuss the potential impact that advances in *wireless* neurotechnology may provide persons with neuromotor disease and insult, but also to explore fundamental basic research avenues enabled by leveraging wireless telemetry in human and nonhuman brain science. Although work in progressing across many labs and in many corners of the world to address the devastating effects of neuromotor disease, here we will focus our discussion on the restoration of ambulatory function after SCI, as it exemplifies how state-of-the-art wireless microelectronic technology is being applied to solve biomedical therapeutic challenges for both reading from and writing into the nervous system. A conceptual schematic of our approach is presented in Fig. 1.

2. Why go wireless?

Interestingly, the computational power and multiscale connections between short (cortical) and long-range (spinal cord) neural circuits that produce movement remain poorly understood. Many of our best insights into neural computation derive from studying how single brain areas enable trained animals to solve simple, highly constrained tasks with fully observable sense data. Tethered and constrained experimental paradigms for nonhuman primates, for example, have enabled computational neuroscientists to develop sophisticated data-driven models of cognitive processing, yet have been limited principally to studying brain activity within an isolated context. To understand the dynamic neural processes that mediate fluid, natural human behavior, such as dexterous manipulation of objects with multiple parallel sensory cues, we must study recurrent neural computations spanning many interacting brain regions during complex and naturalistic behaviors. This cannot be accomplished with today's tethered electronic connections. Such connections not only physically limit the areas of implantation due to their size and cabling mechanics but also functionally limit the behavioral repertoire possible to explore. Decoding strategies for brain-controlled interfaces must coalesce dynamic data entering the body along multiple sensory pathways in order to synthesize a stable decision. In order to achieve this, synergistic advances must be made in the key areas of neurotechnology and computational neuroscience so that researchers can interpret neural dynamics at multiple scales in real time during unconstrained movements applicable to human brain–computer communication.

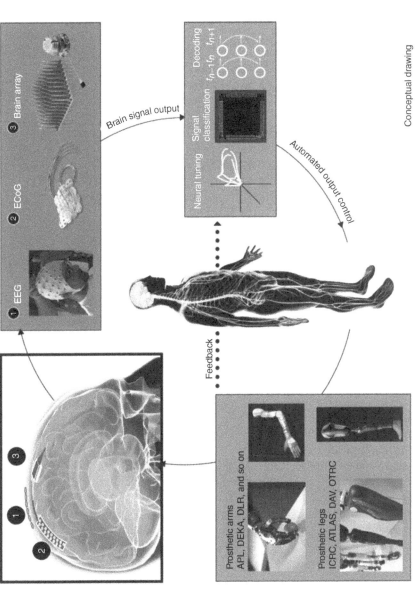

Figure 1. Conceptual illustration of a brain–machine interface technology. Neural data is collected from a variety of neural interfaces, then translated (decoded) into meaningful output control signals through machine learning, and finally used to control a motor prosthetic. (Source: https://commons.wikimedia.org/wiki/File:Brain-Controlled_Prosthetic_Arm_2.jpg. Public domain; https://commons.wikimedia.org/wiki/File:Prosthetic_Limbs_at_Headley_Court_MOD_45157827.jpg. Photo: Cpl Richard Cave RLC (Phot)/MOD).

3. One wireless recording solution used to explore primary motor cortex control of locomotion

Considerable investment and decades of intensive research have brought great insight into how movement[1-5] and proprioception or sensation[6-9] are accomplished in upper limb processing pathways. Such insight has brought about impressive demonstrations of brain interfaces to external effectors replacing upper limb control.[10-13] Conversely, lower limb systems remain the poorly understood cousins of their biologically elevated counterparts. Primate locomotion relies on dynamic interactions between cortical, brainstem, and spinal circuits. While the contribution of primary motor cortex to skilled hand movement has been studied extensively, its role during natural locomotor behaviors remains largely unknown.[14] Early work performed in cat quadrupedal locomotion was focused on active forelimb control and accommodation after external perturbation.[15-19] Only recently have hind limb areas of primary motor cortex recordings been the subject of electrophysiological experiments in rodents,[20] with the anatomical area still under some debate.[21] Perhaps this limited basic research on locomotor centers in the cortex is due to the prevailing dogma that walking is a highly automated process largely organized within the spinal cord, putting into question whether the primate motor cortex participates meaningfully in locomotor control.

While few attempts have been made to access neuronal modulation from cortical regions of nonhuman primates during locomotion,[22] such recordings were performed under highly constrained conditions due to the need for cabled electronics. Recently, a limited data set of spiking activity was collected from premotor cortical areas of a rhesus monkey walking on a treadmill.[23, 24]

In order to move beyond constrained conditions, we designed, built, and deployed a miniaturized wireless high data rate neurosensor platform, shown in Fig. 2. This device leveraged numerous earlier advances from several laboratories[25-27] and integrates custom amplification microelectronic circuitry that amplifies and multiplexes 100 channels of broadband neural signals, transmits data digitally at high sampling rates (20 kSps/channel) up to a 5 m distance via a single-input multiple-output (SIMO) wireless link for extended spatial coverage; and operates on a single one-half AA Li-ion battery continuously for more than 48 hours. The neurosensor weighs only 46.1 g and incorporates three ultralow-power, custom-designed ASICs for signal amplification, packaging, and transmission. The assembled components are protected from mechanical and electrical impact by a static-dissipative, carbon-fiber-reinforced poly-ether-ether-ketone enclosure.[28]

This wireless technology enabled acquisition of full spectrum neuronal population dynamics in untethered freely moving monkeys. In an experiment to more fully characterize hind limb area of primary motor cortex, we implanted a

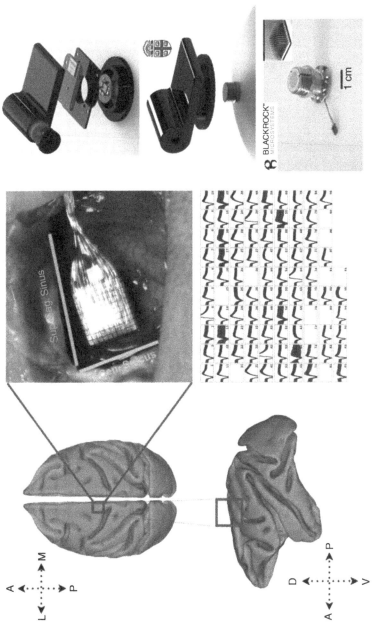

Figure 2. A silicon-based microelectrode array (MEA) was implanted into primary motor cortex of the nonhuman primate brain. A transcutaneous pedestal connection exits the skin for sampling of the signal. Traditionally, a cable is connected to this pedestal. Here, we attach a wireless neurosensor capable of transmitting the high-bandwidth data meters, untethering the subject from large recording electronics. (Source: Reproduced with permission of Blackrock Microsystems).

Figure 3. A wireless neuromotor recording and stimulation platform: neural data is recorded wirelessly and transmitted meters away to a control center, where decoding and interpretation drive outgoing commands for spinal stimulation. In parallel, muscle activity and kinematic movements are recorded for basic research.

96-electrode microelectrode array (MEA) in the hind limb area of the primary motor cortex along with bipolar electrodes into a pair of agonist and antagonist muscles for each joint of the contralateral leg in order to record electromyography (EMG) activity (a total of eight muscles were acquired wirelessly). The EMG activity and neural data were collected in conjunction with whole-body kinematics while the monkeys were walking quadrupedal, without behavioral constraints, on a treadmill across a range of velocities – see Fig. 3. All animals displayed robust and reproducible modulation of motor cortex spiking activity that was distributed across the entire gait cycle. The temporal structure of neuronal ensemble modulation coevolved with speed-dependent changes in joint angles and leg muscle activity patterns.[28] We found that the leg area in primary motor cortex contains information that can be leveraged to build a locomotor prosthetic. Cross-task neural and kinematic data were recorded from five nonhuman primates, underscoring the robustness of such a platform prosthetic in a diverse human population. What is likely, given the results of that study, is that there is at least a shared control by the motor cortex in the production of locomotion, enough to inspire us to propose a platform to bypass a spinal lesion and enable active, dynamic modulation of gait in closed-loop, that is, gait modulation controlled by the subject's intention: a brain–spinal interface.[29]

But how should we modulate gait?

4. Writing into the nervous system with epidural electrical stimulation of spinal circuits effectively modulates gait

Thousands of years ago (and until the 20th century in some countries) humans used the electric eels to treat pain – one of the first documented uses of electricity to deliver therapy. Since then, advances in material science, charge storage, and control of charge delivery have enabled more precise perturbations of the nervous system. The challenge for the next generation of neuroprosthetic treatments is to capitalize on the myriad technologies and methods at our disposal to initiate a productive conversation with the nervous system wherein devices not only deliver therapy but also continuously titrate themselves based on neural states and detected motor impairments.

Let us first take a step back and explore the historical interpretations of spinal dynamics. The spinal sensorimotor infrastructure has traditionally been viewed as an assembly of reflex subsystems and central pattern generators that produce automated and stereotypical motor activity in response to sensory input or descending command.[30] The high degree of automaticity embedded in spinal circuits enables the execution of complex motor behaviors with considerable precision without conscious thought. The spinal brain acts as a smart information-processing interface that integrates dynamic input from sensory ensembles and, on this basis, makes decision on how to continuously adjust motor output in order to meet environmental constraints while maintaining stability. Despite these advanced properties, the markedly depressed state of the spinal cord postinjury prevents the production of standing and walking. Consequently, much effort has been invested in developing paradigms to replace the missing sources of neuromodulation and excitation that are normally delivered to spinal sensorimotor circuits from higher cortical areas to coordinating movement. Electrical stimulation has been the primary strategy used to compensate for the interrupted source of spinal excitation after injury – see Fig. 4. Continuous electrical stimulations applied to the dorsal roots,[31] over the dorsal aspect of the spinal cord,[32] or directly into the ventral horn of lumbar segments, have showed the ability to elicit standing and stepping patterns in animal models and humans with SCI,[33] PD,[34] and MS.[35] In combination with monoamine agonists, epidural electrical stimulation (EES) of lumbosacral circuits has been able to restore full weight-bearing locomotion in rats with complete SCI.[36] This electrochemical neuroprosthesis replaces the missing source of neuromodulation and excitation after the interruption of descending pathways, although the exact mechanisms remain unclear.

Neurotechnology and stimulation protocols are at the early stages of development. Empirical knowledge and visual observations have guided electrode positioning, as well as the selection of electrode configurations and stimulation parameters. Extensive mappings revealed that various locations and stimulation

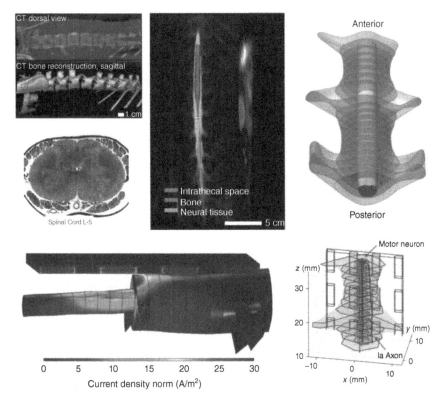

Figure 4. Biophysical models of spinal anatomy enable accurate estimation of electrical stimulation effects on spinal roots and motor reflexes.

profiles are necessary to facilitate standing, stepping, and isolated movements. This manual tuning is inherently impractical and suboptimal. These experiments emphasize the need to establish a mechanistic framework to personalize multisite stimulation algorithms, and develop closed-loop control systems that take full advantage of this paradigm to facilitate movement in motor-impaired subjects.

Integration of lower limb sensorimotor information into generalized control after limb loss or disease stands to drastically improve postinjury success in everyday tasks, for example, walking, control of balance, removal of phantom limb sensation, and "simple" control of the sit to stand transition. Rapid reintegration of stretch, pressure, and texture information into the ongoing proprioceptive dynamics will enable limb replacements that truly integrate with the host. Recent technological advancements have made read and write access to the nervous system possible in untethered and freely moving subjects,[28, 37] leading us to envision a novel, chronic open communication window for sensorimotor prosthesis to restore proprioception from and control of lower limb prostheses.[29]

Often, in order to gain access to the spinal-specific and cortical-specific mechanisms controlling motor function, lesion studies are conducted to remove the contribution of a specific pathway to functional recovery, for example, pyramidal tract lesions, hemisectioning of the spinal cord, and induced stroke in the motor cortex. Experiments conducted in nonhuman primates have significantly contributed to identifying primate-specific mechanisms of recovery after partial SCI. For example, it was shown[38] that the recovery of fine manual skills and locomotion after a cervical hemisection in rhesus monkeys is associated with the extensive sprouting of spinal cord midline crossing corticospinal fibers, which are abundant in primates[39] but rare in rodents.[40] Along the same lines, Isa and coworkers[41] showed that recovery of finger dexterity after a partial cervical SCI in monkeys relies on extensive regions of the contralesional primary motor cortex and bilateral premotor cortex, which have no equivalent in rodents.

However, current nonhuman primate injury-based models for studying the dysfunctional locomotor system limit our ability to observe, quantify, and enhance recovery of function in a few critical ways: (i) after a surgical or impact injury, the kinematic and nervous system state is constantly in flux and competing with inflammatory and compensatory responses; (ii) intrasubject variability cannot be assessed as the lesion only occurs once; (iii) as recovery invariably occurs, an iterative development process cannot be used, hampering the development of restorative interfaces; and (iv) physical lesions have systemic effects, complicating care and experimental progress.[42] These experiments also highlight the limitations of current methodologies of studying recovery in within the spinal system as well as the dynamics of reorganization in the forebrain, and fail to provide a stable platform for the development of restorative neurotechnology.

5. Genetic technology brings a better model to neuroscience

If we imagine a reversible, noninjuring lesion model of the nervous system, many of the physical injury model limitations described above disappear. This problem could be solved if it were possible to *block* the corticospinal or reticulospinal pathways specifically and temporarily. Such a method would abolish inflammatory complications, its reversibility will allow for multiple trials and developmental iterations within the same subject, and specific targeting will greatly reduce the systemic effects of current lesion models. The last decade has produced a toolbox of methods for genetic dissection of the nervous system.[43–45] The combination of both retrograde and anterograde viruses (e.g., double-virus) enables unprecedented neural pathway selectivity and specificity. Control of gene expression in mammalian cells by tetracycline-locked promoters can enable lasting inhibition of many days.[46, 47] Isa and coworkers of the National Institute for Physiological Sciences in Osaka, Japan used these viral-based methods to show temporary shutdown of the propriospinal (PN) connections in the ventral horn between

C6 and T1. Sensitivity of the expressed protein to the antibiotic doxycycline[48] enabled the elaboration of the contribution of PN neurons toward hand dexterity. In order to specifically remove the contributions of supraspinal structures to locomotor pattern generation, we use these methods to block the linkage among *cortical, midbrain, and spinal* targets. We believe this will yield a significant contribution to both basic science methodologies and lay the foundation for the demonstration of a motor restorative interface to gain fundamental knowledge about the nervous system and to use that knowledge to reduce the burden of neurological disease.

6. The wireless bridge for closed-loop control and rehabilitation

Many years of prior research within the scientific community has culminated in the proof-of-concept demonstration that neural information from the motor cortex could be used to drive stimulation signals to the spinal cord. However, there currently exists no dynamically adapting electrical neuromodulation therapy to alleviate or modulate gait function. While recent studies by our group have shown that sufficient data exists to control such modulation, the integration of real-time, closed-loop technologies into a brain–spinal interface platform has not previously been accomplished and is nontrivial. Critically, in order to drive stimulation pulses in meaningful spatiotemporal patterns, we must use a wireless implantable *stimulator*. Fortunately, we found such a solution in industry – a modified commercial implantable pulse generator (or IPG).

These devices were wirelessly connected with a common control state-machine running with complete system latencies near 100 ms from neural and kinematic recording, to decoding of intention, and finally modification of spinal stimulation parameters sufficient for therapeutic effect. The resulting innovation, shown in Fig. 5, is a platform that will soon be evaluated for clinical translation.

The brain–spinal interface platform will also be a resource for the neuroscientific community. Such a system enables untethered perturbation of the nervous system, with simultaneous high fidelity and high-resolution reporting of neural and kinematic correlates. Short-term, single trial use of the BSI technology may temporarily improve gait, but it is unclear what lasting impact closed-loop spinal neuromodulation will have on recovery and compensation of function.

Furthermore, continuous use may expose functionalities previously unexpected. Given the long battery life and untethered nature of the platform, researchers will study subjects using the brain–spinal interface while in their home environment. In addition, neuroscientists will have the ability to record neural data continuously, over extended periods of time from days to weeks. Such recordings enable the evaluation of "prosthetic integration" or how the subject uses the brain–spinal interface beyond primary objective to restore gait. "Prosthetic learning"[49–52] may be revealed as long-term experiments are

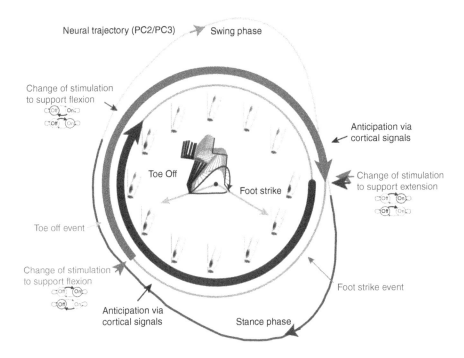

Figure 5. A brain–spinal interface timing diagram illustrating key changes in stimulation profile based on neural activity anticipating gait modification. Neural recordings enabled adjustment for stimulation command transmission delay.

performed. Through advances in neurotechnology, computational neuroscience, and closed-loop interactions between electrophysiological recording and stimulation systems, we hope to advance in a nonincremental way our ability to understand continuous and dynamic changes of neuromotor circuits with unprecedented temporal and spatial breadth. Such advancements will have vast implications on biobehavioral research, neuromotor disease diagnosis, and neurorestorative therapies.

7. Conclusion

This chapter discusses a platform of technologies that push neuroscience into the wireless age with the hope of opening new experimental and therapeutic pathways. As we progress toward therapeutic approaches, we are mindful of the patient population. For example, there have been over 1573 major limb amputations to-date in military operations in Afghanistan and Iraq. In addition, in 2013 alone there were nearly 27,000 veterans with spinal injuries and related disorders cared for by Veterans Affairs. These veterans stand to benefit directly from the research discussed. In addition, millions of people worldwide live with SCI. There currently exists no clinical electrical neuromodulation therapy to alleviate or modulate gait function of an individual after amputation or SCI. Functional electrical stimulation (FES) has been the most productive tool to temporarily restore stance to paralyzed individuals,[53–55] although peripheral nerve stimulation is being studied.[56, 57] To achieve transformational change for these patients, prosthetic technologies must advance beyond tonic stimulation of muscles and into biophysical, model-based multisegment stimulation modulated across time and space. Temporal modulation will integrate more effectively into the ongoing dynamics of remaining lumbar spinal circuits;[58] spatial modulation will enable activation of multiple spinal segments, based on computational models of anatomical activation[59] through EES technology. Engagement and, as needed, direct control by the brain of therapeutic stimulation will enable neural plasticity, important for rehabilitation over long periods, and more natural control limb kinematics. Further, by combining EES with cortical stimulation of the sensory areas (in conjunction to recording from the motor area), one envisions a true internally closed-loop sensorimotor prosthetic system.

Acknowledgments

This work was supported in part by the International Foundation for Research in Paraplegia (IRP Grant P152), the DARPA Young Faculty Award (YFA, Grant D15AP00112), and Brown University.

References

1. J. M. Carmena, M. A. Lebedev, R. E. Crist, *et al.*, "Learning to control a brain–machine interface for reaching and grasping by primates," *PLoS Biol.* **1**, E42 (2003).
2. M. Velliste, S. Perel, M. C. Spalding, A. S. Whitford, and A. B. Schwartz, "Cortical control of a prosthetic arm for self-feeding," *Nature* **453**, 1098–1101 (2008).
3. C. T. Moritz, S. I. Perlmutter, and E. E. Fetz, "Direct control of paralysed muscles by cortical neurons," *Nature* **456**, 639–642 (2008).
4. C. Ethier, E. R. Oby, M. J. Bauman, and L. E. Miller, "Restoration of grasp following paralysis through brain-controlled stimulation of muscles," *Nature* **485**, 368–371 (2012).
5. M. D. Serruya, N. G. Hatsopoulos, L. Paninski, M. R. Fellows, and J. P. Donoghue, "Instant neural control of a movement signal," *Nature* **416**, 141–142 (2002).
6. M. Bergenheim, E. Ribot-Ciscar, and J. P. Roll, "Proprioceptive population coding of two-dimensional limb movements in humans: I. Muscle spindle feedback during spatially oriented movements," *Exp. Brain Res.* **134**, 301–310 (2000).

7. S. N. Baker, M. Chiu, and E. E. Fetz, "Afferent encoding of central oscillations in the monkey arm," *J. Neurophysiol.* **95**, 3904–3910 (2006).

8. G. A. Tabot, J. F. Dammann, J. A. Berg, *et al.*, "Restoring the sense of touch with a prosthetic hand through a brain interface," *Proc. Nat. Acad. Sci.* **110**, 18279–18284 (2013).

9. M. C. Dadarlat, J. E. O'Doherty, and P. N. Sabes, "A learning-based approach to artificial sensory feedback leads to optimal integration," *Nature Neurosci.* **18**, 138–144 (2015).

10. G. R. Muller-Putz, R. Scherer, G. Pfurtscheller, and R. Rupp, "EEG-based neuroprosthesis control: A step towards clinical practice," *Neurosci. Lett.* **382**, 169–174 (2005).

11. L. R. Hochberg, D. Bacher, B. Jarosiewicz, *et al.*, "Reach and grasp by people with tetraplegia using a neurally controlled robotic arm," *Nature* **485**, 372–395 (2012).

12. J. L. Collinger, M. A. Kryger, R. Barbara, *et al.*, "Collaborative approach in the development of high-performance brain–computer interfaces for a neuro-prosthetic arm: Translation from animal models to human control," *Clin. Transl. Sci.* **7**, 52–59 (2014).

13. V. Gilja, C. Pandarinath, C. H. Blabe, *et al.*, "Clinical translation of a high-performance neural prosthesis," *Nature Med.* **21**, 1142–1145 (2015).

14. R. N. Lemon, "Descending pathways in motor control," *Ann. Rev. Neurosci.* **31**, 195–218 (2008).

15. T. Drew, "Motor cortical cell discharge during voluntary gait modification," *Brain Res.* **457**, 181–187 (1988).

16. T. Drew, W. Jiang, and W. Widajewicz, "Contributions of the motor cortex to the control of the hindlimbs during locomotion in the cat," *Brain Res. Rev.* **40**, 178–191 (2002).

17. D. M. Armstrong and T. Drew, "Discharges of pyramidal tract and other motor cortical neurons during locomotion in the cat," *J. Physiol.* **346**, 471–495 (1984).

18. I. N. Beloozerova and M. G. Sirota, "The role of motor cortex in the control of accuracy of locomotor movements in the cat," *J. Physiol.* **461**, 1–25 (1993).

19. I. N. Beloozerova, B. J. Farrell, M. G. Sirota, and B. I. Prilutsky, "Differences in movement mechanics, electromyographic, and motor cortex activity between accurate and nonaccurate stepping," *J. Neurophysiol.* **103**, 2285–2300 (2010).

20. W. Song, and S. F. Giszter, "Adaptation to a cortex-controlled robot attached at the pelvis and engaged during locomotion in rats," *J. Neurosci.* **31**, 3110–3128 (2011).

21. S. B. Frost, M. Iliakova, C. Dunham, *et al.*, "Reliability in the location of hindlimb motor representations in Fischer-344 rats: Laboratory investigation," *J. Neurosurg.* **19**, 248–255 (2013).

22. N. A. Fitzsimmons, M. A. Lebedev, I. D. Peikon, and M. A. Nicolelis, "Extracting kinematic parameters for monkey bipedal walking from cortical neuronal ensemble activity," *Front. Integr. Neurosci.* **3**, article no. 3 (2009).

23. J. D. Foster, P. Nuyujukian, O. Freifeld, *et al.*, "A freely-moving monkey treadmill model," *J. Neural Eng.* **11**, 046020 (2014).
24. D. A. Schwarz, M. A. Lebedev, T. L. Hanson, *et al.*, "Chronic, wireless recordings of large-scale brain activity in freely moving rhesus monkeys," *Nature Methods* **11**, 670–676 (2014).
25. W. R. Patterson, Y. K. Song, C. W. Bull, *et al.*, "A microelec-trode/microelectronic hybrid device for brain implantable neuroprosthesis applications," *IEEE Trans. Biomed. Eng.* **51**, 1845–1853 (2004).
26. Y.-K. Song, W. R. Patterson, C. W. Bull, *et al.*, "A brain implantable microsystem with hybrid RF/IR telemetry for advanced neuroengineering applications," *IEEE Eng. Med. Biol. Soc.* 2007, 445–448 (**2007**).
27. Y.-K. Song, D. A. Borton, S. Park, *et al.*, "Active microelectronic neurosensor arrays for implantable brain communication interfaces," *IEEE Trans. Neural Syst. Rehabil. Eng.* **17**, 339–345 (2009).
28. M. Yin, D. A. Borton, J. Komar, *et al.*, "Wireless neurosensor for full-spectrum electrophysiology recordings during free behavior," *Neuron* **84**, 1170–1182 (2014).
29. D. A. Borton, M. Bonizzato, J. Beauparlant, *et al.*, "Corticospinal neuro-prostheses to restore locomotion after spinal cord injury," *Neurosci. Res.* **78**, 21–29 (2014).
30. S. Grillner, "Biological pattern generation: the cellular and computational logic of networks in motion," *Neuron* **52**, 751–766 (2006).
31. D. Barthelemy, H. Leblond, and S. Rossignol, "Characteristics and mechanisms of locomotion induced by intraspinal microstimulation and dorsal root stimulation in spinal cats," *J. Neurophysiol.* **97**, 1986–2000 (2007).
32. K. Minassian, B. Jilge, F. Rattay, *et al.*, "Stepping-like movements in humans with complete spinal cord injury induced by epidural stimulation of the lumbar cord: electromyographic study of compound muscle action potentials," *Spinal Cord* **42**, 401–416 (2004).
33. V. R. Edgerton, and S. Harkema, "Epidural stimulation of the spinal cord in spinal cord injury: current status and future challenges," *Expert Rev. Neurother.* **11**, 1351–1353 (2011).
34. M. B. Santana, P. Halje, H. Simplicio, *et al.*, "Spinal cord stimulation alleviates motor deficits in a primate model of Parkinson disease," *Neuron* **84**, 716–722 (2014).
35. L. S. Illis, A. E. Oygar, E. M. Sedgwick, and M. A. Awadalla, "Dorsalcolumn stimulation in the rehabilitation of patients with multiple sclerosis," *Lancet* **307**, 1383–1386 (1976).
36. R. van den Brand, J. Heutschi, Q. Barraud, *et al.*, "Restoring voluntary control of locomotion after paralyzing spinal cord injury," *Science* **336**, 1182–1185 (2012).
37. D. A. Borton, M. Yin, J. Aceros, and A. Nurmikko, "An implantable wireless neural interface for recording cortical circuit dynamics in moving primates," *J. Neural Eng.* **10**, 026010 (2013).

38. E. S. Rosenzweig, G. Courtine, D. L. Jindrich, *et al.*, "Extensive spontaneous plasticity of corticospinal projections after primate spinal cord injury," *Nature Neurosci.* **13**, 1505–1512 (2010).

39. E. S. Rosenzweig, J. H. Brock, M. D. Culbertson, *et al.*, "Extensive spinal decussation and bilateral termination of cervical corticospinal projections in rhesus monkeys," *J. Comp. Neurol.* **513**, 151–163 (2009).

40. C. Brosamle and M. E. Schwab, "Cells of origin, course, and termination patterns of the ventral, uncrossed component of the mature rat corticospinal tract," *J. Comp. Neurol.* **386**, 293–303 (1997).

41. Y. Nishimura and T. Isa, "Cortical and subcortical compensatory mechanisms after spinal cord injury in monkeys," *Exp. Neurol.* **235**, 152–161 (2012).

42. L. H. Sekhon and M. G. Fehlings, "Epidemiology, demographics, and pathophysiology of acute spinal cord injury," *Spine* **26**, S2–S12 (2001).

43. G. Nagel, T. Szellas, W. Huhn, *et al.*, "Channelrhodopsin-2, a directly light-gated cation-selective membrane channel," *Proc. Nat. Acad. Sci.* **100**, 13940–13945 (2003).

44. F. Zhang, L.-P. Wang, E. S. Boyden, and K. Deisseroth, "Channelrhodopsin-2 and optical control of excitable cells," *Nature Methods* **3**, 785–792 (2006).

45. I. Diester, M. T. Kaufman, M. Mogri, *et al.*, "An optogenetic toolbox designed for primates," *Nature Neurosci.* **14**, 387–397 (2011).

46. M. Gossen and H. Bujard, "Tight control of gene expression in mammalian cells by tetracycline-responsive promoters," *Proc. Nat. Acad. Sci.* **89**, 5547–5551 (1992).

47. M. Yamamoto, N. Wada, Y. Kitabatake, *et al.*, "Reversible suppression of glutamatergic neurotransmission of cerebellar granule cells *in vivo* by genetically manipulated expression of tetanus neurotoxin light chain," *J. Neurosci.* **23**, 6759–6767 (2003).

48. M. Kinoshita, R. Matsui, S. Kato, *et al.*, "Genetic dissection of the circuit for hand dexterity in primates," *Nature* **487**, 235–238 (2012).

49. J. M. Carmena, "Advances in neuroprosthetic learning and control," *PLoS Biol.* **11**, e1001561 (2013).

50. S. Dangi, A. L. Orsborn, H. G. Moorman, and J. M. Carmena, "Design and analysis of closed-loop decoder adaptation algorithms for brain–machine interfaces," *Neural Comput.* **25**, 1693–1731 (2013).

51. K. Ganguly, and J. M. Carmena, "Emergence of a stable cortical map for neuroprosthetic control," *PLoS Biol.* **7**, e1000153 (2009).

52. K. V. Shenoy and J. M. Carmena, "Combining decoder design and neural adaptation in brain–machine interfaces," *Neuron* **84**, 665–680 (2014).

53. V. K. Mushahwar, P. L. Jacobs, R. A. Normann, R. J. Triolo, and N. Kleitman, "New functional electrical stimulation approaches to standing and walking," *J. Neural Eng.* **4**, S181–S197 (2007).

54. R. J. Triolo, S. N. Bailey, M. E. Miller, L. M. Lombardo, and M. L. Audu, "Effects of stimulating hip and trunk muscles on seated stability, posture, and reach after spinal cord injury," *Arch. Phys. Med. Rehabil.* **94**, 1766–1775 (2013).

55. M. L. Audu, L. M. Lombardo, J. R. Schnellenberger, *et al.*, "A neuropros-thesis for control of seated balance after spinal cord injury," *J. Neuroeng. Rehabil.* **12**, article no. 8 (2015).

56. L. E. Fisher, D. J. Tyler, J. S. Anderson, and R. J. Triolo, "Chronic stability and selectivity of four-contact spiral nerve-cuff electrodes in stimulating the human femoral nerve," *J. Neural Eng.* **6**, 046010 (2009).

57. K. H. Polasek, H. A. Hoyen, M. W. Keith, R. F. Kirsch, and D. J. Tyler, "Stimulation stability and selectivity of chronically implanted multicontact nerve cuff electrodes in the human upper extremity," *IEEE Trans. Neural Syst. Rehabil. Eng.* **17**, 428–437 (2009).

58. N. Wenger, E. M. Moraud, S. Raspopovic, *et al.*, "Closed-loop neuromodu-lation of spinal sensorimotor circuits controls refined locomotion after com-plete spinal cord injury," *Sci. Transl. Med.* **6**, 255ra133 (2014).

59. M. Capogrosso, N. Wenger, S. Raspopovic, *et al.*, "A computational model for epidural electrical stimulation of spinal sensorimotor circuits," *J. Neu-rosci.* **33**, 19326–19340 (2013).

3.3

Nanopower-Integrated Electronics for Energy Harvesting, Conversion, and Management

A. Romani, M. Dini, M. Filippi, M. Tartagni, and E. Sangiorgi
Department of Electrical, Electronic, and Information Engineering "G. Marconi" and Advanced Research Center on Electronic Systems "E. De Castro", University of Bologna, Via Venezia 52, 47521 Cesena FC, Italy

1. Introduction

The increasing interest on pervasive sensor networks and the steady development of electronic devices with low power consumption motivates the research on electronic systems capable of harvesting energy from the surrounding environment. Currently, most energy harvesters can provide in practical cases an output power density of about 10–100 µW/cm^3.[1] In this scenario, mechanical vibrations, thermal gradients, and photovoltaics represent the most promising power sources for supplying portable low-power electronic equipment.

For a successful application, it becomes essential to efficiently convert such low levels of input power. In this context, electronic interfaces based on recent commercial discrete components provide a cost-effective and easily implementable solution. Notable examples are reported in Refs 2–5, where intrinsic current consumptions down to 1 µA or less are achieved. However, in order to keep consumption low, discrete electronics must necessarily implement simplified, and thus less efficient, control schemes. In addition, nonnegligible, power-hungry parasitics are unavoidable when connecting discrete components.

The use of microelectronic technologies allows a series of circuit optimizations paving the way toward the exploitation of sub-microwatt power regimes, thanks to the significant reduction of parasitics and to the efficiency of custom-designed circuit topologies. As a direct consequence, the break-even in the power budget is achieved with smaller transducers or lower input power densities. This promises new applications that until recently have been stymied by the lack of power. Notable examples are reported in Refs 6–10, where several types of energy transducers providing up to a few microwatts are managed by fully custom-integrated circuits.

Future Trends in Microelectronics: Journey into the Unknown, First Edition.
Edited by Serge Luryi, Jimmy Xu, and Alexander Zaslavsky.

This chapter reviews some of the most promising integrated circuits (ICs) for power harvesting, conversion, and management, achieved either by industry or academia, intended for exploiting several different types of energy transducers. A specific case study consisting in a nanopower IC for harvesting power from multiple energy sources is also reviewed. The focus of the chapter is on the advantages conferred by nanoelectronic ICs in this specific field.

2. Commercial ICs for micropower harvesting

Many commercial ICs have been marketed over the years claiming to meet stringent energy-harvesting requirements. However, only very recently did commercial off-the-shelf components reach the ability of dealing with power sources providing input power down to few microwatts. Examples of recent IC implementations are chips such as the Texas Instruments bq25504[11] and the STMicroelectronics SPV1050.[12] Both ICs manage power sources down to few microwatts by implementing internal nanocurrent bias networks and by pursuing trade-offs between intrinsic power consumption and energy conversion efficiency. An ultralow intrinsic consumption is the key for the exploitation of the lowest possible input power densities. Both ICs perform maximum power point tracking (MPPT) by means of the fractional open circuit voltage (FOCV) technique, depicted in Fig. 1. This technique computes the maximum power point (MPP) as a fixed fraction of the open circuit voltage output by the energy transducer. This also means that power conversion is briefly stopped during the refresh of the MPPT. However, this technique offers negligible power consumption and a satisfactory trade-off with energy conversion efficiency. In the aforementioned ICs, the FOCV is in both cases obtained by means of resistive dividers. Large resistances reduce the consumed current but slow down the sampling time of the divided voltage. Another interesting feature of the above energy-harvesting ICs is the capability of performing a battery-less cold start-up. According to their datasheets, the bq25504 and SPV1050 chips can self-start, respectively, from 330 and 550 mV, with a minimum input power of few microwatts.

Excellent performance is also reported for the Cypress S6AE101A,[13] which achieves super-low-power operation with an intrinsic consumption current of only 250 nA and a required start-up power of only 1.2 µW. According to the manufacturer, even the negligible levels of power obtained from compact solar cells under low-brightness environments of approximately 100 lm can be exploited. Unlike the previously mentioned ICs that implement inductor-based switching converters, the S6AE101A transfers power to an output capacitor using built-in switch control, by turning on a power switching circuit for as long as the capacitor voltage is within a preset maximum and minimum range. In case

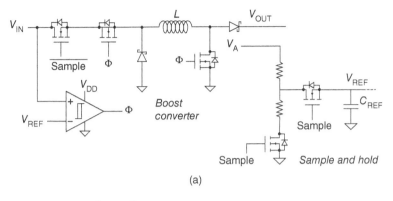

(a)

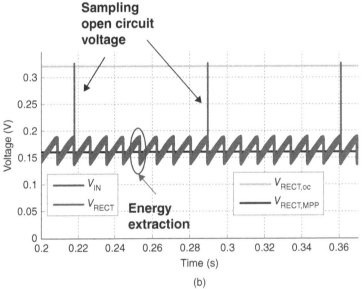

(b)

Figure 1. The FOCV MPPT technique: the basic circuit (a) and reference waveforms (b) during power conversion.

of low power levels, energy can be provided from connected primary batteries for auxiliary power. However, with respect to the previously mentioned ICs, this circuit requires a higher minimum input voltage in the order of 1 V.

Although the performance achieved by recent products is notable, there are still margins for improvement in terms of minimum required power and activation voltage. The following sections highlight recent achievements presented in scientific literature, with a special focus on nanopower-integrated solutions.

3. **State-of-the-art integrated nanocurrent power converters for energy-harvesting applications**

Many recently presented commercial integrated circuits claim to fulfill strict energy-harvesting requirements. However, intrinsic current consumptions on the order of microamperes or less have been reported only recently, largely in the scientific literature rather than commercial products – some recently published nanopower converters are collected in Table 1. These types of devices pave the way toward the exploitation of weaker power sources.

Optimized solutions based on discrete electronic components can reduce the intrinsic power consumption of energy-harvesting circuits. However, the practical baseline power limit of this type of circuits is in the order of tens of microwatts. To cite an example, a power management system for RF energy harvesting has been reported, based on an ultralow-power microcontroller unit (MCU) and an MPPT circuit built with on-chip peripherals.[2] In Ref. 14, an FOCV MPPT technique was implemented with a similar configuration. However, even optimized MCU-based circuits typically consume about $10\,\mu W$, which poses a limit to the possible exploitable sources. Among the circuits with lower consumptions, a self-powered resonant DC/DC converter was reported with an input power lower than $4\,\mu W$ by Adami et al.[15]

However, in order to operate in sub-microwatt regimes, the very low parasitics of circuits implemented in modern microelectronic processes can confer a significant advantage. In fact, several ICs reported in literature achieve even better performance, partly due to the implementation of nanopower techniques. A first example reported in Refs 7, 16 is an integrated power management circuit consuming $330\,nA$ with additional features such as battery charging and cold start-up from $300\,mV$. Similar power consumptions with the capability of managing multiple types of transducers are reported in Refs 8, 17. An integrated DC/DC converter with a low dropout regulator (LDO) is reported to consume in standby about $520\,nA$,[18] with the capability of battery-free self-starting at input voltages down to $200\,mV$. In these cases, the most frequently adopted technique is switching power conversion, and FOCV MPPT is assumed to be a good compromise between converted and consumed power. An alternative technique that exploits a buck switching converter with dynamic on–off time calibration and a regulated output voltage was reported to consume $217\,nW$.[19] However, unlike boost converters, step-down topologies are not suitable for continuous energy storage, because the voltage on the output capacitor is limited. Inductor-less charge pumps[20] with comparable intrinsic consumption and lower activation voltages down to $150\,mV$[21] have also been proposed recently. In general, charge pumps have lower efficiencies than switching converters. Low-threshold MOSFETs allow to decrease the start-up voltage.[22–24] The problem of activation voltage is particularly relevant in thermoelectric energy harvesting, where the voltage available is on the order of tens of millivolts. Lower input voltages have been exploited by implementing step-up oscillators.[25–27]

Work	Type of sources	Features	Quiescent current/power	Minimum input power	Efficiency	Input voltage range
45	RF	MPPT with on–off time regulation	181 nA	NA	<95%	1.2–2.5 V
21	DC	Dynamic body bias, adaptive dead time	<0.5 µW	NA	34%/72% at 0.18/0.45 V	>0.15 V
7	DC	FOCV MPPT, battery charger, cold start-up	330 nA	5 µW	38%/>80% at 0.1/5 V	0.08–3 V (330 mV start-up)
46	Thermoelectric	Variation-tolerant FOCV MPPT, no output voltage regulation, battery required	NA	NA	72%	70–600 mV
6	PV	On-chip PV cell, double-boost converter	NA	1 µW	65%	>0.5 V
47	PV	Reconfigurable circuit power and speed	390 nW	NA	90%	0.9–2 V
8	Thermoelectric, PV, piezoelectric	Single shared inductor with asynchronous arbiter logic	431 nA (9 sources)	3 µW	89.6%	<5.5 V (1.65 V start-up)
18	DC	Asynchronous control logic, fast battery-less start-up, nanopower LDO, active 223 mV start-up circuit	121 nA	935 nW	77.1%	0.074–2.5 V (223 mV start-up)
44	Piezoelectric	SECE with residual charge inversion, fast battery-less start-up	160 nA	300 nW	<85%	<5.5 V (1.65 V start-up)

Table 1. A collection of recent nanopower energy-harvesting circuits.

Among the viable energy transducers for supplying low-power electronic systems, piezoelectric transducers have been extensively studied, and a significant amount of recent scientific literature has focused on various aspects, from transducer physics to power conversion circuits. In the latter context, significant advantages have been introduced by synchronized switch harvesting techniques,[28] that are typically based on the activation of LCR resonant circuits that are synchronized with vibrations. Because of the mainly capacitive output impedance of piezoelectric transducers, the peak voltage and the harvested power are usually increased. In addition, piezoelectric transducers limit converter intrinsic consumption because of very low activation rates and offer significant advantages because of their generally low oscillation frequencies, on the order of tens of Hz for macroscopic devices.

A well-known technique is the synchronized switch harvesting on inductor (SSHI),[28, 29] in which the transducer is connected to a rectifier bridge. The series connection of a switch and an inductor is connected in parallel to the transducer. When a peak-to-peak elongation ends, and the rectifier stops conducting, the switch is closed for a short time for inverting the polarity of voltage on the transducer. This inversion brings the transducer close to the opposite conduction threshold, so the next elongation starts from this favorable offset. The main consequence is that the rectifier, in an ideal case, is always conducting, and charge is continuously transferred to the output. However, since electrical charge is still transferred through the rectifier, the efficiency of the energy transfer depends on the operating point, and can be adversely impacted by very low or very high load currents. This drawback is not present in the synchronous electrical charge extraction (SECE) technique.[5, 30–34] In SECE, which is depicted in Fig. 2, the transducer is left in an open circuit state for most of the time. Only on voltage peaks, a series of switches connects the piezoelectric transducer to an inductor, so that charge is removed and the corresponding energy is stored in the magnetic field of the inductor. When the transducer is fully discharged, the inductor current has reached its maximum value. At that exact moment, switches between the transducer and inductor open, just as a new set of switches connects the charged inductor to a storage capacitor. The current decreases and in a way similar to a boost converter, as shown in Fig. 2. The SECE technique can also be implemented with a flyback topology,[3, 35] in order to reduce losses on freewheeling diodes. However, longer transients occur, and this increases energy losses, especially in case of large capacitive loads. It is worth to mention that the efficiency of SECE does not depend on the output bias, and that removing electrical charge on voltage peaks doubles the peak voltage. Besides this, SSHI[36] and other synchronized-switch harvesters based on multistage converters or transformers, such as DSSH[37] or SSHI-MR[38] have also been reported to provide a better performance.

In this type of interface, one of the main limiting factors in micropower scenarios is the use of diodes in circuit implementations.[28, 39, 40] In general, the voltage drop across diodes wastes significant amounts of power. Negative

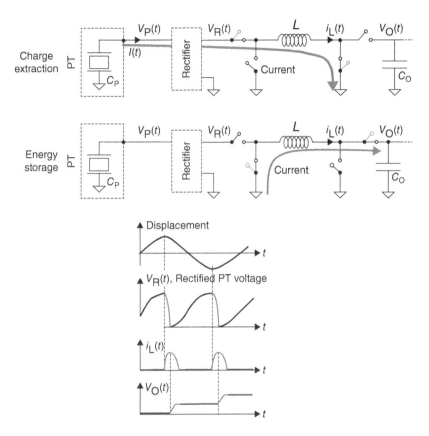

Figure 2. The SECE technique: a basic circuit showing current flow and reference waveforms during power conversion.

voltage converters, based on a cross-coupled MOSFET bridge, represent a better solution for rectifying voltages while permitting bidirectional current flow.[41] Active rectifiers[42, 43] also provide efficient performance at the cost of higher intrinsic power consumption, as do digitally controlled switches.[8, 31] A nanopower-integrated circuit implementing SECE and able to exploit the residual charge left on the transducer before the rectifier, leading to a more favorable voltage offset is reported in Ref. 44.

4. A multisource-integrated energy-harvesting circuit

As a case study, this section presents more insights into a nanocurrent IC able to convert and manage power from multiple and heterogeneous transducers.[8] The IC employs a series of techniques that pursue a favorable trade-off between conversion efficiency and internal power consumption. Its operating principle is based on multichannel buck/boost conversion with a single shared inductor. Since a power converter is likely to operate in discontinuous current conduction mode when working with micropower transducers, the duty cycle of inductor utilization is expected to be very low. This is also true with SECE, where the duty cycle of activation is extremely low, due to the typically low oscillation frequencies of piezoelectric transducers. As a consequence, it is possible to serialize the accesses of multiple sources to the shared inductor by designing a specific arbiter circuit that manages MPPT and prevents concurrent accesses to the inductor.

The block diagram of the IC is shown in Fig. 3. It manages up to nine input channels for heterogeneous types of energy sources, namely, five piezoelectric and four DC sources. Two of the DC input ports are compatible with high voltage (HV) sources with $1 \leq V_{DC0} \leq 5$ V, and two ports are optimized for low voltage (LV) sources, typically with $100 \leq V_{DC0} \leq 1$ V. Such high number of input ports was included in the design for showing the degree of scalability of the technique and the still very low associated power consumption. Compatible transducers include piezoelectric devices, thermoelectric generators, rectennas, and photovoltaic cells. The converter core manages all the input channels and utilizes a single external inductor. The inductor value should be carefully determined, according to the associated switching speeds: values on the order of tens of microhenries require high switching rates with high consumption, whereas values in millihenries range require relatively large inductors and low switching rates.

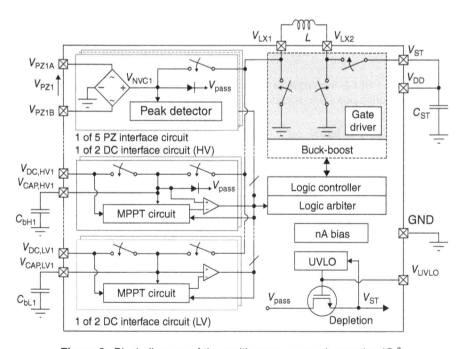

Figure 3. Block diagram of the multisource energy-harvesting IC.[8]

A passive path through a depletion-mode MOSFET makes it possible to initially charge the storage capacitor. When the voltage stored on C_{ST} is sufficient to turn on the active circuits, the passive path is cut off by pulling low its gate signal. A 16 nA under voltage lockout circuit (UVLO in Fig. 3) manages this phase and enables active conversion as well.

When the system is fully self-powered and battery-less, the V_{DD} input is connected to the output node V_{ST}, so that the converter supplies itself with the harvested energy, leading to a fully autonomous and self-powered solution. The core of the system is a fully asynchronous logic controller and arbiter that manage power conversion through the converter core from the individual sources. An FOCV MPPT is applied for DC sources, whereas piezoelectric sources are managed with an SECE circuit including local peak detection.

The IC has been designed in a 0.32 μm microelectronic process.

Figure 4 shows in more detail the circuits used for managing piezoelectric sources. The piezoelectric voltage is first rectified by a negative voltage converter with a MOSFET bridge for performing zero-voltage-drop rectification. When the IC operates in active mode, additional parallel-connected switches are used to fully discharge the piezoelectric source during the SECE cycle. The latter block is shown in the bottom part of Fig. 4 on the left. In order to issue piezoelectric conversion requests to the logic controller, a peak detector is implemented with a comparator. Since the circuit has no stable voltage supply, and the transducers have varying voltages, a specialized circuit block (denoted as HiPick in Fig. 4) was designed to select the highest available voltage to properly drive MOSFET switches. The nanopower comparator is shown at the top of Fig. 4 on the left,

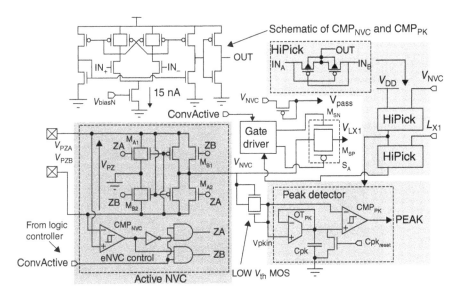

Figure 4. Detailed view of circuits for managing the piezoelectric transducers.[8]

and has a nominal bias current of 16 nA. Comparators were frequently used in the design, and where fast response times were necessary, the bias current was dynamically increased only during critical phases of operation, for example, during single energy conversion cycles.

The FOCV MPPT circuits are depicted in Fig. 5. Switched capacitors are used to set the FOCV at the 50% for linear sources and 75% for photovoltaic sources. Every eight conversion cycles, the MPP is refreshed by enabling the SAMPLE signal. An input buffer capacitor allows for continuous operation and low switching frequencies. During power conversion, a comparator with a dynamic 7× boost of bias current is used to maintain the source in the desired range around the MPP, with a hysteresis of ±28 mV.

The power converter core is shown in Fig. 6. Conversion timing is controlled by means of zero-voltage switching (ZVS) and zero-current switching (ZCS) circuits. Also in this block, the comparators are dynamically biased with an increased current in the active phase in order to ensure zero idle static power and a fast ~800 ns response of comparators during energy conversion. As a final design choice, the arbiter and the logic controller are implemented with

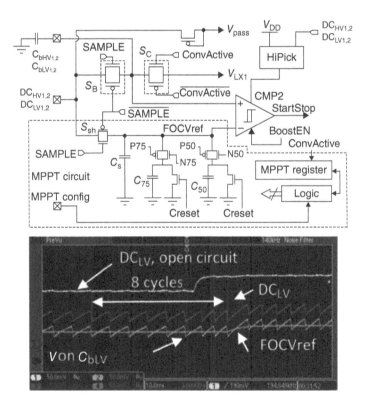

Figure 5. The FOCV MPPT circuit for DC sources (a) and experimental voltage waveforms (b) acquired during chip operation.[8]

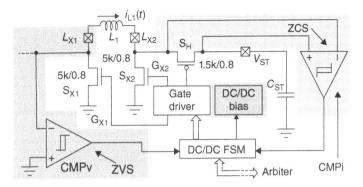

Figure 6. The buck/boost converter core.[8]

asynchronous logic circuits. This spares the designers the need of generating and distributing clocks in the IC, with all the associated dynamic energy consumption.

The IC has a measured quiescent current of 360 nA with the V_{DD} input biased at 3 V. The peak efficiency for piezoelectric input ports is 82%, while DC ports achieve up to 79%. Among the most important metrics, the energy consumed by the IC during a single energy transfer is very low, and was measured to be less than 2% of the energy input during a single conversion cycle with a 50 µW source. This low value indicates that the main energy losses are due to switch resistances rather than circuit consumption. The switches in the power converter could be made larger, but the current sizing was designed to keep the intrinsic consumption as low as possible. A top-view micrograph of the IC is shown in Fig. 7.

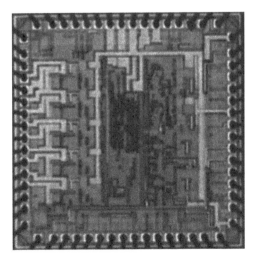

Figure 7. Micrograph of the IC. The total die area is 4.6 mm².

5. Conclusion

Energy harvesting is an exciting field of research that is experiencing continuous progress. The increasing performance of energy transducers and the decreasing power requirements of power management ICs is bridging the gap toward the implementation of disruptive application. As a matter of fact, the microwatts barrier has already been broken, and current state-of-the-art ICs consume only a few hundred nanowatts. Such ultralow consumption allows power conversion and management circuits to achieve a positive power budget even when relying on weak power sources in harsh conditions. In this context, several commercial ICs for power management with excellent performance are available on the market. However, the potential of energy harvesting and of fully autonomous nodes is still far from being fully utilized. Shifting conventional design paradigms are encouraging designers to implement energy-aware techniques. A full energy optimization of all system components, including CPUs, sensors, and wireless interfaces, is expected to further contribute to the deployment of energy-autonomous nodes.

Acknowledgments

This work was supported in part by the European Community under the ENIAC Nanoelectronics Framework Program (Call ENIAC-2012-2, grant agreement Lab4MEMS no. 325622-2); by the 7th Framework Program (grant agreement NANOFUNCTION no. 257375); and by the Italian Ministry of Instruction, Research and University (MIUR), within the framework of the national project PRIN 2011 GRETA.

References

1. R. J. M. Vullers, R. Van Schaijk, I. Doms, C. Van Hoof, and R. Mertens, "Micropower energy harvesting," *Solid State Electron.* **53**, 684–693 (2009).
2. A. Dolgov, R. Zane, and Z. Popovic, "Power management system for online low power RF energy harvesting optimization," *IEEE Trans. Circ. Syst. I* **57**, 1802–1811 (2010).
3. T. Paing and R. Zane, "Design and optimization of an adaptive non-linear piezoelectric energy harvester," *Proc. 26th Annual IEEE Appl. Power Electronics Conf. (APEC)* (2011), pp. 412–418.
4. S. Boisseau, P. Gasnier, M. Gallardo, and G. Despesse, "Self-starting power management circuits for piezoelectric and electret-based electrostatic mechanical energy harvesters," *J. Phys. Conf. Series* **476**, 012080 (2013).
5. A. Romani, M. Filippi, M. Dini, and M. Tartagni, "A sub-µA standby current synchronous electric charge extractor for piezoelectric energy harvesting," *ACM J. Emerg. Technol. Comput. Syst.* **12**, article no. 7 (2015).

6. E. Dallago, A. L. Barnabei, A. Liberale, P. Malcovati, and G. Venchi, "An interface circuit for low-voltage low-current energy harvesting systems," *IEEE Trans. Power Electron.* **30**, 1411–1420 (2015).

7. K. Kadirvel, Y. Ramadass, U. Lyles, *et al.*, "A 330 nA energy-harvesting charger with battery management for solar and thermoelectric energy harvesting," *Tech. Dig. ISSCC* (2012), pp. 106–108.

8. M. Dini, A. Romani, M. Filippi, V. Bottarel, G. Ricotti, and M. Tartagni, "A nanocurrent power management IC for multiple heterogeneous energy harvesting sources," *IEEE Trans. Power Electron.* **30**, 5665–5680 (2015).

9. E. E. Aktakka and K. Najafi, "A micro inertial energy harvesting platform with self-supplied power management circuit for autonomous wireless sensor nodes," *IEEE J. Solid-State Circ.* **49**, 2017–2029 (2014).

10. S. Bandyopadhyay, P. P. Mercier, A. C. Lysaght, K. M. Stankovic, and A. Chandrakasan, "23.2 A 1.1 nW energy harvesting system with 544 pW quiescent power for next-generation implants," *Tech. Dig. ISSCC* (2014), pp. 396–397.

11. Texas Instruments, www.ti.com/product/bq25504 (accessed on Apr. 23, 2016).

12. STMicroelectronics, www.st.com/spv1050 (accessed on Apr. 23, 2016).

13. Cypress, www.spansion.com/downloads/S6AE101A_DS405-00026-E.pdf (accessed on Apr. 23, 2016).

14. A. Costanzo, A. Romani, D. Masotti, N. Arbizzani, and V. Rizzoli, "RF/baseband co-design of switching receivers for multiband microwave energy harvesting," *Sens. Actuators A* **179**, 158–168 (2012).

15. S.-E. Adami, V. Marian, N. Degrenne, C. Vollaire, B. Allard, and F. Costa, "Self-powered ultra-low power DC–DC converter for RF energy harvesting," *IEEE Faible Tension Faible Consommation (FTFC) Conf.* (2012), pp. 1–4.

16. Y. Ramadass, "A 330 nA charger and battery management IC for solar and thermoelectric energy harvesting," Presented at *3rd Intern. Workshop Power SoC* (2012), paper 5.6.

17. S. Bandyopadhyay and A. Chandrakasan, "Platform architecture for solar, thermal, and vibration energy combining with MPPT and single inductor," *IEEE J. Solid-State Circ.* **47**, 2199–2215 (2012).

18. M. Dini, A. Romani, M. Filippi, and M. Tartagni, "A nano-current power management IC for low voltage energy harvesting," *IEEE Trans. Power Electron.*, **31**, 4292–4304 (2015).

19. T.-C. Huang, C.-Y. Hsieh, Y.-Y. Yang, *et al.*, "A battery-free 217 nW static control power buck converter for wireless RF energy harvesting with α-calibrated dynamic on/off time and adaptive phase lead control," *IEEE J. Solid-State Circ.* **47**, 852–862 (2012).

20. T. Tanzawa, "An optimum design for integrated switched-capacitor Dickson charge pump multipliers with area power balance," *IEEE Trans. Power Electron.* **29**, 534–538 (2014).

21. J. Kim, P. K. T. Mok, and C. Kim, "A 0.15 V input energy harvesting charge pump with dynamic body biasing and adaptive dead-time for efficiency improvement," *IEEE J. Solid-State Circ.* **50**, 414–425 (2015).

22. A. Richelli, S. Comensoli, and Z. M. Kovacs-Vajna, "A DC/DC boosting technique and power management for ultralow-voltage energy harvesting applications," *IEEE Trans. Ind. Electron.* **59**, 2701–2708 (2012).

23. G. Bassi, L. Colalongo, A. Richelli, and Z. M. Kovacs-Vajna, "A 150 mV–1.2 V fully-integrated DC–DC converter for thermal energy harvesting," *Proc. Intern. Symp. Power Electronics, Electrical Drives, Automation Motion (SPEEDAM)* (2012), pp. 331–334.

24. L. Colalongo, Z. M. Kovacs-Vajna, A. Richelli, and G. Bassi, "100 mV–1.2 V fully-integrated DC–DC converters for thermal energy harvesting," *IET Power Electron.* **6**, 1151–1156 (2013).

25. A. Richelli, L. Colalongo, and Z. M. Kovacs-Vajna, "A 30 mV–2.5 V DC/DC converter for energy harvesting," *J. Low Power Electron.* **11**, 190–195 (2015).

26. E. Macrelli, A. Romani, R. P. Paganelli, A. Camarda, and M. Tartagni, "Design of low-voltage integrated step-up oscillators with microtransformers for energy harvesting applications," *IEEE Trans. Circ. Syst. I* **62**, 1747–1756 (2015).

27. A. Camarda, A. Romani, E. Macrelli, and M. Tartagni, "A 32 mV/69 mV input voltage booster based on a piezoelectric transformer for energy harvesting applications," *Sens. Actuators A* **232**, 341–352 (2015).

28. E. Lefeuvre, A. Badel, C. Richard, L. Petit, and D. Guyomar, "A comparison between several vibration-powered piezoelectric generators for standalone systems," *Sens. Actuators A* **126**, 405–416 (2006).

29. Y. C. Shu, I. C. Lien, and W.-J. Wu, "An improved analysis of the SSHI interface in piezoelectric energy harvesting," *Smart Mater. Struct.* **16**, 2253–2264 (2007).

30. A. Romani, M. Filippi, and M. Tartagni, "Micropower design of a fully autonomous energy harvesting circuit for arrays of piezoelectric transducers," *IEEE Trans. Power Electron.* **29**, 729–739 (2014).

31. A. Romani, C. Tamburini, A. Golfarelli, *et al.*, "Dynamic switching conversion for piezoelectric energy harvesting systems," *Proc. IEEE Sensors Conf.* (2008), pp. 689–692.

32. A. Romani, R. P. Paganelli, E. Sangiorgi, and M. Tartagni, "Joint modeling of piezoelectric transducers and power conversion circuits for energy harvesting applications," *IEEE Sens. J.* **13**, 916–925 (2013).

33. M. Dini, M. Filippi, M. Tartagni, and A. Romani, "A nano-power power management IC for piezoelectric energy harvesting applications," *Proc. 9th Conf. PhD Research Microelectronics Electronics (PRIME)* (2013), pp. 269–272.

34. M. Dini, M. Filippi, A. Romani, V. Bottarel, G. Ricotti, and M. Tartagni, "A nano-power energy harvesting IC for arrays of piezoelectric transducers," *Proc. SPIE* **8763**, 87631O (2013).

35. P. Gasnier, J. Willemin, S. Boisseau, *et al.*, "An autonomous piezoelectric energy harvesting IC based on a synchronous multi-shot technique," *IEEE J. Solid-State Circ.* **49**, 1561–1570 (2014).

36. D. Guyomar, A. Badel, E. Lefeuvre, and C. Richard, "Toward energy harvesting using active materials and conversion improvement by nonlinear processing," *IEEE Trans. Ultrason. Ferroelectr. Freq. Control* **52**, 584–594 (2005).

37. M. Lallart, L. Garbuio, L. Petit, C. Richard, and D. Guyomar, "Double synchronized switch harvesting (DSSH): A new energy harvesting scheme for efficient energy extraction," *IEEE Trans. Ultrason. Ferroelectr. Freq. Control* **55**, 2119–2130 (2008).

38. L. Garbuio, M. Lallart, D. Guyomar, C. Richard, and D. Audigier, "Mechanical energy harvester with ultralow threshold rectification based on SSHI nonlinear technique," *IEEE Trans. Ind. Electron.* **56**, 1048–1056 (2009).

39. Y. Ramadass and A. Chandrakasan, "An efficient piezoelectric energy harvesting interface circuit using a bias-flip rectifier and shared inductor," *IEEE J. Solid-State Circ.* **45**, 189–204 (2012).

40. N. Krihely and S. Ben-Yaakov, "Self-contained resonant rectifier for piezoelectric sources under variable mechanical excitation," *IEEE Trans. Power Electron.* **26**, 612–621 (2011).

41. C. Peters, J. Handwerker, D. Maurath, and Y. Manoli, "An ultra-low-voltage active rectifier for energy harvesting applications," *IEEE Intern. Symp. Circ. Syst. (ISCAS)* (2010), pp. 889–892.

42. C. Peters, J. Handwerker, D. Maurath, and Y. Manoli, "A sub-500 mV highly efficient active rectifier for energy harvesting applications," *IEEE Trans. Circ. Syst. I* **58**, 1542–1550 (2011).

43. Y. Sun, N. H. Hieu, C.-J. Jeong, and S.-G. Lee, "An integrated high-performance active rectifier for piezoelectric vibration energy harvesting systems," *IEEE Trans. Power Electron.* **27**, 623–627 (2012).

44. M. Dini, A. Romani, M. Filippi, and M. Tartagni, "A nanopower synchronous charge extractor IC for low-voltage piezoelectric energy harvesting with residual charge inversion," *IEEE Trans. Power Electron.* **31**, 1263–1274 (2016).

45. T. Huang, C. Hsieh, Y. Yang, *et al.*, "A battery-free 217 nW static control power buck converter for wireless RF energy harvesting with calibrated dynamic on/off time and adaptive phase lead control," *IEEE J. Solid-State Circ.* **47**, 852–862 (2012).

46. J. Kim and C. Kim, "A DC–DC boost converter with variation-tolerant MPPT technique and efficient ZCS circuit for thermoelectric energy harvesting applications," *IEEE Trans. Power Electron.* **28**, 3827–3833 (2013).

47. N. K. Pour, F. Krummenacher, and M. Kayal, "A reconfigurable micro power solar energy harvester for ultra-low power autonomous microsystems," in *IEEE Intern. Symp. Circ. Syst. (ISCAS)* (2013), pp. 33–36.

3.4
Will Composite Nanomaterials Replace Piezoelectric Thin Films for Energy Transduction Applications?

R. Tao, G. Ardila, R. Hinchet, A. Michard, L. Montès, and M. Mouis
IMEP-LAHC/Minatec, CNRS-Grenoble INP, UJF, 38016 Grenoble, France

1. Introduction

The piezoelectric effect is the conversion of mechanical strain into electrical energy (direct effect) or, conversely, the production of mechanical deformation by an input electric field (converse effect). It is widely used in applications such as lighters, speakers, medical ultrasound transducers, precise positioning systems, and many others, using mainly bulk materials.[1] At a smaller scale, piezoelectric thin films are used in applications ranging from MEMS/NEMS actuators to resonators.[2] Other applications include mechanical sensors and energy harvesters used typically for wireless sensor networks (WSNs), with the objective of monitoring human health, environment, or structures such as airplanes or buildings.[3] The integration of energy harvesters in WSN systems, in particular, helps to develop battery-less autonomous systems or, in other cases, eliminate complex wiring.

The most used piezoelectric materials in these applications are PZT and AlN thin films.[2] At the nanoscale, ZnO and GaN in the shape of nanowires (NWs) are the most studied materials because they are relatively easy to fabricate[4] and because of their enhanced electromechanical properties: higher flexibility and higher piezoelectric coefficients compared to their thin film counterparts.[5]

Piezoelectric NWs exhibit the so-called piezotronic effect,[6] comprised of the coupling between piezoelectric and semiconducting properties, which can be used for such applications as mechanical sensors,[7, 8] transistors,[6] programmable memories,[9] optoelectronic devices,[10] and chemical sensors.[11] Further, such NWs can be integrated with insulators to create composite materials.[12, 13]

After a brief review of thin film piezoelectric materials, their fabrication techniques, and typical applications, we present the experimental and theoretical study on individual piezoelectric NWs, with a focus on ZnO and GaN, and their particular properties. Then, we discuss the integration of such nanowires into

Future Trends in Microelectronics: Journey into the Unknown, First Edition.
Edited by Serge Luryi, Jimmy Xu, and Alexander Zaslavsky.
© 2016 John Wiley & Sons, Inc. Published 2016 by John Wiley & Sons, Inc.

composite materials for sensing and energy-harvesting applications, including a theoretical study on the improvement of their performance.

2. Thin film piezoelectric materials and applications

Usually, MEMS devices utilizing piezoelectric materials exploit two particular response modes, which are identified by subscripts (33 and 31) indicating the axis of the applied field and the produced strain or stress.[14] These are associated with piezoelectric coefficients relating the electric field and the strain (d) or the stress (e). For instance d_{31} relates the in-plane strain to the out-of-plane electric field and d_{33} relates the out-of-plane strain to the electric field in the same direction.

Other than the piezoelectric coefficients, it is often useful to consider the energy conversion efficiency defined by the expression:

$$k_{ij}^2 = \frac{d_{ij}^2}{\varepsilon_r \varepsilon_0 s_{ij}}, \tag{1}$$

where subscripts i and j represent the relative directions, ε_0 is the permittivity of free space, ε_r is the relative dielectric constant, and s_{ij} is the compliance or the inverse of the stiffness. This expression shows that in addition to large piezoelectric coefficients, it is also important to have a large stiffness and low dielectric constant. Thus, materials with high piezoelectric coefficients are very useful in actuation applications, but materials with high conversion efficiencies are also of interest for sensing and energy harvesting, even if they do not have large piezoelectric coefficients.[15]

In order to exploit the piezoelectric effect, typical MEMS applications use piezoelectric materials integrated into bendable structures such as beams, bridges, or membranes. Many applications can be cited using such structures: mirror arrays for video projection, microphones, RF resonators, and filters for mobile applications, accelerometers and ink-jet printing heads.

The applicability of piezoelectric materials in emerging fields depend on several factors:[1] (i) the ability to deposit piezoelectric materials at low temperatures to ensure compatibility with CMOS technologies or polymers; (ii) the ability to integrate them into non-Si substrates, such as metal foils and polymers; and (iii) the ability to deposit piezoelectric materials over topologies for 3D sensors or actuators.

• *Thin film piezoelectric materials*

At the MEMS scale, many piezoelectric materials have been studied and successfully integrated: compound semiconductors such as nitrides like AlN or oxides like ZnO, ferroelectrics such as $Pb(Zr_xTi_{1-x})O_3$ (PZT), and polymers such as polyvinylidene fluoride (PVDF).[16] In the case of polycrystalline ferroelectrics, poling at a certain temperature (100–200 °C) is needed to reorient the polar axis, in which case the piezoelectric effect is averaged over all the grains of the layer, reducing the piezoelectric properties compared to perfectly aligned domains. This is not the case in nonferroelectric materials such as ZnO, GaN, or AlN, where the fabrication process provides the alignment of the polar directions.[17] This is an advantage in the case of 1D structures such as NWs, which are usually highly crystalline.[6]

The thin film materials with the best piezoelectric properties are ferroelectric such as PZT and very recently PMN-PT, which could present piezoelectric properties about 5–10 times those of PZT, but only if very specific conditions are satisfied in the deposition process.[18] On the other hand, materials with lower piezoelectric properties such as AlN, ZnO, or GaN are also very attractive for their lower dielectric constants, which increase the energy conversion efficiency, as in Eq. (1).

Several techniques can be used to deposit thin film piezoelectric materials. Lead-based materials such as PZT can be deposited using metal-organic chemical vapor deposition (MOCVD), sputtering and chemical solution deposition, including sol–gel techniques.[17] Thicknesses below 2 μm can typically be achieved while preserving high piezoelectric properties.[19] In the case of PMN-PT, it can be deposited by sputtering, achieving thicknesses below 4 μm. These fabrication techniques usually require high temperature (~650–750 °C) that may not be compatible with some applications, such as CMOS with Al metallization that cannot withstand temperatures above 400 °C. Some exceptions can be cited where hydrothermal methods (<200 °C) have been used to deposit thin layers (<500 nm) of PZT,[20] although repeated processing must be used to reach a thickness of several tens of micrometers and the performance is inferior to bulk materials.[1] On the other hand, AlN and ZnO thin films (<1 μm) are typically deposited by low-temperature sputtering (<400 °C), simplifying integration and compatibility with CMOS technologies, but the piezoelectric properties of such films are low compared to PZT.[14] Finally, PVDF has higher piezoelectric properties than nonferroelectric materials, but its lower stiffness results in lower energy conversion efficiency.[21]

Fabricating composite materials using piezoelectric NWs could be a way to improve their performance, as well as compatibility with multiple substrates and CMOS technologies. Nanowires with 50 μm length have been reported,[22] meaning that composite materials can be designed without thickness restrictions.

3. Individual ZnO and GaN piezoelectric nanowires: experiments and simulations

Semiconducting piezoelectric NWs, such as ZnO and GaN, have been investigated due to their high piezoelectric coefficients and mature growth methods.[23] Many modeling and experimental approaches have been considered to characterize their mechanical, electrical, and electromechanical properties.

- *Mechanical properties of ZnO and GaN piezoelectric NWs*

Molecular dynamic (MD) modeling[24] and finite element method (FEM)[25] have been used to determine the mechanical properties of ZnO and GaN NWs. At the same time, different loading modes, including tension, compression, and bending, are applied to the NW using MEMS, *in situ* SEM/TEM, AFM, cantilever *in situ* SEM/TEM, and other experimental techniques, in order to obtain experimental mechanical properties.

Agrawal *et al.* studied the elasticity size effects in ZnO NWs,[26] where they calculated Young's modulus of NWs with diameters ranging from 5 to 20 nm by first principle methods. Results showed that as the diameter decreased, Young's modulus increased from 169 to 194 GPa, which was consistent with the size effect reported by Chen *et al.*[27] They also measured ZnO NWs with diameters ranging from 20.4 to 412.9 nm by applying a uniaxial tensile load inside a TEM. Young's modulus was found to increase rapidly as the diameter went below 80 nm. Bernal *et al.* characterized experimentally *c*- and *a*-axes of GaN NWs using a MEMS-based *in situ* TEM uniaxial testing technique.[24] Unlike ZnO NWs, GaN NWs with diameters larger than 30–40 nm presented an elastic modulus similar to bulk values. However, due to the uncertainties of boundary conditions, surface status, and calibration deviation of different methods, the mechanical characterization results in literature exhibit significant scatter for both ZnO and GaN NWs.

- *Electrical properties of ZnO and GaN piezoelectric NWs*

Yang *et al.* reported measurements of the dielectric constant of single pencil-like ZnO nanowires using the scanning conductance microscopy (SCM) technique.[28] Their results showed that the dielectric constant of ZnO NWs had significant size dependence within the experimental manipulation range, and it was much smaller than bulk value (see Fig. 1).

- *Piezoelectricity study of ZnO and GaN piezoelectric NWs*

Piezoelectricity is a linear electromechanical coupling, which manifests the interconversion between electric and mechanical energy. Piezoelectric constants of piezoelectric NWs are commonly measured by applying tensile loading or lateral bending of the material with simultaneous measurement of generated charge or electric potential. For example, Minary-Jolandan *et al.* presented a method based on scanning probe microscopy (SPM) to probe the 3D piezoelectric matrix of individual piezoelectric NWs.[29]

Since the experimental characterization of the coupling properties still faces challenges, such as the manipulation of individual NWs, the contact resistance between the sample and the measurement tools, and the sensitivity of output electrical signals, computational simulations can also contribute to the understanding of this effect. Thus, from first principles calculations, Xiang *et al.*

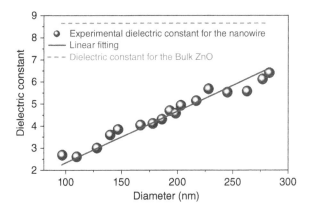

Figure 1. Diameter dependence of the dielectric constant in a single pencil-like ZnO NW. (Source: Yang et al.[28] Reproduced with permission of the American Chemical Society).

demonstrated that piezoelectric constants of ZnO NWs increase monotonically as the radius decreases.[30] This size dependence effect was studied later in both ZnO and GaN NWs with radius ranging from 0.6 to 2.4 nm by Espinosa and coworkers.[31] They reported that GaN NWs manifest a greater size dependence than ZnO NWs and that their piezoelectric constants can be improved by two orders of magnitude compared to the bulk values if the NW diameter can be reduced to under 1 nm.

The piezoelectric behavior of individual NWs was also investigated by FEM simulations. Wang and coworkers first proposed a continuum model for simulating the electrostatic potential in a laterally bent ZnO NW with 50 nm diameter and 600 nm length.[32] The voltage drop created across the cross section of the NW was around ±0.3 V when the bending force was 80 nN.

Initial simulations considered ZnO and GaN as insulators; however, they are semiconductors with a relatively wide bandgap. They are doped during the synthesis by defects and impurities, either intentionally for achieving certain functionality or through accidental doping due to the growth mechanism.[33] Gao and Wang investigated the behavior of free charge carriers in a bent piezoelectric n-type ZnO NW under thermodynamic equilibrium conditions by inducing donors into the model.[34] They demonstrated that the potential generated by NWs is screened by the charge carriers.

4. Piezoelectric composite materials using nanowires

Individual NWs provide typically very low power (~1 pW).[35] To increase the energy generated, NWs are usually integrated either laterally[36] or vertically[37] on the substrate to form piezoelectric composite materials. Considering large-scale

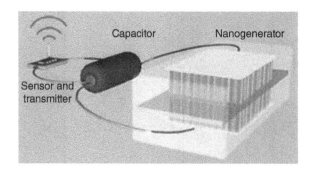

Figure 2. Nanogenerator composed of a five-layer flexible plate with PMMA-surrounded ZnO NWs grown vertically on both sides of a polymer substrate and electrodes deposited on both top and bottom. (Source: Hu *et al.*[37] Adapted with permission of the American Chemical Society).

fabrication and commercial application, vertically integrated nanogenerators (VINGs) including a composite piezoelectric layer appear to be an attractive option.

The VING structure was first introduced by Wang and his group[37] to harvest bending mechanical energy. Initial VING designs employed ZnO NWs and were based on a five-layer flexible plate, illustrated in Fig. 2. Typically, VING composites are fabricated on Si wafers to work with compressive pressures[13] or on flexible substrates such as metallic foil[38] or even paper[39] to work with bending forces.

To study the performance of the VING structure, we have conducted computational simulations on ZnO NW-based nanocomposites in both compression and flexion modes. Matrix materials and ZnO properties were varied to compare with conventional thin film generators.

• *VING working under compressive pressure*

In compression mode, the core function part of the simplified model (named NG cell for nanogenerator cell) is one NW surrounded by matrix material, where the input pressure on the top surface is equal to the one on the entire structure (see Fig. 3). Symmetry boundary conditions are applied at the lateral surfaces on the assumption that identical NG cells surrounding the target one. In this case, the NW is immersed in PMMA, which protects it from the electrical leakage or short circuits. Then, top and bottom surfaces are defined as electrodes to harvest the generated electrostatic energy. When the device is compressed, part of the input mechanical energy is stored inside the core piezoelectric NW, and then it is converted into electric energy through direct piezoelectric effect. Finally, the electric energy is driven out by the external circuit (not shown).

The compressive pressure is considered to be a constant (1 MPa). The initial cell size is $100 \times 100 \times 700$ nm, with NW radius $r = 25$ nm, and length $L = 600$ nm.

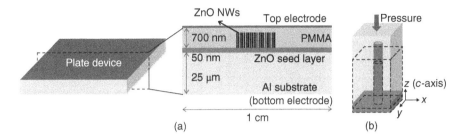

Figure 3. Scheme of a composite piezoelectric material on a metallic substrate (a) and the NG cell in compression mode (b).[38]

The size is varied by changing the NW diameter/cell width ratio, representing the density of the NWs. The electrical and piezoelectric properties of ZnO are varied respectively to study their influence. As mentioned above, the mechanical properties are not known with great certainty, so we do not consider the variation of ZnO elastic moduli in the current simulations.

Figure 4(a) shows the absolute value of the output potential varying with the previously defined ratio. The thin film model generates ~10 mV. When the model of the composite NG cell is considered, the potential increases by a factor of 8 (at ratio = 0.4) even though the NW is assumed to have the same properties as a thin film.[40–42] The enhancement is due to the soft PMMA matrix, which concentrates the strain inside ZnO. Besides, because of the 3D dielectric losses and the deviation of the strain field around the NWs, a smaller size ratio improves the storage of mechanical energy but reduces the electrical energy stored.

In contrast, a high size ratio increases the capacity to store electrical energy in the NG, at the expense of reduced mechanical energy storage (see Fig. 4(b)). The trade-off of these effects forms a potential curve with a peak at ratio = 0.4. Since the dielectric constant decreases with the NW radius,[28] the electric energy loss through the top insulating layer is smaller in NG cell.[13] As the ratio increases, the dielectric constant becomes the major factor that influences the output potential. For the cell with nanoscale piezoelectric coefficients,[8] the potential curve follows the trend of the thin film cell but increases by roughly a factor of 2. In summary, the potential of an NG cell with nanowire properties has been enhanced by 22 times compared to the thin film model and by ~2.4 times compared to the NG cell with thin film properties. When we calculate the electric energy from the potential and the equivalent capacitance of this structure, we find the maximum shifts to higher NW density because of larger capacitance, as shown in Fig. 4(c).

• *VING working under bending force*

Similar to the compression case, the model only considers the unit cell at the central zone of the device (Fig. 5) with appropriate boundary conditions for fast FEM modeling. When the device is bent as a doubly clamped plate, the individual

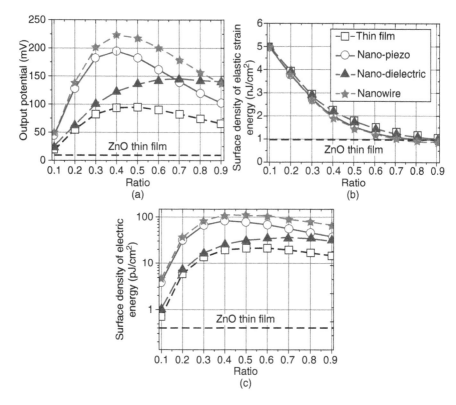

Figure 4. Absolute value of (a) output potential, (b) surface density of elastic strain energy, and (c) electric energy of NG cells working in compression mode using ZnO thin film properties, nano dielectric and piezoelectric constants, and NW properties, respectively. Results using a ZnO thin layer were calculated as reference.

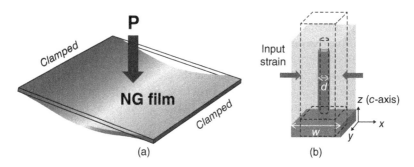

Figure 5. Schematic picture of NG membrane (a) and NG cell (b) in flexion mode.[38]

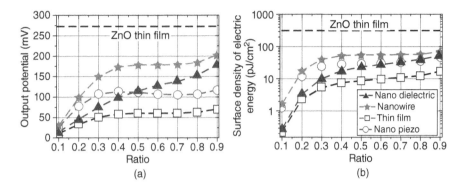

Figure 6. Absolute value of (a) potential and (b) surface density of electric energy of NG cells working in flexion mode using ZnO thin film properties, nano dielectric constant, nano piezoelectric constant, and NW properties, respectively. Results using a thin ZnO layer were calculated as reference.

NW is in compression or extension (depending on the position of the NW on the plate) and is elongated or shortened in the c-axis direction.[43] The input strain on the NG cell is calculated from a 2D model of the plate bent by a hydrostatic pressure (100 Pa) chosen to work in the linear mechanical regime.

Figure 6(a) shows the output potential of the NG cell with thin film properties, which increases with the size ratio. But NG cells with nano properties show different behavior. An NG cell with only nano piezoelectric coefficients only generates a higher potential at a size ratio around 0.4. On the other hand, an NG cell with a nano dielectric constant manifests a monotonically increasing potential with size ratio. Since the nano dielectric constant of ZnO is smaller than PMMA, the electric energy loss decreases as the volume fraction occupied by ZnO increases – see Fig. 6(b). Combining these two nanoproperties, the NG cell shows a similar changing trend compared to a thin film cell but with an increase by a factor of 3.5.

Although using nano properties improves the energy conversion and potential generation of VING in flexion mode, the performances are lower than ZnO thin film generators based on the simulation results. The performance of this composite material can be further improved by using different matrix materials. Figure 7 displays the simulation results of NG cells using PMMA, SiO_2, Si_3N_4, and Al_2O_3 as matrix material, respectively, as well as the comparison with a ZnO thin film.

Given their higher rigidity, SiO_2, Si_3N_4, or Al_2O_3 matrices enhance the input strain of the NG cell when the entire NG membrane is still bent by 100 Pa pressure (see Fig. 7(b)). Also, given their larger relative permittivity, they reduce the energy loss passing through the top insulating layer, as shown in Fig. 7(c). As a result, the potential generated is higher than a thin ZnO film starting from a ratio ~0.3–0.5, and reaches a peak value that is 1.5–2.5 times larger than the reference

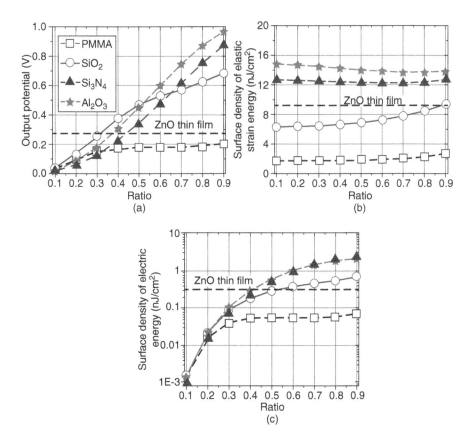

Figure 7. Absolute value of (a) potential, (b) surface density of elastic strain energy, and (c) surface density of electric energy of NG cells working in flexion mode using PMMA, SiO$_2$, Si$_3$N$_4$, and Al$_2$O$_3$ as matrix material, respectively. Results using a thin ZnO layer were calculated as reference.

ZnO thin film of Fig. 7(a). Since the improvement is largely due to the increase of input strain and the reduction of electric energy loss, potential and electric energy both increase with the size ratio.

- *Effect on the alignment of the NWs*

The models described in the preceding sections consider composite materials based on vertically aligned NWs. Although perfectly aligned NWs have been successfully fabricated,[44] to the best of our knowledge all mechanical transducers reported to date have employed slightly inclined piezoelectric NWs.[13] In this section, we evaluate the effect of the inclination of the NWs on the electric potential generated by a single cell.

The reference composite structure is formed by ZnO NWs (600 nm long) immersed on PMMA with a 100 nm-thick top insulating layer of Si_3N_4, which we previously found to be optimal compared to PMMA only. A single cell is evaluated including an increasing number of NWs having different inclination angles with respect to the ideal vertical case, starting from 1 NW up to 64 NWs. The FEM simulations take into account the fact that the NWs are grown along their c-axis, which is important to define the piezoelectric and mechanical properties; this has been done by defining these properties with respect to a rotated Cartesian axis aligned with each NW. All simulations in this section assume ZnO to have thin film piezoelectric properties.

Figure 8(a) shows the absolute value of the potential generated from the individual composite cell containing a single inclined NW as a function of its inclination. Simulations were carried out either taking into account the Cartesian axis rotation to correct the c-axis or without the axis rotation. When the inclination angle is small (below ~5°), the curves are very close to each other and neglecting the c-axis correction does not produce appreciable error (maximum error of 2% below 5° and 8% below 12°). At higher inclination angles the error increases, reaching a maximum of ~85%. An optimal inclination angle can also be observed when the inclination angle is close to 5°, increasing slightly the absolute electric potential at the top electrode. Then the potential is greatly reduced reaching a reduction of 50% for an inclination of ~30°. A comparison of the results with and without c-axis correction for a specific angle is depicted in Fig 8(b) and (c).

When two NWs are included in a single cell, several possibilities can be evaluated. To facilitate the study, we first considered an inclination in the xz-plane only. A series of simulations was conducted inclining only one of the two NWs, as shown in Fig. 9(a). The absolute value of the electric potential at the top electrode

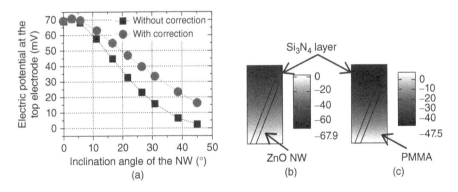

(a)   (b)   (c)

Figure 8. FEM simulation results for a VING composite cell containing a single ZnO NW versus the inclination angle from the vertical, with and without c-axis correction: (a) absolute value of the piezoelectric potential generated; electric potential (mV) inside the cell with (b) and without (c) c-axis correction.

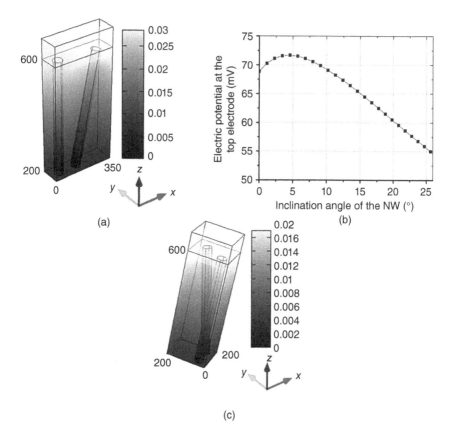

Figure 9. FEM simulations of a single VING cell with two NWs under compression with inclination angles on the xz-plane: (a) displacement (nm) of one cell including one vertical NW and one NW inclined at ~11°; (b) electric potential at the top electrode in function of the inclination angle of one NW; (c) displacement (nm) of one cell with one vertical NW and one NW inclined at ~11° in the yz-plane.

as a function of inclination is shown in Fig. 9(b). In general, the results are similar to the previews case with a single NW, with an optimal inclination close to 5°. This same behavior is observed if the inclination is taken to lie in the yz-plane (see Fig. 9(c)). The simulation results shows no clear trend when both NWs are inclined at small angles between 0° and 12° in the xz-plane. In some cases the electric potential is slightly increased (increase of ~1%) compared to the ideal situation (~69 mV), and in some cases the potential is markedly reduced (reduction of ~17%).

   In the case of four inclined NWs inside a cell, for different scenarios with inclination angles below 12° we observe a systematic reduction of the electric potential, ranging from 1% to 13% of the maximum. In order to verify this trend,

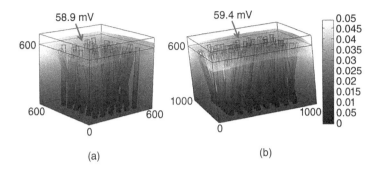

Figure 10. FEM simulations results of the displacement (nm) of different single cells containing many NWs under compression with inclination angles below 12°. The absolute electric potential at the top electrode is presented for a single cell with (a) 36 and (b) 64 ZnO NWs.

cells including 36 and 64 ZnO NWs have been simulated at inclinations up to 12°, as shown in Fig. 10. The absolute value of the electric potential at the top electrode in both cases is close to 59 mV, representing a reduction of ~14% from the ideal case. We expect that these simulations are closer to experimental reality because of the greater number of NWs in actual devices. Since the multi-NW simulations do not include c-axis correction, the results for the generated electric potential are likely underestimated by ~8%.

These results show that the approach of considering a small number of NWs in a composite cell is only applicable if the NWs are truly vertical, providing the maximum electric potential. In the more realistic case of slightly inclined NWs, a large number of NWs is required in the cell to obtain consistent results. The integration of slightly inclined piezoelectric NWs can reduce the electric potential generated by as much of 14% compared perfectly vertical NWs.

5. Conclusion

Semiconducting piezoelectric NWs show significant potential for application in electronic and electromechanical sensors and energy harvesters. In particular, these nanostructures can be used to build composite piezoelectric materials (NWs in a dielectric matrix) that could offer several advantages when integrated vertically. First, NWs of various lengths (in particular, ZnO NWs up to several tens of micrometers long) have been reported, thus opening the possibility to fabricate composite layers thicker than standard thin films (<4 μm). Second, low-temperature fabrication process (<100 °C) make this technology compatible with CMOS devices and with different substrates, such as Si, polymers, plastics, metal foils, and even paper. Third, properties such as piezoelectricity, flexibility, and dielectric constant can be improved in NWs, thereby improving the performance of NW-containing composite materials. Finally, FEM simulations show that composites can provide better performance (more voltage or electric energy generated) compared to piezoelectric thin films of the same thickness, and that performance can be improved by operating in compression or flexion modes by the right choice of the dielectric matrix and NW density. On the other hand, some effects could reduce the performance of these composites: for instance the nonverticality of the integrated NWs and the screening by mobile charges or fixed ionized dopants.

Acknowledgments

This work has been partly supported by the French ministry (ANR COSCOF) and the regional research and development program NANO2017.

References

1. P. Muralt, R. G. Polcawich, and S. Trolier-McKinstry, "Piezoelectric thin films for sensors, actuators, and energy harvesting," *MRS Bull.* **34**, 658–664 (2009).
2. C.-B. Eom and S. Trolier-McKinstry, "Thin-film piezoelectric MEMS," *MRS Bull.* **37**, 1007–1017 (2012).
3. W. J. Choi, Y. Jeon, J.-H. Jeong, R. Sood, and S. G. Kim, "Energy harvesting MEMS device based on thin film piezoelectric cantilevers," *J. Electroceram.* **17**, 543–548 (2006).
4. Z. Guo, C. Andreazza-Vignolle, P. Andreazza, *et al.*, "Tuning the growth of ZnO nanowires," *Phys. B: Condens. Matter* **406**, 2200–2205 (2011).
5. H. D. Espinosa, R. A. Bernal and M. Minary-Jolandan, "A review of mechanical and electromechanical properties of piezoelectric nanowires," *Adv. Mater.* **24**, 4656–4675 (2012).
6. X. Wen, W. Wu, C. Pan, Y. Hu, Q. Yang, Z. L. Wang, "Development and progress in piezotronics," *Nano Energy* **14**, 276–295 (2015).
7. Y. S. Zhou, R. Hinchet, Y. Yang, *et al.*, "Nano-Newton transverse force sensor using a vertical GaN nanowire based on the piezotronic effect," *Adv. Mater.* **25**, 883–888 (2013).
8. R. Hinchet, J. Ferreira, J. Keraudy, *et al.*, "Scaling rules of piezoelectric nanowires in view of sensor and energy harvester integration," *Tech. Dig. IEDM* (2012), p. 6.2.1.
9. W. Wu and Z. L. Wang, "Piezotronic nanowire-based resistive switches as programmable electromechanical memories," *Nano Lett.* **11**, 2779–2785 (2011).
10. X. Wen, W. Wu, and Z. L. Wang, "Effective piezo-phototronic enhancement of solar cell performance by tuning material properties," *Nano Energy* **2**, 1093–1100 (2013).

11. R. M. Yu, C. F. Pan, J. Chen, G. Zhu, and Z. L. Wang, "Enhanced performance of a ZnO nanowire-based self-powered glucose sensor by piezotronic effect," *Adv. Funct. Mater.* **23**, 5868–5874 (2013).

12. S. Lee, R. Hinchet, Y. Lee, *et al.*, "Ultrathin nanogenerators as self-powered/active skin sensors for tracking eye ball motion," *Adv. Funct. Mater.* **24**, 1163–1168 (2014).

13. R. Hinchet, S. Lee, G. Ardila, L. Montes, M. Mouis, Z. L. Wang, "Performance optimization of vertical nanowire-based piezoelectric nanogenerators," *Adv. Funct. Mater.* **24**, 971–977 (2014).

14. P. Muralt, "Piezoelectric thin films for MEMS," *Integr. Ferroelectr.* **17**, 297–307 (1997).

15. C. Baur, D. J. Apo, D. Maurya, S. Priya, and W. Voit, "Advances in piezoelectric polymer composites for vibrational energy harvesting," chapter in: L. Li, W. Wong-Ng, and J. Sharp, eds., *Polymer Composites for Energy Harvesting, Conversion, and Storage*, Washington, DC: American Chemical Society, 2014, pp. 1–27.

16. K. A. Cook-Chennault, N. Thamby, and A. S. Sastry, "Powering MEMS portable devices – A review of non-regenerative and regenerative power supply systems with special emphasis on piezoelectric energy harvesting systems," *Smart Mater. Struct.* **17**, 043001 (2008).

17. P. Muralt, "Recent progress in materials issues for piezoelectric MEMS," *J. Am. Ceram. Soc.* **91**, 1385–1396 (2008).

18. S. H. Baek, M. S. Rzchowski, and V. A. Aksyuk, "Giant piezoelectricity in PMN-PT thin films: Beyond PZT," *MRS Bull.* **37**, 1022–1029 (2012).

19. D. Ambika, V. Kumar, K. Tomioka, and I. Kanno, "Deposition of PZT thin films with {001}, {110}, and {111} crystallographic orientations and their transverse piezoelectric characteristics," *Adv. Mater. Lett.* **3**, 102–106 (2012).

20. T. Morita, Y. Wagatsuma, Y. Cho, H. Morioka, H. Funakubo, and N. Setter, "Ferroelectric properties of an epitaxial lead zirconate titanate thin film deposited by a hydrothermal method below the Curie temperature," *Appl. Phys. Lett.* **84**, 5094–5096 (2004).

21. G. Ardila, R. Hinchet, M. Mouis, and L. Montès, "Scaling prospects in mechanical energy harvesting using piezoelectric nanostructures," *Dig. IEEE Intern. Semicond. Conf. Dresden-Grenoble (ISCDG)* (2012), p. 75.

22. G. Zhu, R. Yang, S. Wang and Z. L. Wang, "Flexible high-output nanogenerator based on lateral ZnO nanowire array," *Nano Lett.* **10**, 3151–3155 (2010).

23. Y. W. Heo, D. P. Norton, L. C. Tien, *et al.*, "ZnO nanowire growth and devices," *Mater. Sci. Eng. Rep.* **47**, 1–47 (2004).

24. R. A. Bernal, R. Agrawal, B. Peng, *et al.*, "Effects of growth orientation and diameter on the elasticity of GaN nanowires," *Nano Lett.* **11**, 548–555 (2011).

25. A. Mitrushchenkov, G. Chambaud, J. Yvonnet, and Q.-C. He, "Towards an elastic model of wurtzite AlN nanowires," *Nanotechnology* **21**, 255702 (2010).

26. R. Agrawal, B. Peng, E. E. Gdoutos, and H. D. Espinosa, "Elasticity size effects in ZnO nanowires – A combined experimental–computational approach," *Nano Lett.* **8**, 3668–3674 (2008).
27. C. Chen, Y. Shi, Y. Zhang, J. Zhu, and Y. Yan, "Size dependence of Young's modulus in ZnO nanowires," *Phys. Rev. Lett.* **96**, 075505 (2006).
28. Y. Yang, W. Guo, X. Wang, Z. Wang, J. Qi, and Y. Zhang, "Size dependence of dielectric constant in a single pencil-like ZnO nanowire," *Nano Lett.* **12**, 1919–1922 (2012).
29. M. Minary-Jolandan, R. A. Bernal, I. Kuljanishvili, V. Parpoil, and H. D. Espinosa, "Individual GaN nanowires exhibit strong piezoelectricity in 3D," *Nano Lett.* **12**, 970–976 (2012).
30. H. J. Xiang, J. Yang, J. G. Hou, and Q. Zhu, "Piezoelectricity in ZnO nanowires: A first-principles study," *Appl. Phys. Lett.* **89**, 223111 (2006).
31. R. Agrawal and H. D. Espinosa, "Giant piezoelectric size effects in zinc oxide and gallium nitride nanowires. A first principles investigation," *Nano Lett.* **11**, 786–790 (2011).
32. Y. Gao and Z. L. Wang, "Electrostatic potential in a bent piezoelectric nanowire. The fundamental theory of nanogenerator and nanopiezotronics," *Nano Lett.* **7**, 2499–2505 (2007).
33. J. Gao, X. Zhang, Y. Sun, Q. Zhao, and D. Yu, "Compensation mechanism in N-doped ZnO nanowires," *Nanotechnology* **21**, 245703 (2010).
34. Y. Gao and Z. L. Wang, "Equilibrium potential of free charge carriers in a bent piezoelectric semiconductive nanowire," *Nano Lett.* **9**, 1103–1100 (2009).
35. C. Sun, J. Shi, and X. Wang, "Fundamental study of mechanical energy harvesting using piezoelectric nanostructures," *J. Appl. Phys.* **108**, 034309 (2010).
36. R. Yang, Y. Qin, C. Li, L. Dai, and Z. L. Wang, "Characteristics of output voltage and current of integrated nanogenerators," *Appl. Phys. Lett.* **94**, 022905 (2009).
37. Y. Hu, Y. Zhang, C. Xu, L. Lin, R. L. Snyder, and Z. L. Wang, "Self-powered system with wireless data transmission," *Nano Lett.* **11**, 2572–2577 (2011).
38. R. Tao, R. Hinchet, G. Ardila, L. Montès, M. Mouis, and A. D. Discription, "FEM modeling of vertically integrated nanogenerators in compression and flexion modes," *Proc. 10th Conf. PhD Research Microelectronics Electronics (PRIME)*, Grenoble (2014).
39. A. Manekkathodi, M.-Y. Lu, C. W. Wang, and L.-J. Chen, "Direct growth of aligned zinc oxide nanorods on paper substrates for low-cost flexible electronics," *Adv. Mater.* **22**, 4059–4063 (2010).
40. T. B. Bateman, "Elastic moduli of single crystal zinc oxide," *J. Appl. Phys.* **33**, 3309–3312 (1962).
41. G. Carlotti, G. Socino, A. Petri, and E. Verona, "Acoustic investigation of the elastic properties of ZnO films," *Appl. Phys. Lett.* **51**, 1889–1891 (1987).

42. N. Ashkenov, B. Mbenkum, C. Bundesmann, *et al.*, "Infrared dielectric functions and phonon modes of high-quality ZnO films," *J. Appl. Phys.* **93**, 126–133 (2003).
43. R. Tao, R. Hinchet, G. Ardila, and M. Mouis, "Evaluation of vertical integrated nanogenerator performances in flexion," *J. Phys. Conf. Series* **476**, 012006 (2013).
44. V. Consonni, E. Sarigiannidou, E. Appert, *et al.*, "Selective area growth of well-ordered ZnO nanowire arrays with controllable polarity," *ACS Nano* **8**, 4761–4770 (2014).

3.5

New Generation of Vertical-Cavity Surface-Emitting Lasers for Optical Interconnects

N. Ledentsov Jr, V. A. Shchukin, N. N. Ledentsov, and J.-R. Kropp
VI Systems GmbH, Hardenbergstraße 7, Berlin 10623, Germany

S. Burger and F. Schmidt
Zuse Institute Berlin (ZIB), Takustraße 7, Berlin 14195, Germany and JCMwave GmbH, Bolivarallee 22, Berlin 14050, Germany

1. Introduction

As silicon downscaling continues, the pitch size is gradually decreasing. The number of transistors per chip, consequently, further increases. Presently, the major IC manufacturers are planning to market 10 nm technology in 2015 with a further upgrade to 7 nm anticipated in 2017.[1, 2]

With the growing number of transistors per chip and upgrades in the architecture, processor productivity continues to approximately double every year, increasing demands on the processor communication bandwidth. Consequently, the speed of the input/output (I/O) ports must also increase. Until recently, Moore's Law for data communications predicted that a fourfold increase in the I/O speed would be needed every 4–5 years. Thus far, this trend has been generally maintained in major communication standards. According to the IEEE Ethernet Roadmap, the single I/O bitrate should approach 100 Gb/s by 2017. Indeed the presently active IEEE 400G Ethernet Task Force for the related standard recently agreed on serial single channel bit data rate of 400 Gb/s in short distance communications.[3] The aggregated transmission rate is to reach only 400G by 2017, but the variety of standard combinations and evolution of mid-board transceivers will allow customized solutions to reach 1 Tb/s or higher. However, at I/O speeds of well above 10 Gb/s, copper cables and connectors are becoming too energy consuming, bulky, and susceptible to electromagnetic pollution. These factors make copper links difficult even inside the box. Thus,

Future Trends in Microelectronics: Journey into the Unknown, First Edition.
Edited by Serge Luryi, Jimmy Xu, and Alexander Zaslavsky.
© 2016 John Wiley & Sons, Inc. Published 2016 by John Wiley & Sons, Inc.

massive deployment of optical interconnects is already starting in advanced areas, such as high-performance servers with 60,000 optical links per single rack.[4]

Beyond supercomputers and data centers, even consumer applications are beginning to require unprecedented data transmission rates. Recently, 4K and 5K displays became broadly available and the data traffic demands increased up to 80–100 Gb/s per screen for the highest image quality specs (color tones, frames per second). Next we will have 8K and volumetric displays, with IMAX-like bandwidth requirements of up to 175 Gb/s. The Thunderbolt 3 interface from Apple due in 2015 is already at 40 Gb/s. At higher data bitrates, the use of copper interfaces steadily shrinks while the numbers of optical links explode to tens of millions per single supercomputer, as a result of the increased bandwidth demand.

In this chapter, we address requirements for vertical-cavity surface-emitting lasers (VCSELs), one of the major devices for data communication to meet the bandwidth demand. Single-mode (SM) operation makes it possible to overcome effects related to significant spectral dispersion of the multimode fiber (MMF) standardized for the 840–860 nm wavelength range.

2. VCSEL requirements

To match the needs of the new generation of optical links, VCSELs have to meet a number of key requirements. They need to provide very high data transmission speeds at high energy efficiency, allowing close packaging into parallel links run by multichannel driver electronics. They also need to ensure data transmission over the necessary lengths of fiber. As the critical distance-limiting factor in modern fiber chromatic dispersion, VCSELs must provide an ultranarrow emission spectrum and low wavelength chirp under signal modulation.

Several designs have been proposed for high-temperature, energy-efficient VCSEL operation.[5–7] These include antiwaveguiding cavity design with AlAs-rich core, increased optical confinement, engineering of the density of states by quantum well (QW) strain engineering, thick oxide-confined apertures, and superlattice barriers aimed at prevention of the leakage of nonequilibrium carriers.[6]

This progress, for example, enabled serial digital data transmission up to ~50 Gb/s without using advanced electronics.[7] Even higher bitrates can be achieved with equalization schemes, at a cost of extra energy consumption and more ancillary electronics.

Today's technology is based on current-modulated VCSELs, which still show a high potential for future technology generations. Electrooptically modulated VCSELs represent a potential alternative. In such devices, an additional "shutter" section based on the electrooptic effect extends the optical modulation bandwidth beyond 35 GHz and the electrical bandwidth beyond 60 GHz, putting data transmission at 10 Gb/s within reach.

Special VCSEL designs may also allow uncooled wavelength division multiplexing (WDM) within the narrow 840–860 nm spectral range of low modal dispersion in the standard MMFs. Complete temperature stability of the VCSEL can be achieved by using the passive cavity concept. The gain medium is placed in the bottom part of the bottom semiconductor distributed Bragg reflector (DBR), while the upper part of the bottom DBR, the cavity region and the top DBR are all made of dielectric materials. Due to the fairly weak or zero dependence of the refractive index on temperature for certain combinations of dielectric materials, temperature stabilized VCSEL operation without cooling becomes possible. Furthermore, the dielectric DBRs and cavity extend a simple VCSEL technology with a high optical confinement factor and good heat conductivity even to materials normally less suitable for VCSEL fabrication, for example, for InP-based 1300–1550 nm VCSELs. This extends the range of VCSEL applications.

All VCSEL applications in the datacom space require data transmission over long sections of MMF, because the size of modern data centers is growing continuously. On the other hand, even at 25 Gb/s, the transmission distance for multimode VCSELs shrinks to 20–100 m only, depending on the particular standard and application. Single-mode low wavelength chirp VCSELs can be used to overcome this problem. Such devices allow $25G \times 12$ parallel transmission over 1 km of standard MMF, whereas arrays of conventional high-speed multimode VCSELs can hardly exceed 150 m distances.[8]

WDM based on the diffraction-induced angle separation of the wavelengths also requires single-mode VCSELs to ensure a stable, reproducible far-field pattern under high-speed modulation. The same requirement applies to VCSELs mated to multicore fibers with reduced diameters and close spacing of the core regions.

Single-mode VCSEL operation can be achieved in devices with very small aperture sizes (~2–3 µm). However, such devices show fairly low power and high series resistance. Industrial applications generally require VCSELs with moderate oxide apertures (5–6 µm) capable of producing 2–3 mW of output power at moderate current densities <20 kA/cm^2 over a broad temperature range. The differential resistance should be kept below 100 Ω to match standard driver electronics.

Progress in the field of high-speed VCSEL devices has been quite rapid. Very recently, single-mode (SM) VCSELs demonstrated 54 Gb/s transmission over 1 km of OM4 MMF under nonreturn to zero (NRZ) modulation[9] and 100 Gb/s over 300 m of MMF using multitone transmission.[10]

Several approaches are possible for producing SM VCSELs. For example, surface patterning of the device adjacent to the aperture boundary can create higher scattering loss for high-order modes, favoring the fundamental mode in the lasing process.[11] Alternatives include increasing the thickness of the optical cavity to reduce the impact of the oxide aperture on mode confinement and placing thin oxide aperture layers in the mode position of the longitudinal optical mode. Unfortunately, these approaches either require high process precision to align the sizes of the related features or suffer from the unstable far field, fairly low optical confinement factor in the gain medium and thus a low speed and high capacitance. The leaky VCSEL concept can allow SM operation without any adjustments in the oxide-confined VCSEL technology and does not require sacrificing the basic VCSEL parameters.[12] Such a device is the main topic of the present chapter.

3.   Optical leakage

The optical leakage concept allows the realization of high-power SM VCSELs, which can be used in multiple fields such as illumination, gas sensing, gesture recognition, data transmission, and other fields.[6, 13] There is a growing interest in the development of leaky-mode VCSELs.[14, 15]

The leaky VCSEL concept, as compared to other SM VCSEL approaches, provides a possibility to couple several devices through the leaky emission into a coherent VCSEL array to achieve increased brightness, realize two-dimensional beam steering,[14] or on-chip lateral integration with other types of devices such as monitor photodetectors, slow light waveguides, and 2D optical logic gates for optical computing.[16] Electrooptically controlled leakage can be applied for VCSEL modulation at ultrahigh speed.[17, 18]

Leaky-mode VCSELs can be fabricated using different techniques. In one approach, VCSEL mesas are formed by etching with subsequent overgrowth using a material with a higher refractive index.[19, 20] The VCSEL cavity modes can leak out into the high refractive index material. As the high-order transverse modes have a higher intensity at the boundary of the aperture, they exhibit a much higher leakage loss than the fundamental mode. Another approach is based on etching the top DBR over the aperture region, resulting in a locally shorter effective cavity length that allows the leakage of cavity modes into the surrounding parts of the structure. A similar concept can be realized by thinning the cavity layer in the aperture region prior to the DBR deposition or using photonic crystal patterning of the VCSEL surface, filtering out undesirable high-order modes from the aperture region.[21]

Recently, it was shown that leaky VCSELs can be realized by a proper epitaxial design of an oxide-confined device without any need for surface patterning or etching and overgrowth.[15] As a result, leaky VCSELs can be fabricated in a process that is fully compatible with the standard oxide-confined VCSEL technology without additional processing steps.[6]

In spite of active research in the field of leaky VCSELs, no studies of the impact of the leakage effect on the far-field pattern of the device have been reported so far. This is in a strict contrast to leaky-design edge-emitting lasers, where the leakage effect is clearly manifested by a specific emission tilted away from the normal to the facet of the device.[22] Such emission should exist in leaky-mode VCSELs and may serve as a fingerprint of this design, as was recently proposed.[23]

In the present chapter, we investigate oxide-confined, leaky-design VCSELs and observe the characteristic leakage-effect-induced features in the vertical far-field pattern of the device. No such emission is detected in conventional oxide-confined VCSELs. A perfect match is found between the near and far fields, in agreement with cold-cavity, three-dimensional (3D) modeling. Clearly resolved narrow emission lobes at high tilt angles are shown to be a unique property of the leaky-mode device. The leaky emission provides a fingerprint and a quantitative measure of the optical leakage in the lateral direction and represents a powerful tool for proper design of the related vertical microcavity devices, their coherent assemblies, and on-chip combinations with other devices.

4. Experiment

Epitaxial wafers for oxide-confined 850 nm VCSELs were grown by MOCVD in an industrial multiwafer reactor. The epitaxial design, optimized for ultrahigh-speed operation, included conventional AlGaAs DBRs with graded compositional profiles, an AlGaAs-based microcavity, and an active region composed of five compressively strained InGaAs quantum wells (QWs).

The VCSEL wafer was fabricated in a high-frequency contact pad configuration with BCB planarization and was tested without soldering to a heat sink. The design and the basic properties of the oxide-confined leaky VCSELs are presented in Refs 12, 15.

A CCD camera was used to evaluate the near-field pattern. Far-field measurements were performed over the full 180° range with 2° resolution by using a Si detector.

• *Benchmarking to conventional VCSELs*

For benchmarking, we first studied conventional double aperture VCSEL device similar to the one described in Ref. 6.

Double oxide aperture and a $3\lambda/2$ cavity were applied both to the reference and the leaky VCSELs. A characteristic feature of the leaky VCSEL is either single-mode operation at moderate aperture diameters (~5 μm) or a multimode spectrum at larger ~8–10 μm apertures but at a strongly reduced overall spectral width as compared to the nonleaky VCSEL design at similar aperture values.[15, 24] Modeling of the near and far fields of conventional VCSELs has a long history and is well understood.[6, 12, 15, 23, 24] Oxide-confined leaky VCSELs were predicted to have a specific emission at large tilt angles with a narrow angular width.[23] This feature originates from the leaky emission in the direction outside of the aperture. As no confinement for this emission is provided by the aperture diameter, the diffraction-related broadening is suppressed and the feature in the far field can be narrow.

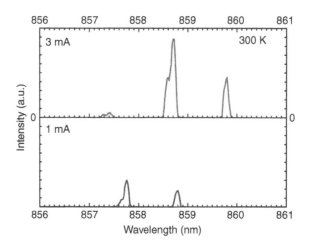

Figure 1. Characteristic electroluminescence spectrum of conventional benchmark VCSEL at 1 mA (bottom) and 3 mA (top). The wavelength spacing between the modes in the spectrum (1.2 nm) points to the aperture diameter of ~4 μm.

In Fig. 1, we show electroluminescence spectra of a conventional double oxide aperture $3\lambda/2$ VCSEL at room temperature as a function of current. It follows from Fig. 1 that the conventional VCSEL is heavily multimode even at low currents in spite of the narrow ~4 μm diameter aperture, as revealed by the wavelength splitting between the fundamental and the excited modes of 1.2 nm.

Figure 2(a) shows the far-field diagrams of the same VCSEL as a function of current. At low current, a signal background having a cosine intensity distribution is observed due to the spontaneous emission, which saturates when the lasing

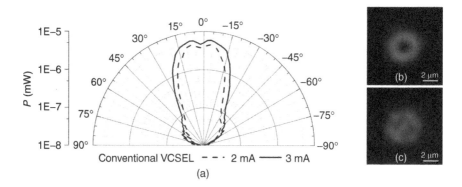

Figure 2. (a) Far field of the conventional VCSEL with double oxide aperture at different currents; corresponding CCD camera images are also shown at 1 mA (b) and 4 mA (c).

evolves. Lasing is revealed in the far field by appearance of the tilted overlapping features (at ~7° in the angular space) representing high-order transverse mode(s). Measurements made with a CCD camera (sometimes called "near-field" images) clearly reveal multimode behavior as well (see Fig. 2(b) and (c)).

• *Leaky VCSELs*

A very different behavior is observed for the leaky VCSEL. Even the device with a mode spacing of ~0.85 nm and an aperture diameter of ~5 μm shows predominantly single-mode lasing up to ~4 mA (see Fig. 3).

The far-field pattern emitted by the device above threshold confirms the single-mode behavior. At low intensity, spontaneous emission can be revealed. At currents below 4 mA, the excited mode is not manifested in the spectra. Figure 4 presents the far field of the leaky VCSEL with double oxide aperture as a function of current. At low current, we find the spontaneous emission background with a cosine intensity distribution. Lasing is revealed by appearance of a single-lobe structure in the angular space at currents above threshold.

As follows from Fig. 4, the leaky VCSEL device of the leaky design is single mode up to 4 mA. Electroluminescence spectra of the leaky VCSEL at larger aperture size of ~5.5 μm and smaller mode spacing of ~0.75 nm are shown in Fig. 5 for completeness.

The CCD camera images and the far-field diagrams of the device are shown in Fig. 6. The transformation in the near-field pattern agrees with the appearance of the tilted lobes in the far-field pattern at tilt angles ~7°. The spectra are still narrower both in wavelength and angular space as compared to those of the conventional nonleaky VCSEL design of Figs 1 and 2 used for benchmarking.

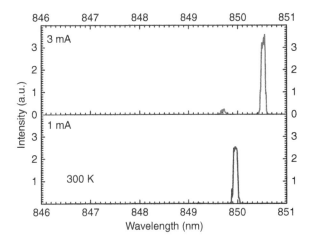

Figure 3. Characteristic electroluminescence spectra of leaky quasi-SM device at 1 mA (bottom) and 3 mA (top). The transverse mode spacing is 0.85 nm.

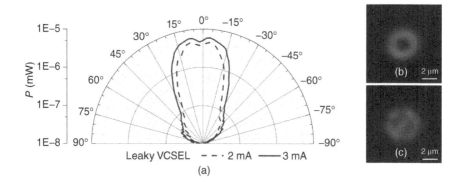

Figure 4. (a) Far field of the leaky VCSEL with double oxide aperture; corresponding CCD camera images are shown at 1 mA (b) and 4 mA (c).

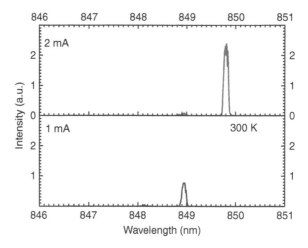

Figure 5. Characteristic electroluminescence spectra of the device at 1 and 2 mA. Mode spacing is 0.75 nm (aperture ~5.5 μm).

The most interesting phenomenon is the emergence of an angularly narrow emission at large tilt angles ~35°. The full width at half maximum of the emission is only ~2°, much narrower than the angular width of the fundamental and excited lobes that are broadened by diffraction due to the small oxide aperture.

5. Simulation

To understand the experimental results, we now turn to numerical modeling of the optical modes in the VCSEL structure in the cold cavity approach.

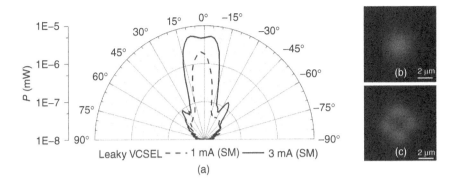

Figure 6. Far-field pattern of the leaky VCSEL with double oxide aperture at 5.5 μm aperture diameter. Note that at higher currents a high-order transverse mode evolves. This mode is accompanied by an appearance of an angular narrow emission at large ~35° tilt angle. The CCD camera images are taken at 1 mA (b) and 4 mA (c).

Our 3D simulations of electromagnetic fields were performed with the JCM Wave finite element software package based on full vector Maxwell's equations. Cylindrical symmetry was applied,[25] allowing a 2D solution to represent the entire 3D picture. Refractive indices of materials for the simulation were taken from Ref. 26.

- *Leaky design*

The electromagnetic field was modeled for in-plane and cross-section intensity distributions, as shown in Fig. 7. For clarity in transverse images, we simulated the optical field transformation up to 10 μm above the surface of the VCSEL.

In Fig. 7, the bright areas represent the maximum of intensity, and dark areas represent the minimum of intensity (see the logarithmic intensity scale). In the fundamental (HE11) mode, the maximum intensity is located in the center of the VCSEL and lasing emission is visible above the surface as a homogeneous column in the center. Simulation also picks up a small in-plane leakage parallel to the surface and even weaker leakage of the light into the air at some tilt angle.

The first excited mode (HE21), on the other hand, has the maximum of intensity not in the center but closer to the boundary of the oxide aperture, at a radius $r = 1.5$ μm.

As expected, since the intensity maximum of the higher order lateral mode is located closer to the boundary of the oxide layers, leading to far greater in-plane leakage. Consequently, the strongly tilted emission in the air that is a manifestation of the leakage process is much stronger for the high-order transverse mode.

Based on this simulation, the leakage emission has an angle of 35°. This result corresponds well to the 1D simulation of the optical reflectance spectra of the structure in oxidized and nonoxidized regions and with the experimental results of Fig. 7.[15]

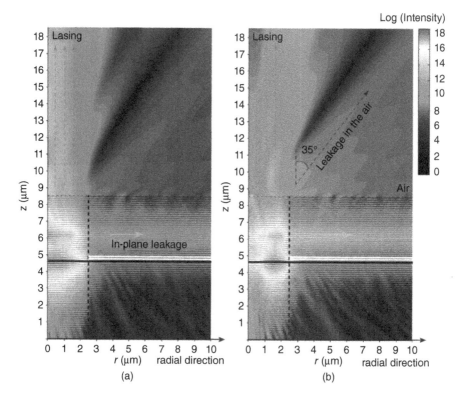

Figure 7. Simulated electric field intensity distribution, plotted on a logarithmic scale, in a device with 5 μm aperture: (a) fundamental HE11 mode ($\lambda = 843.3$ nm, lifetime $\tau = 3.7$ ps); (b) first excited mode HE21 ($\lambda = 841.4$ nm, lifetime $\tau = 1.4$ ps). Black dashed line represents the radius of the oxide-confined aperture, white lines represent the oxide layers, and thick black line indicates the active region of the device. Note slight bright stripe parallel to the surface characterizing the in-plane leakage effect and the related evolution of the tilted beam in the air.

In Refs 15, 24, the optical power reflectance (OPR) spectra of the same leaky VCSEL structure was simulated by the transfer matrix method applied separately to a multilayer all-semiconductor structure in the nonoxidized core region and in the oxidized periphery region of the device. The OR spectrum of the nonoxidized region at normal incidence contains a reflectivity dip at the lasing wavelength 850 nm, which matches a dip in the OR spectrum in the oxidized region calculated for propagation of light at a tilt angle of ~10° in the semiconductor. Thus, the VCSEL mode of the nonoxidized cavity is in resonance with a tilted optical mode of the oxidized region. The angle of ~10° in the semiconductor transfers, upon refraction, into an angle of ~33° in air, which agrees well both with the 3D simulation and experiment. The discrepancy in the angle between 1D and 3D

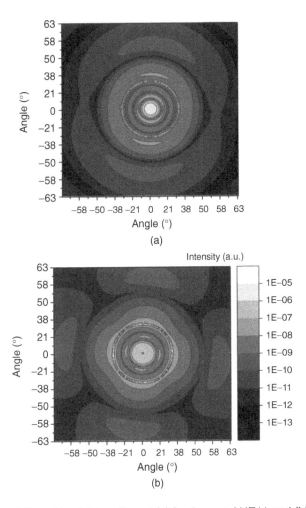

Figure 8. Simulation of far-field profiles of (a) fundamental HE11 and (b) first excited HE21 modes. Circle at ~35° is the fingerprint of the lateral leakage effect.

simulation can be due to the additional lateral confinement taken into account in the 3D simulation.

Figure 8 shows the simulated far-field profiles of the fundamental and the first excited modes, whereas Fig. 9 compares the experimentally measured and the simulated far-field profiles.

Since the experimental setup has an angular resolution of 2°, the observed leakage features are broadened compared to simulation. Furthermore, the leakage lobes in the simulation are narrower compared to the experiment because the simulation assumes an effectively infinite surface area for the leakage, since optical field propagation is limited only by the gradual decay of the optical field

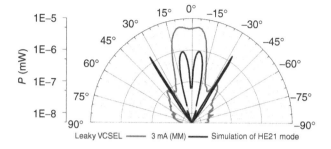

Figure 9. Comparison of experimental and simulated far-field profiles of the first excited HE21 mode.

intensity due to attenuation. In the real device, the area is also limited by the contact metal ring opening of only 11 μm in diameter that diffracts the leaky emission.

The simulated integral intensity of the leakage-related lobes is comparable to the one found in the experiment within the experimental resolution. This is an important observation, proving that in spite of all the uncertainties – for example, strain fields in and around the oxide layers, tapering of the oxide aperture caused by the side oxidation of the graded $Al_xGa_{1-x}As$ layers, free carrier absorption, and other effects – it is the leakage that mainly determines the emission pattern.

- *Leaky design with oxide relief*

The leakage effect is easier to control when the difference in the refractive index between the semiconductor inner region and the oxide outer region of the aperture-forming layer is maximized. Selective etching of the oxide layer provides an extra tool to control and enhance the leakage effect. Thus, we model VCSELs based on the oxide relief concept[27, 28] and evaluate the impact on the leakage effect. In the simple case, we just replace the oxide apertures with air gaps.

Influence of oxide relief on conventional VCSELs. First we look at a conventional VCSEL design, similar to the one described in Figs 1 and 2. As mentioned earlier, no directed leakage is observed for this type of device, which is confirmed by far-field modeling. When oxide relief is applied to the VCSEL, narrow tilted lobes appear in the modeling of the far-field profile at ~60°. Far-field profiles before and after oxide relief are shown in Fig. 10.

The modal lifetimes are compared in Fig. 11. The lifetimes of the first excited modes are relatively shorter than the fundamental mode in structures with oxide relief. The difference is particularly strong for 6–9 μm aperture sizes. This promises increased mode selectivity and improved range of single-mode operation.

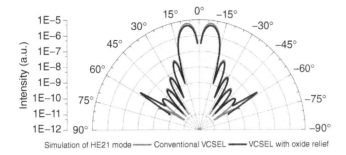

Figure 10. Comparison of simulated far-field profiles of the first excited HE21 mode of conventional VCSEL design with and without oxide relief. Note that characteristic narrow "leaky" features appear only in the structure with oxide-relief apertures.

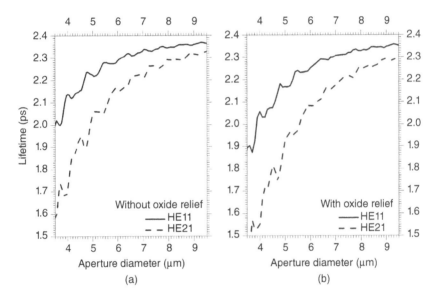

Figure 11. Lifetimes of the fundamental HE11 and first excited HE21 modes of conventional VCSEL design with and without oxide relief.

Influence of oxide relief on leaky VCSELs. Now we turn to applying oxide relief to the leaky VCSELs. Simulated electric field intensity distribution in a device with 5 μm aperture shows that after applying the oxide relief, strong field intensity inside the air gaps can be seen. In the simulation of the far-field profiles, we observe an increase in the intensity and a slight decrease in the angle of leakage emission – see Fig. 12.

Comparison of the lifetimes of the fundamental and first excited modes presented in Fig. 13 shows that the structure with oxide relief has decreased lifetimes, but there seems to be no significant improvement in the mode selectivity.

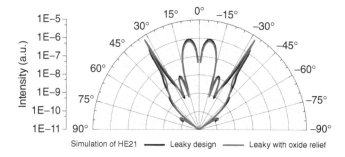

Figure 12. Comparison of simulated far-field profiles of the first excited mode (HE21) of leaky VCSEL design with and without oxide relief.

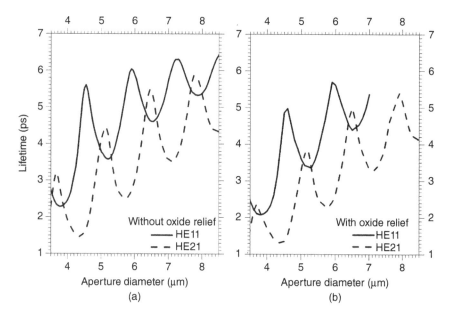

Figure 13. Calculated wavelengths lifetimes of the fundamental HE11 and first excited HE21 modes of leaky VCSEL design with and without oxide relief.

Oscillatory behavior of the mode lifetimes in the leaky structures is evident in both cases, with the reasons of the effect discussed in Ref. 16.

Our modeling shows that the leakage effect can be further improved in leaky VCSELs by employing the oxide relief technique, but the impact is weaker than in otherwise nonleaky-design, oxide-confined VCSELs.

Our modeling approach can be further extended to oxide-relief VCSELs with locally Zn-intermixed DBRs.[29] It can also be extended to VCSELs with different types of surface relief,[30] photonic crystals, and coherent arrays of leaky VCSELs.[31, 32]

6. Conclusion

To conclude, we have considered requirements for modern VCSELs used in data communications. Single-mode VCSEL technologies at high data rates of up to 100 Gb/s are needed for future communication networks. As any deployed VCSEL technology should meet reliability of standard oxide-confined VCSELs, the leaky design approach appears particularly promising. We studied both theoretically and experimentally the leakage-related effects in oxide-confined VCSELs applying 3D vector modeling of the optical field and evaluated spectral, near- and far-field properties of the device. In line with the theory, we experimentally observed the leakage-induced emission revealed as narrow tilted lobes in the far field of the VCSEL. This observation confirms the validity of the leaky VCSEL concept, allows for better understanding of the device properties such as possible strain and heat gradients, oxide layer tapering, in-plane light scattering, and absorption. Our work enables the engineering of advanced devices and photonic-integrated circuits with targeted design of oxide apertures or air gaps through quantitative evaluation of the leaky emission.

Acknowledgments

The authors acknowledge support by ADDAPT project of the FP7 Program of the European Union under Grant Agreement No. 619197.

References

1. W. M. Holt, "Moore's law: A path forward," *Proc. Intern. Solid-State Circ. Conf. (ISSCC)* (2016), p. 1.
2. Taiwan Semiconductor Manufacturing Company, "Future R&D plans," see www.tsmc.com/ (accessed on Apr. 21, 2016).
3. Documents of IEEE P802.3bs 400 Gb/s Ethernet Task Force, (2015), see www.ieee802.org/3/bs/index.html (accessed on Apr. 21, 2016).
4. C. Nitta, R. Proietti, Y. Yin, *et al.*, "Leveraging AWGR-based optical packet switches to reduce latency in petascale computing systems," technical report CSE-2013-68, University of California–Davis, 2012.
5. N. N. Ledentsov and J. Lott, "New-generation vertically emitting lasers as a key factor in the computer communication era," *Phys.-Usp.* **54**, 853–858 (2011).
6. N. N. Ledentsov, "Ultrafast nanophotonic devices for optical inter-connects," chapter in: S. Luryi, J. M. Xu, and A. Zaslavsky, eds., *Future Trends in Microelectronics: Into the Cross Currents*, New York: Wiley, 2013, pp. 43–48.
7. N. N. Ledentsov, J.-R. Kropp, V. A. Shchukin, *et al.*, "High-speed mod-ulation, wavelength, and mode control in vertical-cavity surface-emitting lasers," *Proc. SPIE* **9381**, 93810F (2015).

8. J.-R. Kropp, V. A. Shchukin, N. N. Ledentsov, *et al.*, "850 nm single mode VCSEL-based 25G×16 transmitter/receiver boards for parallel signal transmission over 600 m of mm fiber," *Proc. SPIE* **9390**, 93900C (2015).

9. G. Stepniak, A. Lewandowski, J.-R. Kropp, *et al.*, "54 Gbit/s OOK transmission using single-mode VCSEL up to 2.2 km MMF," *Electron. Lett.* **52**, 633–635 (2016).

10. B. Wu, X. Zhou, N. Ledentsov, and J. Luo, "Towards 100 Gb/s serial optical links over 300 m of multimode fibre using single transverse mode 850 nm VCSEL," *Asia Commun. Photonics Conf. (ACP)* (2015), paper Asu4C.3.

11. E. Haglund, Å. Haglund, J. S. Gustavsson, *et al.*, "Reducing the spectral width of high speed oxide confined VCSELs using an integrated mode filter," *Proc. SPIE* **8276**, 82760L (2012).

12. V. Kalosha, N. N. Ledentsov, D. Bimberg, *et al.*, "Design considerations for large-aperture single-mode oxide-confined vertical-cavity surface-emitting lasers," *Appl. Phys. Lett.* **101**, 071117 (2012).

13. M. Grabherr, "New applications boost VCSEL quantities: Recent developments at Philips," *Proc. SPIE* **9381**, 938102 (2015).

14. C. Lei and K. D. Choquette, eds., "Vertical-cavity surface-emitting lasers X," *Proc. SPIE* **6132**, 61320J (2006).

15. V. A. Shchukin, N. N. Ledentsov, J.-R. Kropp, *et al.*, "Single-mode vertical cavity surface emitting laser via oxide-aperture-engineering of leakage of high-order transverse modes," *IEEE J. Quantum Electron.* **50**, 990–995 (2014).

16. D. K. Serkland, K. M. Geib, G. M. Peake, *et al.*, "Final report on LDRD project: Leaky-mode VCSELs for photonic logic circuits," *U.S. Department of Energy, SAND* 7118 (2005).

17. V. A. Shchukin, N. N. Ledentsov, J. A. Lott, *et al.*, "Ultra high-speed electro-optically modulated VCSELs: Modeling and experimental results," *Proc. SPIE* **6889**, 68890H (2008).

18. A. Paraskevopoulos, H.-J. Hensel, W.-D. Molzow, *et al.*, "Ultra-high-bandwidth (>35 GHz) electrooptically-modulated VCSEL," *Proc. Optical Fiber Commun. Conf. (OFC/NFOEC)* (2006), pp. 1–3.

19. Y. A. Wu, G. S. Li, R. F. Nabiev, *et al.*, "Single-mode, passive antiguide vertical cavity surface emitting laser," *IEEE J. Sel. Top. Quantum Electron.* **1**, 629–637 (1995).

20. L. Bao, N. H. Kim, L. J. Mawst, *et al.*, "Near-diffraction-limited coherent emission from large aperture antiguided vertical-cavity surface-emitting laser arrays," *Appl. Phys. Lett.* **84**, 320–322 (2004).

21. L. Bao, N. H. Kim, L. J. Mawst, *et al.*, "Single-mode emission from vertical-cavity surface-emitting lasers with low-index defects," *IEEE Photonics Technol. Lett.* **19**, 239–241 (2007).

22. D. R. Scifres, W. Streifer, and R. D. Burnham, "Leaky wave room temperature double heterostructure GaAs/GaAlAs diode laser," *Appl. Phys. Lett.* **29**, 23–25 (1976).

23. V. P. Kalosha, D. Bimberg, and N. N. Ledentsov, "Leakage-assisted transverse mode selection in vertical-cavity surface-emitting lasers with thick large-diameter oxide apertures," *IEEE J. Quantum Electron.* **49**, 1034–1039 (2013).

24. V. A. Shchukin, N. N. Ledentsov, J.-R. Kropp, *et al.*, "Engineering of optical modes in vertical-cavity microresonators by aperture placement: Applications to single-mode and near-field lasers," *Proc. SPIE* **9381**, 93810V (2015).

25. M. Rozova, J. Pomplun, L. Zschiedrich, *et al.*, "3D finite element simulation of optical modes in VCSELs," *Proc. SPIE* **8255**, 82550K (2012).

26. S. Adachi, "GaAs, AlAs, and $Al_xGa_{1-x}As$: Material parameters for use in research and device applications," *J. Appl. Phys.* **58**, R1–R29 (1985).

27. J.-W. Shi, W. C. Weng, F. M. Kuo, *et al.*, "Oxide-relief vertical-cavity surface-emitting lasers with extremely high data-rate/power-dissipation ratios," *Proc. Optical Fiber Commun. Conf. (OFC/NFOEC)* (2011), pp. 1–3.

28. J.-W. Shi, C. C. Wei, J. J. Chen, *et al.*, "850-nm Zn-diffusion vertical-cavity surface-emitting lasers with oxide-relief structure for high-speed and energy-efficient optical interconnects from very-short to medium (2 km) reaches," *Proc. SPIE* **9381**, 93810E (2015).

29. J.-W. Shi, C. C. Wei, J. J. Chen, *et al.*, "High-performance Zn-diffusion 850-nm vertical-cavity surface emitting lasers with strained InAlGaAs multiple quantum wells," *IEEE Photonics J.* **2**, 960–966 (2010).

30. J. S. Gustavsson, Å. Haglund, J. Bengtsson, *et al.*, "Dynamic behavior of fundamental-mode stabilized VCSELs using shallow surface relief," *IEEE J. Quantum Electron.* **40**, 607–619 (2004).

31. D. Zhou and L. J. Mawst, "Two-dimensional phase-locked antiguided vertical-cavity surface-emitting laser arrays," *Appl. Phys. Lett.* **77**, 2307–2309 (2000).

32. D. F. Siriani, S. P. Carney, and K. D. Choquette, "Coherence of leaky-mode vertical-cavity surface-emitting laser arrays," *IEEE J. Quantum Electron.* **47**, 672–675 (2011).

3.6

Reconfigurable Infrared Photodetector Based on Asymmetrically Doped Double Quantum Wells for Multicolor and Remote Temperature Sensing

X. Zhang, V. Mitin, G. Thomain, T. Yore, and Y. Li
Department of Electrical Engineering, SUNY at Buffalo, Buffalo NY 14260, USA

J. K. Choi
Memory R&D Division, SK Hynix, Icheon-si Gyeonggi-do 467-701, South Korea

K. Sablon and A. Sergeev
U.S. Army Research Laboratory, Adelphi MD 20783, USA

1. Introduction

Multicolor IR detectors are providing new solutions in physical, chemical, and biological sensing and imaging.[1, 2] Multispectral sensing makes it possible to increase detection sensitivity, to improve object identification and discrimination capabilities, and to measure the absolute temperature of the object regardless of its emissivity and geometry.[3–5] Numerous practical applications of such detectors include defense and commercial technologies, such as night vision, low visibility navigation, monitoring of industrial high-temperature processes, noncontact temperature imaging, target detection and tracking, and remote earth observations.

Quantum well infrared photodetectors (QWIPs) are a well-established technology with numerous possibilities to control electron levels and to manage photoelectron processes.[6, 7] In recent years, significant efforts were devoted to development of multispectral QWIPs, particularly for temperature sensing. Initial studies were focused on multistack packaging of single-frequency QWIPs with multiple electrical terminals, which allow for independent bias of each QWIP and detection of multiple wavelengths.[8–12] Results of Ref. 10 have demonstrated that the temperature of the object may be determined with high accuracy by monitoring the ratio(s) of photocurrents generated simultaneously

Future Trends in Microelectronics: Journey into the Unknown, First Edition.
Edited by Serge Luryi, Jimmy Xu, and Alexander Zaslavsky.
© 2016 John Wiley & Sons, Inc. Published 2016 by John Wiley & Sons, Inc.

in two or more single-frequency QWIPs. In particular, the two-color detectors operating at 10 K have demonstrated high sensitivity of the photocurrent ratio to the object temperature. Employing the corrugated structure that provides high wide-band coupling between IR radiation and electrons in QWs, the authors of Ref. [10] show that the relative photoresponse ratio changes by ~0.4% per 1 °C.[10] Unfortunately, the multiterminal design of multicolor QWIPs leads to significant technological challenges in the device processing, which limits scalability of this technology. Various approaches to improve packaging and biasing of single-frequency QWIPs in multicolor detectors have been proposed and investigated.[13] However, difficulties in detector multiplexing and related technological challenges put significant limitations on development of multicolor detectors on the base of traditional single-frequency QWIPs.

Emerging reconfigurable nanostructures provide novel sensing opportunities, such as multiplexing capabilities, high portability, and scalability. Bias-induced charge redistribution between selectively doped nanoblocks, such as quantum wells or dots, strongly changes the absorption spectra of these nanoblocks and the relevant photoelectron processes.[14–16] Advanced reconfigurable QWIP structures may substitute a stack of single-frequency QWIPs. Nanostructures with double quantum wells (DQWs) are very promising for multicolor sensing. They demonstrate strong double-peak photoresponse, where the peak positions and their magnitudes may be controlled by the bias voltage via the charge redistribution between DQWs.[17–19]

In this chapter, we study the effects of DQW asymmetric doping on the parameters of IR detector, its spectral selectivity, and bias tunability. Strong doping is expected to increase the bias-induced asymmetry in electron filling of the double wells in the DQW unit. The filling asymmetry, in turn, should increase a difference between the main peak and the second peak. We also evaluate the applicability of our devices to remote temperature sensing. With this goal, we study photoresponse as a function of emitter temperature in the 300–800 °C range. Our results show that due to high spectral sensitivity to the biasing, the relative photoresponse, that is, the ratio of photocurrents at two bias voltages, exhibits a substantial dependence on the emitter temperature. Continuous variation of spectral characteristics with the bias voltage provides fast collection of large amounts of the relative photoresponse data and, in this way, improves accuracy in determination of the emitter temperature.[20] High bias-tunable spectral selectivity of asymmetrically doped DQWs makes this nanomaterial very attractive for precise thermometric measurements.

2. Fabrication of DQWIP with asymmetrical doping

Asymmetrically doped double-well QWIP structures were grown on semi-insulating GaAs wafer to permit backside illumination, as illustrated in Fig. 1. The growth sequence started with a 500 nm undoped GaAs buffer layer, and was

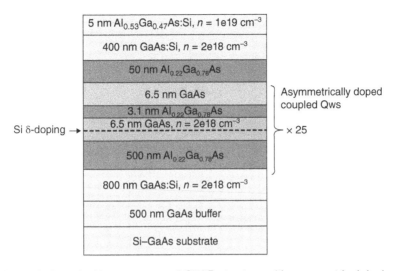

| 5 nm Al$_{0.53}$Ga$_{0.47}$As:Si, n = 1e19 cm^{-3} |
| 400 nm GaAs:Si, n = 2e18 cm^{-3} |
| 50 nm Al$_{0.22}$Ga$_{0.78}$As |
| 6.5 nm GaAs |
| 3.1 nm Al$_{0.22}$Ga$_{0.78}$As |
| 6.5 nm GaAs, n = 2e18 cm^{-3} |
| 500 nm Al$_{0.22}$Ga$_{0.78}$As |
| 800 nm GaAs:Si, n = 2e18 cm^{-3} |
| 500 nm GaAs buffer |
| Si–GaAs substrate |

Asymmetrically doped coupled Qws

Si δ-doping →

× 25

Figure 1. The double-quantum well QWIP structure with asymmetrical doping.

followed by an 800 nm heavily doped ($N_D = 2 \times 10^{18}$ cm^{-3}) GaAs contact layer, 25 stages of the detection unit, 400 nm layer of GaAs doped 2×10^{18} cm^{-3} as a contact layer, and finally 5 nm InGaAs layer doped to 10^{19} cm^{-3}. One DQW unit is composed of a 6.5 nm GaAs layer doped by Si with sheet density 5×10^{11} cm^{-2} and 6.5 nm layer of undoped GaAs separated by a 3.1 nm-thick Al$_{0.2}$Ga$_{0.8}$As barrier. The DQW units are separated by 50 nm Al$_{0.2}$Ga$_{0.8}$As barriers.

Grown structures were processed by standard photolithography, wet chemical etching, and metallization techniques. The first lithography step defined the 100×100 µm^2 mesa structures. The mesa structures were etched with sulfuric acid (H$_2$SO$_4$:H$_2$O$_2$:H$_2$O = 3:10:450) at a rate of 1.5 nm/s. The etch rate was calibrated by atomic force microscopy. The second lithography step was performed for the metal deposition. Prior to the deposition, the native oxide layer was removed by an ammonium hydroxide solution (NH$_4$OH:H$_2$O = 10:300). Ohmic contacts made with Ni/Ge/Au/Ni/Au metallization (2/14/28/40/100 nm layer thicknesses) were deposited by e-beam evaporation and underwent rapid thermal annealing in N$_2$ ambient for 40 s at 430 °C. The resulting Ohmic contact to each device had an average resistance about 30 Ω.

3. Optoelectronic characterization of DQWIPs

To characterize QWIP devices, we measured dark current, photocurrent, and spectral characteristics of the photoresponse. These data were used to calculate the device responsivity and activation energy. The measurement setups are shown schematically in Fig. 2.

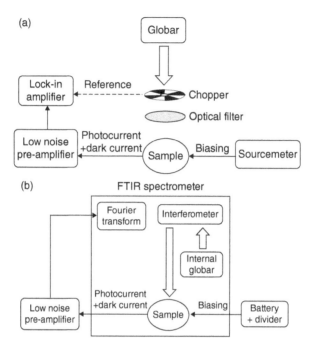

Figure 2. The measurement setups: (a) dark current and photocurrent; (b) spectral characteristics of photoresponse.

The device was mounted inside the cryostat with a KBr window, which has 90% transmissivity in the mid-IR range. A Ge filter was placed between the cryostat and globar to rule out the absorption by the interband transitions. The sample temperature was measured and controlled by Cryocon temperature controller. The current–voltage curves were registered by the Keithley 2602 source meter with accuracy of 10^{-7} A/cm^2. To enhance IR absorption by our DQWs we use the following two approaches. In the first scheme, the radiation is incident on the 45°-angle facet of substrate. In the second scheme we employ the diffuse reflector, which was fabricated by polishing the substrate backside with a 3 μm pad. Both approaches provide substantial coupling of IR radiation to electrons localized in DQWs.

The measured results are presented in Fig. 3. It is known that for a single quantum well device, the dark current at positive bias is usually higher than that at negative bias of the same absolute value. This is explained by the dopant migration in the growth direction.[21] However, as seen in Fig. 3(a), the dark current in our devices is practically symmetrical with respect to bias. Thus, in our case, the dopant migration is essentially compensated by the asymmetrical doping in our QWIP structure.

The activation energy was determined by an Arrhenius-type temperature dependence fit of the dark current: $I_D \sim \exp(-E_A/kT)$. As shown in Fig. 3,

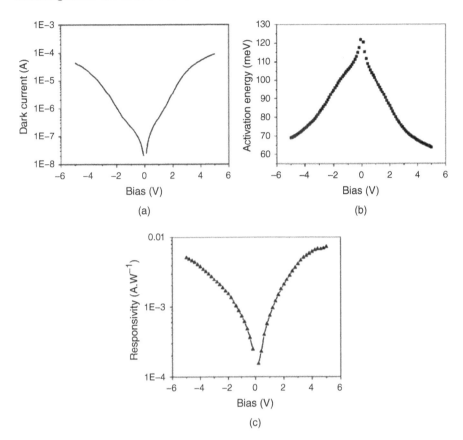

Figure 3. Electro-optical characteristics of DQWIP devices at 80 K: (a) dark current; (b) activation energy; and (c) responsivity.

the activation energy looks symmetric with respect to the bias polarity and demonstrates significant dependence on the absolute value of bias. The obtained value of activation energy (120 mV) is in very good agreement with the numerical modeling of the DQW band structure at zero bias.[22] The activation energy strongly decreases with increasing bias due to Poole–Frenkel and Fowler–Nordheim effects.

The responsivity R was determined as a ratio of the photocurrent to the IR power, selected by the 2–20 μm Ge filter from 1000 °C globar radiation. The power incident on the device was measured by the optical power meter. The results in Fig. 3(c) show that the responsivity increases with the bias and achieves ~10 mA/W at 5 V.

The spectral selectivity and bias-tunability of DQW structures at 80 K are reported in Fig. 4 (the spectral characteristics of photoresponse at 20 K were studied in our previous publication[22]). Two spectral peaks can be distinguished at

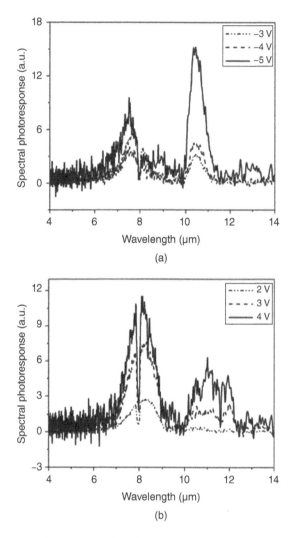

Figure 4. The spectral characteristics of photoresponse at 80 K: (a) negative bias of −3, −4, and −5 V; (b) positive bias of 2, 3, and 4 V.

negative and positive bias, as shown in Fig. 4(a) and (b) respectively. The spectral peaks 7.5 and 10.3 μm at negative bias shift to 8.3 and 11.1 μm at positive bias. This indicates that the spectral characteristics of photoresponse strongly depend on the polarity of bias voltage.

As expected, the asymmetrical doping enhances spectral asymmetry of the photoresponse. As shown in Fig. 4(a), the 7.5 μm spectral peak dominates at −3 V, while 10.3 μm dominates at higher bias of −5 V. In Fig. 4(b) at positive bias, this effect is not as evident, but we still can see the switch of the spectral peak at high bias ($V > 2\,\mathrm{V}$). Thus, the spectral characteristics of photoresponse substantially depend not only on polarity of the bias voltage but also on the absolute value of the applied bias.

The detailed study of photoelectron processes[22] shows that the bias-induced spectral selectivity is determined by the strong asymmetry in electron filling of QWs in the double-well unit and by effective charge redistribution over the unit under bias. Local fields modify both electron wave functions and matrix elements of electron transitions. The electron energy levels are shifted due to the Stark effect. Electric field also modifies tunneling processes due to the Poole–Frenkel and Fowler–Nordheim effects.

Summarizing this subsection, we would like to highlight that asymmetric doping significantly enhances the spectral selectivity and its bias tunability. Further improvements may be reached at higher doping levels.

4. Temperature sensing

After characterizing the multicolor capabilities of DQWIPs, we investigated the applicability of our detectors to remote temperature sensing. As an IR emitter, we employ a blackbody radiator with precisely controlled temperature in the 300–1000 °C range. In our measurements, the detector was located 15 cm away from the aperture of the blackbody. The DQWIP photocurrent as a function of voltage bias at different temperatures of the blackbody source is presented in Fig. 5. As seen, our DQWIP devices work best below 700 °C.

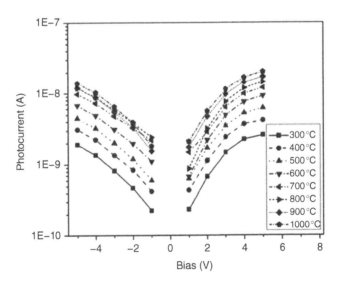

Figure 5. The photocurrent of DQWIP illuminated by blackbody radiation with source temperature in the 300–1000 °C range.

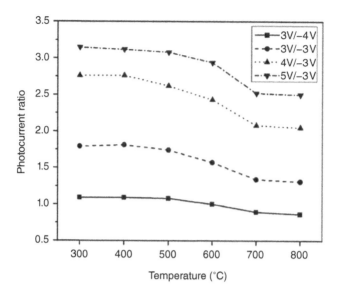

Figure 6. Temperature dependence of the photocurrent ratio under various combinations of bias values.

For quantitative description of the temperature sensing, we studied the temperature dependence of photocurrent ratio at four combinations of bias voltages: 3/−4, 3/−3, 4/−3, and 5/−3 V. The corresponding results are presented in Fig. 6. The photocurrent ratio shows the strongest temperature dependence in the 600–700 °C range. In this range, the sensitivity of the photocurrent ratio $R^{-1}(dR/dT)$ is ~0.2% per 1 °C. Thus, the bias-tunable DQWIP operating at nitrogen temperatures shows the temperature sensitivity just two times smaller than the complex multistack QWIP devices operating at helium temperatures.[10]

5. Conclusion

Multicolor DQW devices with asymmetrical doping of double wells in the DQW unit were fabricated and characterized. We found that the asymmetrical doping may compensate the effect of dopant migration in the growth direction. As a result, the dark current and responsivity of our devices are only weakly dependent on the bias polarity. Doping strongly enhances spectral selectivity and spectral tunability by the bias voltage because specific selective electron filling of quantum wells in the double-well unit increases the photocurrent from one of the wells and suppresses the contribution of the other. At 80 K, the two spectral peaks at 7.5 and 10.3 μm shift to 8.3 and 11.1 μm when moderate voltage bias changes from negative to positive. Also, at high bias voltages spectral peaks may be shifted by the bias. Our results demonstrate feasibility of DQWIPs for remote temperature sensing. Ratios of photocurrents at different bias voltages show significant dependence on the object temperature in the 300–700 °C range. The corresponding temperature sensitivity for two-color sensing (up to 0.2% per 1 °C) at nitrogen temperatures is comparable with the sensitivity of detectors based on multistack of two single-frequency QWIPs and operating at helium temperatures.[10] Reconfigurable multicolor detection allows for high accuracy in temperature measurements via statistical analysis of the photoresponse as a function of bias voltage. The sensing with DQWIPs may be extended to higher temperatures of IR emitters via appropriate bandgap engineering of the double wells.

Acknowledgments

This work was supported by the US Air Force (#FA9550-10-1-391); A. Sergeev acknowledges NRC support.

References

1. A. Rogalski, J. Antoszewski, and L. Faraone, "Third-generation infrared photodetector arrays," *J. Appl. Phys.* **105**, 091101 (2009).
2. M. Henini and M. Razeghi, *Handbook of Infrared Detection Technologies*, Oxford: Elsevier, 2002.
3. W. Minkina and S. Dudzik, *Infrared Thermography: Errors and Uncertainties*, New York: Wiley, 2009.
4. M. Vollmer and K.-P. Möllmann, *Infrared Thermal Imaging: Fundamentals, Research and Applications,* New York: Wiley-VCH, 2010.
5. X. P. V. Maldague, *Nondestructive Evaluation of Materials by Infrared Thermography*, London: Springer-Verlag, 1997.
6. K. K. Choi, *The Physics of Quantum Well Infrared Photodetectors*, River Edge, NJ: World Scientific, 1997.
7. H. Schneider and H. C. Liu, *Quantum Well Infrared Photodetectors: Physics and Applications*, New York: Springer, 2007.
8. K. L. Tsai, K. H. Chang, C. P. Lee, K. F. Huang, J. S. Tsang, and H. R. Chen, "Two-color infrared photodetector using GaAs/AlGaAs and strained InGaAs/AlGaAs multiquantum wells," *Appl. Phys. Lett.* **62**, 3504–3506 (1993).
9. M. Z. Tidrow, K. K. Choi, A. J. DeAnni, W. H. Chang, and S. P. Svensson, "Grating coupled multicolor quantum well infrared photodetectors," *Appl. Phys. Lett.* **67**, 1800–1802 (1995).
10. C. J. Chen, K. K. Choi, W. H. Chang, and D. C. Tsui, "Two-color corrugated quantum-well infrared photodetector for remote temperature sensing," *Appl. Phys. Lett.* **72**, 7–9 (1998).
11. A. Majumdar, K. K. Choi, L. P. Rokhinson, and D. C. Tsui, "Towards a voltage tunable two-color quantum-well infrared photodetector," *Appl. Phys. Lett.* **80**, 538–540 (2002).

12. A. Kock, E. Gornik, G. Abstreiter, G. Bohm, M. Walther, and G. Weimann, "Double wavelength selective GaAs/AlGaAs infrared detector device," *Appl. Phys. Lett.* **60**, 2011–2013 (1992).

13. H. C. Liu, J. Li, J. R. Thompson, Z. R. Wasilewski, M. Buchanan, and J. R. Simmons, "Multicolor voltage-tunable quantum-well infrared photodetector," *IEEE Electron Device Lett.* **14**, 566–568 (1993).

14. K. Sablon, A. Sergeev, N. Vagidov, A. Antipov, J. W. Little, and V. Mitin, "Effective harvesting, detection, and conversion of IR radiation due to quantum dots with built-in charge," *Nanoscale Res. Lett.* **6**, article no. 584 (2011).

15. A. Sergeev, N. Vagidov, V. Mitin, and K. Sablon, "Charged quantum dots for high efficiency photovoltaics and IR sensing," chapter in: S. Luryi, J. Xu, and A. Zaslavsky, eds, *Future Trends in Microelectronics: Frontiers and Innovations*, Hoboken, NJ: Wiley–IEEE Press, 2013, pp. 244–253.

16. K. A. Sablon, A. Sergeev, N. Vagidov, J. W. Little, and V. Mitin, "Effects of quantum dot charging on photoelectron processes and solar cell characteristics," *Solar Energy Mater. Solar Cells* **117**, 638–644 (2013).

17. Z. An, T. Ueda, J.-C. Chen, S. Komiyama, and K. Hirakawa, "A sensitive double quantum well infrared phototransistor," *J. Appl. Phys.* **100**, 044509 (2006).

18. T. Ueda, Z. An, K. Hiakawa, and S. Komiyama, "Charge-sensitive infrared phototransistors: Characterization by an all-cryogenic spectrometer," *J. Appl. Phys.* **103**, 093109 (2008).

19. T. Ueda, Z. An, K. Hirakawa, and S. Komiyama, "Temperature dependence of the performance of charge-sensitive infrared phototransistors," *J. Appl. Phys.* **105**, 064517 (2009).

20. Ü. Sakoğlu, J. S. Tyo, M. M. Hayat, S. Raghavan, and S. Krishna, "Spectrally adaptive infrared photodetectors with bias-tunable quantum dots," *J. Opt. Soc. Am. B* **21**, 7–17 (2004).

21. H. C. Liu, Z. R. Wasilewski, M. Buchanan, and H. Chu, "Segregation of Si δ doping in GaAs–AlGaAs quantum wells and the cause of the asymmetry in the current–voltage characteristics of intersubband infrared detectors," *Appl. Phys. Lett.* **63**, 761–763 (1993).

22. J. K. Choi, N. Vagidov, A. Sergeev, S. Kalchmair, G. Strasser, F. T. Vasko, and V. V. Mitin, "Asymmetrically doped GaAs/AlGaAs double-quantum-well structure for voltage-tunable IR detection," *Japan. J. Appl. Phys.* **51**, 074004 (2012).

3.7

Tunable Photonic Molecules for Spectral Engineering in Dense Photonic Integration

M. C. M. M. Souza, G. F. M. Rezende, A. A. G. von Zuben, G. S. Wiederhecker, and N. C. Frateschi
"Gleb Wataghin" Physics Institute, University of Campinas, Campinas, SP 13083-859, Brazil

L. A. M. Barea
Department of Electrical Engineering, UFSCAR, São Carlos, SP 13565-905, Brazil

1. Introduction

After two decades of continuous progress, photonic integration has proved its indisputable role as an enabling technology. It is poised to address the high-performance demands of future computing systems, pushing the limits of ultrafast optical data transfer and processing while complying with tight power budgets and drastic footprint constraints. In addition, unprecedented photonic-enabled capabilities have been responsible for the substantial progress achieved in emerging areas such as sensing,[1, 2] lab-on-a-chip,[3] and integrated microwave photonics.[4]

The successful deployment of photonic-based solutions across different fields depends on the ability to fully control the spectral response of the building blocks of photonic circuitry. The basic functionalities required for most applications (light sources, optical modulators, filters, delay lines, detectors, etc.) are now available in a variety of designs and platforms,[5, 6] but the challenge remains to realize these functionalities with devices allowing for flexible and reconfigurable spectral control. A microring resonator,[2, 7] for instance, while presenting a certain level of design freedom and spectral tunability, has its filtering and power enhancement properties dictated by its free spectral range (FSR), linewidth, and extinction ratio (ER), which are difficult to control actively.

In this chapter, we present some recent advances in realizing novel devices with flexible spectral response using systems of coupled microring resonators, or

Future Trends in Microelectronics: Journey into the Unknown, First Edition.
Edited by Serge Luryi, Jimmy Xu, and Alexander Zaslavsky.
© 2016 John Wiley & Sons, Inc. Published 2016 by John Wiley & Sons, Inc.

photonic molecules.[8, 9] The design flexibility of such devices offers extra degrees of freedom to perform spectral engineering using standard tuning techniques, such as microheaters and *pn* junctions, and allows to overcome trade-offs faced by single microrings in some applications.

2.   Photonic molecules and their spectral features

When multiple microring resonators are coupled to form photonic molecules, the resulting transmission spectrum will depend on the number of resonators and the way they couple to each other. Still, general trends can be qualitatively explained by considering the optical modes of uncoupled (bare) resonators and their degeneracy breaking due to mutual coupling.[9, 10]

The transmission spectrum of a single microring resonator, shown in Fig. 1(a), features a set of resonances that are separated by the FSR. The resonator FSR is inversely proportional to the microring length. The FWHM resonance linewidths are set by the net optical losses, and the resonance extinction ratio (ER) is dictated by the balance between intrinsic losses and the coupling between the microring and the bus waveguide.[2, 7]

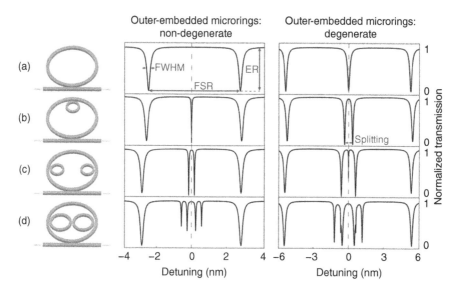

Figure 1. Calculated transmission spectrum of a single microring resonator (a) and different photonic molecule designs (b–d) for both nondegenerate and degenerate conditions between outer and embedded rings. Multiple mode-splitting dips appear in the transmission spectrum depending on the number of coupled resonators and their degeneracy. The transmission plots are shown as a function of detuning from the embedded microring resonances.

When a second distinct microring is coupled to the first one, as in Fig. 1(b), the two rings may have different (nondegenerate) or identical (degenerate) resonance wavelengths. In the nondegenerate case, the resonances of the primary (outer) ring remain unchanged and a new resonance notch appears due to the secondary (embedded) ring. When the two rings are degenerate, however, their mutual coupling may break their degeneracy through a mode-splitting that is proportional to their coupling strength.

If another embedded ring identical to the first one is introduced, as in Fig. 1(c), the coupling-induced mode-splitting will produce a doublet when the embedded and outer rings are nondegenerate. Further, a triplet with a very sharp central notch appears if all the three rings are resonant. Note that although the embedded microrings are not directly coupled to each other, they do interact through the outer microring and this interaction is strong enough to induce a substantial splitting between their bare optical modes.[9]

When the two embedded microrings couple directly to each other, as in Fig. 1(d), the situation is quite different, as the excitation of counterpropagating modes will occur whenever the embedded rings are resonant.[10] When only the embedded rings are resonant, a quadruplet appears in the transmission spectrum due to the coupling of four degenerate modes (two counterpropagating modes of each embedded microring). When the outer and embedded rings are degenerate, a total of six modes are coupled and a sextuplet may appear.

Regardless of the coupling configuration between microrings, the coupled system is no longer adequately described by the optical modes of the individual resonators. Instead, a good base to describe coupled rings is formed by "supermodes" of the coupled system. In other words, the optical power will be spatially distributed through the resonators in a particular manner for each one of the split resonances.[11]

The distinct amplitude and phase relations for the electric field localized in the three microrings of Fig. 1(c) are described in Fig. 2, using 2D finite-difference time-domain (FDTD) simulations for the triplet case.[9] For the two outermost resonances, labeled (i) and (iii) in Fig. 2(a), the optical mode is distributed through the three rings. The field amplitude within each ring is identical in (i) and (iii), whereas the relative phase between the fields of the outer and embedded rings is antisymmetric in (i) and symmetric in (iii). For the central resonance (ii), on the other hand, the field is localized almost completely within the embedded rings. Such confinement toward the embedded rings reduces the mode decay rate to the bus waveguide, yielding a higher total quality factor for this particular resonance.

The spectral and spatial distribution of the optical supermodes can be readily observed experimentally (see Fig. 2(b)). A triplet with enhanced optical quality factor for the central resonance is shown in the experimental transmission spectrum. The spatial distribution of the optical modes can be indirectly observed through the scattered light shown in the infrared microscopy images for each resonance. The fabricated device, shown in the inset of Fig. 2(b), was realized on an silicon-on-insulator (SOI) platform with outer and embedded ring radii of

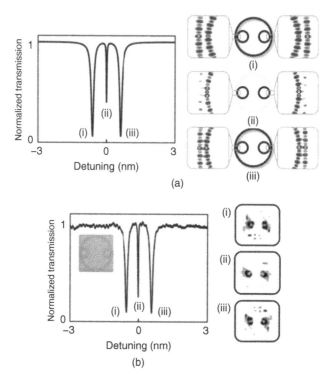

Figure 2. Spatial distribution of optical supermodes: (a) 2D FDTD simulations show the electric field distribution in each one of the resonances of the triplet; (b) transmission spectrum of an SOI photonic molecule (inset) around a triplet and the corresponding scattering power in each resonance as captured by an infrared camera. The triplet is centered at 1580 nm.[9]

20 and 5 μm, respectively, and 200-nm gaps between microrings and between the outer microring and the bus waveguide. The resonances were scanned using the quasi-TE mode of the silicon waveguide with 220×450 nm cross section.

The diversified spatial distribution of supermodes in photonic molecules can be harnessed as a powerful tool in practical implementations. Selectively changing the refractive index at different points of the coupled system may affect supermodes differently and represents an interesting feature to be explored in sensing applications. In addition, the supermodes do not always overlap with each other, providing freedom to engineer mode-selective couplings to distinct bus waveguides.[12]

3. Coupling-controlled mode-splitting: GHz-operation on a tight footprint

The interdependence between resonance spacing, footprint, and linewidth of a single microring resonator may undermine device performance in applications requiring closely spaced resonances on the GHz range. Closely spaced resonances in a single microring are determined by the FSR and can only be achieved with very large devices, compromising footprint and integration density. For example, to achieve resonance spacing of 50 GHz with a single microring in SOI, the device should have a radius of 225 μm. The finesse is also compromised by a large microring size, affecting low-power operation for functionalities based on nonlinear processes.

Using photonic molecules, closely spaced resonances and high finesse can be obtained simultaneously exploiting mode-splitting in compact devices. Splitting of tens of GHz can be easily achieved using typical SOI microrings with radius ranging from five to a few tens of micrometers and gap distances of a few hundreds of nanometers. Compared to a single microring with equivalent FSR, this represents a 100-fold footprint reduction.[9, 10, 13]

Integrated microwave photonics is one of the important areas that can hugely benefit from obtaining resonance spacing of GHz in compact devices.[4] One example of functionality that can be easily implemented using a simple photonic molecule with a doublet in the transmission spectrum is a compact integrated photonic microwave notch filter, illustrated in Fig. 3.[13] The device will filter the radio frequency (RF) sidebands of a modulated optical signal when the carrier is tuned to the central wavelength of the split resonance and the RF sidebands are resonant with the doublets. The coupled-cavity implementation also raises the possibility of achieving a tunable microwave notch filter if the coupling between the resonators can be controlled.

Optical signal-processing functionalities, such as wavelength multicasting, can also be considerably enhanced by the coexistence of closely spaced channels and high field enhancement in a small device. Wavelength multicasting consists of replicating the signal from one wavelength channel into multiple different channels and is used to improve traffic management in DWDM networks. Ideally, one should be able to achieve conversion bandwidths of a few THz, covering the main communication bands, as well as to utilize close DWDM channels, separated by tens of GHz. Using the design of Fig. 1(d), we recently demonstrated low-power wavelength conversion covering 8.5 THz with converted channels separated by 50 GHz in a $40 \times 40 \, \mu m^2$ device (see Fig. 4).[10] Exploiting the free-carrier dispersion (FCD) in silicon,[14] an optical control signal located near the resonance at $\lambda_C = 1548$ nm carrying a signal with 1 mW of peak power was simultaneously converted to four adjacent channels located near each of the four resonances (λ_{1-4}) of a quadruplet around 1617 nm.

4. Reconfigurable spectral control

In our previous discussion on the general spectral features of different photonic molecule designs, we presented their transmission spectra with respect to the degeneracy condition between embedded and outer rings. The transition between

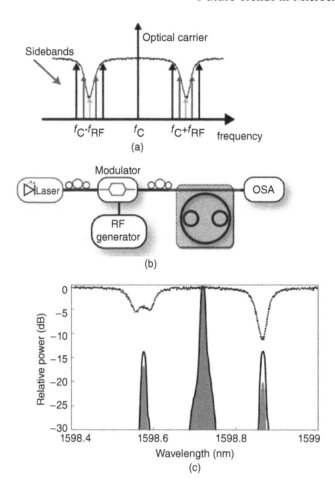

Figure 3. Compact microwave photonic notch filter. (a) When the optical carrier is tuned to the center of a split duplet, sidebands with RF frequencies matching the splitting will be filtered. (b) Experimental setup and (c) spectrum of a modulated signal before (solid line) and after (filled curve) the filter (top curve).[13]

these two conditions can be easily achieved using an integrated microheater to shift the outer ring resonances, as illustrated in Fig. 5. As the outer ring resonance approaches the identical embedded rings resonances, a clear avoided crossing (anticrossing) is observed for all the designs, a typical feature of coupled systems.

The control of the coupled modes around an anticrossing can be used to fine-tune the ER of resonances; see transmission traces in Fig. 5(a). Such ER tuning can be used to optimize the efficiency of power-dependent functionalities in passive integrated structures, such as parametric frequency comb generation.[15] In addition to ER tuning, the spectral anticrossing can be used for local mode dispersion compensation.[16, 17]

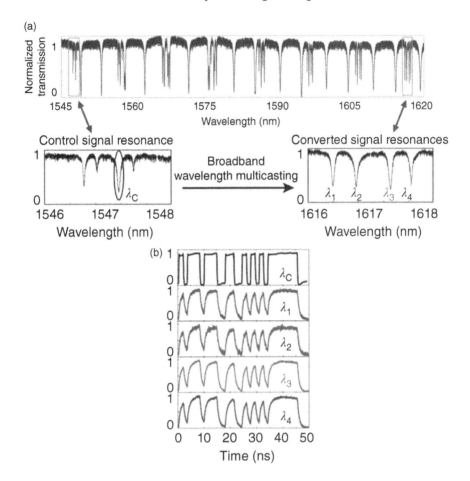

Figure 4. Wavelength multicasting using the coupled design of Fig. 1(d): (a) broad-band transmission spectrum of the device and detail of the resonances used for the conversion of a signal at 1548 nm to four adjacent channels separated by 50 GHz and centered around 1617 nm; (b) input λ_C and converted (λ_{1-4}) waveforms for a PRBS signal with 1 mW of peak power at λ_C.[10]

While Fig. 5 only shows the spectral evolution when the outer microring res-onances are shifted, a variety of spectral shapes can also be obtained by actuating the microheaters of the embedded microrings.

5. Toward reconfigurable mode-splitting control

The dynamic control of individual microrings in a photonic molecule design enables novel spectral control capabilities not fully deployed thus far, such as the active control of mode-splitting in a passive monolithically integrated platform.

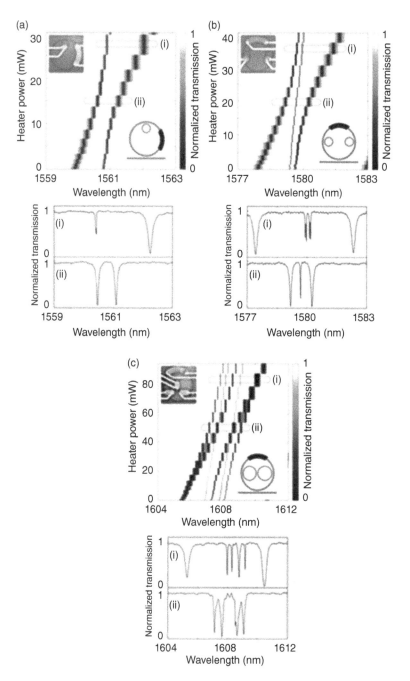

Figure 5. Reconfigurable spectral control: evolution of the transmission spectrum of photonic molecules with (a) one, (b) two uncoupled, and (c) two coupled embedded microrings as the outer microring resonances are shifted using a microheater. Insets in upper panels show the fabricated devices and schematics highlighting the actuated microheater (black). Bottom panels show the transmission traces for selected heater powers indicated in the upper plots by (i) and (ii).

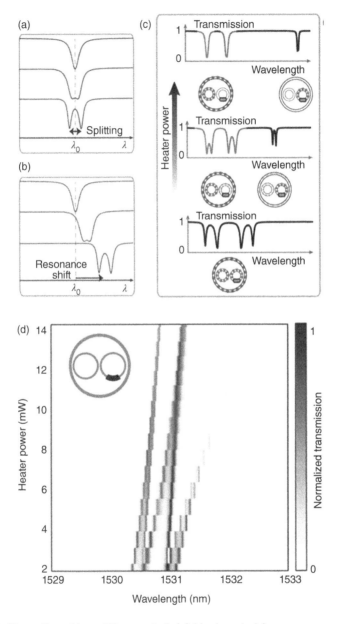

Figure 6. Reconfigurable splitting control: (a) ideal control (no resonance shift) and (b) actual control using MZI couplers (significant resonance shift). (c) The proposed scheme consists of detuning one embedded ring with respect to the other to control the CW–CCW mode coupling in the system. (d) Experimental splitting control from 60 GHz (central splitting) down to a single notch resonance. The small resonance shift is only due to thermal crosstalk. Inset: Schematic of the device highlighting the actuated microheater (black).[18]

Dynamical control of mode-splitting in monolithically integrated platforms is highly desirable but challenging.[18] The traditional solution based on a tunable Mach–Zehnder interferometer as the coupling element between resonators[19, 20] works only partially, since it also affects the optical phase of the resonant mode itself. As a result, instead of splitting control with fixed resonance positions, illustrated schematically in Fig. 6 (a), the obtained splitting control comes with significant resonance shifts, as in Fig. 6(b).

A splitting control mechanism with markedly suppressed resonance shifts has been recently demonstrated in our group exploiting photonic molecules.[18] Using the design of Fig. 1(d), the detuning between the embedded rings is used to control the coupling between clockwise (CW) and counterclockwise (CCW) modes. Starting with all three rings degenerate, one of the embedded rings is detuned with respect to the other two rings by means of a microheater (see Fig. 6(c)). As the detuning between embedded rings increases, the CW–CCW coupling is reduced.

Particularly, for those modes that remain confined in the two unaffected rings, the CW–CCW coupling reduction occurs with no change in their resonance condition and thus no resonance shift. The experimental results of Fig. 6(d) confirm the principle of reconfigurable splitting control from 60 GHz (central splitting) down to a single notch resonance. A small resonance shift is observed, which is only due to thermal crosstalk effects.[18] Such effects can be eliminated using distinct dispersion control mechanisms, such as carrier effects or electro-optic effects. In addition, these mechanisms would enable low-power, ultrafast splitting modulation.

6. Conclusion

In this chapter, we discussed some of the features that make photonic molecules such valuable tools in the quest for photonic circuits with flexible spectral response and enhanced performance. We presented some basic designs and discussed their spectral features, spatial mode distributions, and dynamic spectral control. A few examples illustrated the advantages of using photonic molecules to overcome the drawbacks of single resonant devices and to enable novel mechanisms of performance optimization and spectral control, such as extinction ratio tuning and reconfigurable mode-splitting control. Photonic molecules might enable further advances in sensing, optical signal processing, microwave photonics, and nonlinear nanophotonics.

Acknowledgments

This work was partially supported by CNPq, CAPES, CePOF (05/51689-2), FOTONICOM (08/57857-2), and FAPESP (2012/17765-7, 2014/04748-2).

References

1. V. M. N. Passaro, C. de Tullio, B. Troia, M. La Notte, G. Giannoccaro, and F. De Leonardis, "Recent advances in integrated photonic sensors," *Sensors (Switzerland)* **12**, 15558–15598 (2012).
2. S. Feng, T. Lei, H. Chen, H. Cai, X. Luo, and A. W. Poon, "Silicon photonics: From a microresonator perspective," *Laser Photonics Rev.* **6**, 145–177 (2012).
3. J. Vila-Planas, E. Fernández-Rosas, B. Ibarlucea, *et al.*, "Cell analysis using a multiple internal reflection photonic lab-on-a-chip," *Nature Photonics* **6**, 1642–1655 (2011).
4. D. Marpaung, C. Roeloffzen, R. R. Heideman, A. Leinse, S. Sales, and J. J. Capmany, "Integrated microwave photonics," *Laser Photonics Rev.* **7**, 506–538 (2013).
5. B. Jalali and S. Fathpour, "Silicon photonics," *J. Lightwave Technol.* **24**, 4600–4615 (2006).
6. M. J. R. Heck, J. F. Bauters, M. L. Davenport, *et al.*, "Hybrid silicon photonic integrated circuit technology," *IEEE J. Sel. Top. Quantum Electron.* **19**, article no. 6387568 (2013).
7. W. Bogaerts, P. De Heyn, T. Van Vaerenbergh, *et al.*, "Silicon microring resonators," *Laser Photonics Rev.* **6**, 47–73 (2012).
8. S. V. Boriskina, "Photonic molecules and spectral engineering," chapter in: I. Chremmos, O. Schwelb, and N. Uzunoglu, eds., *Photonic Microresonator Research and Applications*, New York: Springer, 2010, pp. 393–421.
9. L. A. M. Barea, F. Vallini, G. F. M. Rezende, and N. C. Frateschi, "Spectral engineering with CMOS compatible SOI photonic molecules," *IEEE Photonics J.* **5**, article no. 6657691 (2013).
10. M. C. M. M. Souza, L. A. M. Barea, F. Vallini, G. F. M. Rezende, G. S. Wiederhecker, and N. C. Frateschi, "Embedded coupled microrings with high-finesse and close-spaced resonances for optical signal processing," *Opt. Express* **22**, 10430–10438 (2014).
11. C. Schmidt, M. Liebsch, A. Klein, *et al.*, "Near-field mapping of optical eigenstates in coupled disk microresonators," *Phys. Rev. A* **85**, 033827 (2012).
12. X. Zeng and M. A. Popovic, "Design of triply-resonant microphotonic parametric oscillators based on Kerr nonlinearity," *Opt. Express* **22**, 15837–15867 (2014).
13. L. A. M. Barea, F. Vallini, P. F. Jarschel, and N. C. Frateschi, "Silicon technology compatible photonic molecules for compact optical signal processing," *Appl. Phys. Lett.* **103**, 201102 (2013).
14. Q. Xu, V. R. Almeida, and M. Lipson, "Micrometer-scale all-optical wave-length converter on silicon," *Opt. Lett.* **30**, 2733–2735 (2005).
15. S. A. Miller, Y. Okawachi, S. Ramelow, *et al.*, "Tunable frequency combs based on dual microring resonators," *Opt. Express* **23**, 21527–21540 (2015).
16. C. M. Gentry, X. Zeng, and M. A. Popovic, "Tunable coupled-mode dispersion compensation and its application to on-chip resonant four-wave mixing," *Opt. Lett.* **39**, 5689–5692 (2014).

17. X. Lu, S. Rogers, W. C. Jiang, and Q. Lin, "Selective engineering of cavity resonance for frequency matching in optical parametric processes," *Appl. Phys. Lett.* **105**, 151104 (2014).

18. M. C. M. M. Souza, G. F. M. Rezende, L. A. M. Barea, A. A. G. von Zuben, G. S. Wiederhecker, and N. C. Frateschi, "Spectral engineering with coupled microcavities: Active control of resonant mode-splitting," *Opt. Lett.* **40**, 3332–3335 (2015).

19. A. H. Atabaki, B. Momeni, A. Eftekhar, E. S. Hosseini, S. Yegnanarayanan, and A. Adibi, "Tuning of resonance-spacing in a traveling-wave resonator device," *Opt. Express* **18**, 9447–9455 (2010).

20. X. Sun, L. Zhou, J. Xie, Z. Zou, L. Lu, H. Zhu, X. Li, and J. Chen, "Investigation of coupling tuning in self-coupled optical waveguide resonators," *IEEE Photonics Technol. Lett.* **25**, 936–939 (2013).

Index

Future Trends in Microelectronics: Journey into the Unknown, First Edition.
Edited by Serge Luryi, Jimmy Xu, and Alexander Zaslavsky.
© 2016 John Wiley & Sons, Inc. Published 2016 by John Wiley & Sons, Inc.